Spatiotemporal Data Analysis

Spatiotemporal Data Analysis

Gidon Eshel

PRINCETON UNIVERSITY PRESS
PRINCETON AND OXFORD

Published by Princeton University Press, 41 William Street, Princeton, New Jersey 08540
In the United Kingdom: Princeton University Press, 6 Oxford Street, Woodstock, Oxfordshire OX20 1TW
press.princeton.edu

Library of Congress Cataloging-in-Publication Data

Eshel, Gidon, 1958–
Spatiotemporal data analysis / Gidon Eshel.
p. cm.
Includes bibliographical references and index.
ISBN 978-0-691-12891-7 (hardback)
1. Spatial analysis (Statistics) I. Title.
QA278.2.E84 2011
519.5'36—dc23
2011032275

British Library Cataloging-in-Publication Data is available

This book has been composed in Minion Pro

Printed on acid-free paper. ∞

Printed in the United States of America

10 9 8 7 6 5 4 3 2 1

To Laura, Adam, and Laila, with much love and deep thanks.

Contents

Preface

THIS BOOK IS ABOUT analyzing multidimensional data sets. It strives to be an introductory level, technically accessible, yet reasonably comprehensive practical guide to the topic as it arises in diverse scientific contexts and disciplines.

While there are nearly countless contexts and disciplines giving rise to data whose analysis this book addresses, your data must meet one criterion for this book to optimally answer practical challenges your data may present. This criterion is that the data possess a meaningful, well-posed, covariance matrix, as described in later sections. The main corollary of this criterion is that the data must depend on at least one coordinate along which order is important. Following tradition, I often refer to this coordinate as "time," but this is just a shorthand for a coordinate along which it is meaningful to speak of "further" or "closer," "earlier" or "later." As such, this coordinate may just as well be a particular space dimension, because a location 50 km due north of your own is twice as far as a location 25 km due north of you, and half as far as another location 100 km to the north. If your data set does not meet this criterion, many techniques this book presents may still be applicable to your data, but with a nontraditional interpretation of the results. If your data are of the scalar type (i.e., if they depend only on that "time" coordinate), you may use this book, but your problem is addressed more thoroughly by time-series analysis texts.

The data sets for which the techniques of this book are most applicable and the analysis of which this book covers most straightforwardly are vector time series. The system's state at any given time point is a group of values, arranged by convention as a column. The available time points, column vectors, are arranged side by side, with time progressing orderly from left to right.

I developed this book from class notes I have written over the years while teaching data analysis at both the University of Chicago and Bard College. I have always pitched it at the senior undergraduate–beginning graduate level. Over the years, I had students from astronomy and astrophysics, ecology and evolution, geophysics, meteorology, oceanography, computer science, psychology, and neuroscience. Since they had widely varied mathematical backgrounds, I have tended to devote the first third of the course to mathematical priming, particularly linear algebra. The first part of this book is devoted to this task. The course's latter two-thirds have been focused on data analysis, using examples from all the above disciplines. This is the focus of this book's second part. Combining creatively several elements of each of this book's two parts in a modular manner dictated by students' backgrounds and term length, instructors can design many successful, self-contained, and consistent courses.

It is also extremely easy to duplicate examples given throughout this book in order to set up new examples expressly chosen for the makeup and interests of particular classes. The book's final chapter provides some sample homework, suggested exams, and solutions to some of those.

In this book, whenever possible I describe operations using conventional algebraic notation and manipulations. At the same time, applied mathematics can sometimes fall prey to idiosyncratic or nonuniversal notation, leading to ambiguity. To minimize this, I sometimes introduce explicit code segments and describe their operations. Following no smaller precedence than the canonical standard bearer of applied numerics, Numerical Recipes,[1] I use an explicit language, without which ambiguity may creep in anew. All underlying code is written in Matlab or its free counterpart, Octave. Almost always, the code is written using primitive operators that employ no more than basic linear algebra. Sometimes, in the name of pedagogy and code succinctness, I use higher-level functions (e.g., `svd`, where the font used is reserved for code and machine variables), but the operations of those functions can always be immediately understood with complete clarity from their names. Often, I deliberately sacrifice numerical efficiency in favor of clarity and ease of deciphering the code workings. In some cases, especially in the final chapter (homework assignments and sample exams), the code is also not the most general it can be, again to further ease understanding.

In my subjective view, Matlab/Octave are the most natural environments to perform data analysis (R[2] is a close free contender) and small-scale modeling (unless the scope of the problem at hand renders numerical efficiency the deciding factor, and even then there are ways to use those languages to develop, test, and debug the code, while executing it more efficiently as a native executable). This book is not an introduction to those languages, and I assume the reader possesses basic working knowledge of them (although I made every effort to comment extensively each presented code segment). Excellent web resources abound introducing and explaining those languages in great detail. Two that stand out in quality and lucidity, and are thus natural starting points for the interested, uninitiated reader, are the Mathworks general web site[3] and the Matlab documentation therein,[4] and the Octave documentation.[5]

Multidimensional data analysis almost universally boils down to linear algebra. Unfortunately, thorough treatment of this important, broad, and wonderful topic is beyond the scope of this book, whose main focus is practical data analysis. In Part 1, I therefore introduce just a few absolutely essential and

[1] www.nrbook.com/nr3/.

[2] www.r-project.org/.

[3] www.mathworks.com.

[4] www.mathworks.com/help/matlab/.

[5] www.gnu.org/software/octave/doc/interpreter/.

salient ideas. To learn more, I can think of no better entry-level introduction to the subject than Strang's.[6] Over the years, I have also found Strang's slightly more formal counterpart by Noble and Daniel[7] useful.

Generalizing this point, I tried my best to make the book as self-contained as possible. Indeed, the book's initial chapters are at an introductory level appropriate for college sophomores and juniors of any technical field. At the same time, the book's main objective is data analysis, and linear algebra is a means, not the end. Because of this, and book length limitations, the discussion of some relatively advanced topics is somewhat abbreviated and not fully self-contained. In addition, in some sections (e.g., 9.3.1), some minimal knowledge of real analysis, multivariate calculus, and partial differentiation is assumed. Thus, some latter chapters are best appreciated by a reader for whom this book is not the first encounter with linear algebra and related topics and probably some data analysis as well.

Throughout this book, I treat data arrays as real. This assumption entails loss of generality; many results derived with this assumption require some additional, mostly straightforward, algebraic gymnastics to apply to the general case of complex arrays. Despite this loss of generality, this is a reasonable assumption as nearly all physically realizable and practically observed data, are in fact, most naturally represented by real numbers.

In writing this book, I obviously tried my best to get everything right. However, when I fail (on notation, math, or language and clarity, which surely happened)—please let me know (`geshel@gmail.com`) by pointing out clearly where and how I erred or deviated from the agreed upon conventions.

[6] Strang, G. (1988) *Linear Algebra and Its Applications*, 3rd ed., Harcourt Brace Jovanovich, San Diego, 520 pp., ISBN-13: 978-0155510050.

[7] Noble, B. and J. W. Daniel (1987) *Applied Linear Algebra*, 3rd ed., Prentice Hall, Englewood Cliffs, NJ, 521 pp., ISBN-13: 978-0130412607.

Acknowledgments

WRITING THIS BOOK has been on and off my docket since my first year of graduate school; there are actually small sections of the book I wrote as notes to myself while taking a linear algebra class in my first graduate school semester. My first acknowledgment thus goes to the person who first instilled the love of linear algebra in me, the person who brilliantly taught that class in the applied physics program at Columbia, Lorenzo Polvani. Lorenzo, your Italian lilt has often blissfully internally accompanied my calculations ever since!

Helping me negotiate the Columbia graduate admission's process was the first in a never-ending series of kind, caring acts directed at me by my mentor and friend, Mark Cane. Mark's help and sagacious counsel took too many forms, too many times, to recount here, but for his brilliant, generous scientific guidance and for his warmth, wisdom, humor, and care I am eternally grateful for my good fortune of having met, let alone befriended, Mark.

While at Columbia, I was tirelessly taught algebra, modeling, and data analysis by one of the mightiest brains I have ever encountered, that belonging to Benno Blumenthal. For those who know Benno, the preceding is an understatement. For the rest—I just wish you too could talk shop with Benno; there is nothing quite like it.

Around the same time, I was privileged to meet Mark's close friend, Ed Sarachik. Ed first tried, unpersuasively, to hide behind a curmudgeonly veneer, but was quickly exposed as a brilliant, generous, and supportive mentor, who shaped the way I have viewed some of the topics covered in this book ever since.

As a postdoc at Harvard University, I was fortunate to find another mentor/friend gem, Brian Farrell. The consummate outsider by choice, Brian is Mark's opposite in some ways. Yet just like Mark, to me Brian has always been loyal, generous, and supportive, a true friend. Our shared fascination with the outdoors and fitness has made for excellent glue, but it was Brian's brilliant and enthusiastic, colorful yet crisp teaching of dynamical systems and predictability that shaped my thinking indelibly. I would like to believe that some of Brian's spirit of eternal rigorous curiosity has rubbed off on me and is evident in the following pages.

Through the Brian/Harvard connection, I met two additional incredible teachers and mentors, Petros J. Ioannou and Eli Tziperman, whose teaching is evident throughout this book (Petros also generously reviewed section 9.7 of the book), and for whose generous friendship I am deeply thankful. At Woods Hole and then Chicago, Ray Schmidt and Doug McAyeal were also inspiring mentors whose teaching is strewn about throughout this book.

My good friend and one time modeling colleague, David Archer, was the matchmaker of my job at Chicago and an able teacher by example of the formidable power of understated, almost Haiku-like shear intellectual force. While I have never mastered David's understatement, and probably never will, I appreciate David's friendship and scientific teaching very much. While at Chicago, the paragon of lucidity, Larry Grossman, was also a great teacher of beautifully articulated rigor. I hope the wisdom of Larry's teachings and his boyish enthusiasm for planetary puzzles is at least faintly evident in the following pages.

I thank, deeply and sincerely, editor Ingrid Gnerlich and the board and technical staff at Princeton University Press for their able, friendly handling of my manuscript and for their superhuman patience with my many delays. I also thank University of Maryland's Michael Evans and Dartmouth's Dan Rockmore for patiently reading this long manuscript and making countless excellent suggestions that improved it significantly.

And, finally, the strictly personal. A special debt of gratitude goes to Pam Martin, a caring, supportive friend in trying times; Pam's friendship is not something I will or can ever forget. My sisters' families in Tel Aviv are a crucial element of my thinking and being, for which I am always in their debt. And to my most unusual parents for their love and teaching while on an early life of unparalleled explorations, of the maritime, literary, and experiential varieties. Whether or not a nomadic early life is good for the young I leave to the pros; it was most certainly entirely unique, and it without a doubt made me who I am.

PART 1

Foundations

ONE

Introduction and Motivation

BEFORE YOU START working your way through this book, you may ask yourself—Why analyze data? This is an important, basic question, and it has several compelling answers.

The simplest need for data analysis arises most naturally in disciplines addressing phenomena that are, in all likelihood, inherently nondeterministic (e.g., feelings and psychology or stock market behavior). Since such fields of knowledge are not governed by known fundamental equations, the only way to generalize disparate observations into expanded knowledge is to analyze those observations. In addition, in such fields predictions are entirely dependent on empirical models of the types discussed in chapter 9 that contain parameters not fundamentally constrained by theory. Finding these models' numerical values most suitable for a particular application is another important role of data analysis.

A more general rationale for analyzing data stems from the complementary relationship of empirical and theoretical science and dominates contexts and disciplines in which the studied phenomena have, at least in principle, fully knowable and usable fundamental governing dynamics (see chapter 7). In these contexts, best exemplified by physics, theory and observations both vie for the helm. Indeed, throughout the history of physics, theoretical predictions of yet unobserved phenomena and empirical observations of yet theoretically unexplained ones have alternately fixed physics' ropes.[1] When theory leads, its predictions must be tested against experimental or observational data. When empiricism is at the helm, coherent, reproducible knowledge is systematically and carefully gleaned from noisy, messy observations. At the core of both, of course, is data analysis.

Empiricism's biggest triumph, affording it (ever so fleetingly) the leadership role, arises when novel data analysis-based knowledge—fully acquired and processed—proves at odds with relevant existing theories (i.e., equations previously thought to govern the studied phenomenon fail to explain and reproduce the new observations). In such cases, relatively rare but game changing,

[1] As beautifully described in Feuer, L. S. (1989) *Einstein and the Generations of Science*, 2nd ed., Transaction, 390 pp., ISBN-10: 0878558993, ISBN-13: 978-0878558995, and also, with different emphasis, in Kragh, H. (2002) *Quantum Generations: A History of Physics in the Twentieth Century*, Princeton University Press, Princeton, NJ, 512 pp., ISBN13: 978-0-691-09552-3.

the need for a new theory becomes apparent.[2] When a new theory emerges, it either generalizes existing ones (rendering previously reigning equations a limiting special case, as in, e.g., Newtonian vs. relativistic gravity), or introduces an entirely new set of equations. In either case, at the root of the progress thus achieved is data analysis.

Once a new theory matures and its equation set becomes complete and closed, one of its uses is model-mediated predictions. In this application of theory, another rationale for data analysis sometimes emerges. It involves phenomena (e.g., fluid turbulence) for which governing equations may exist in principle, but their applications to most realistic situations is impossibly complex and high-dimensional. Such phenomena can thus be reasonably characterized as fundamentally deterministic yet practically stochastic. As such, practical research and modeling of such phenomena fall into the first category above, that addressing inherently nondeterministic phenomena, in which better mechanistic understanding requires better data and better data analysis.

Data analysis is thus essential for scientific progress. But is the level of algebraic rigor characteristic of some of this book's chapters necessary? After all, in some cases we can use some off-the-shelf spreadsheet-type black box for some rudimentary data analysis without any algebraic foundation. How you answer this question is a subjective matter. My view is that while in a few cases some progress can be made without substantial understanding of the underlying algebraic machinery and assumptions, such analyses are inherently dead ends in that they can be neither generalized nor extended beyond the very narrow, specific question they address. To seriously contribute to any of the progress routes described above, in the modular, expandable manner required for your work to potentially serve as the foundation of subsequent analyses, there is no alternative to thorough, deep knowledge of the underlying linear algebra.

[2] Possibly the most prominent examples of this route (see Feuer's book) are the early development of relativity partly in an effort to explain the Michelson-Morley experiment, and the emergence of quantum mechanics for explaining blackbody radiation observations.

TWO

Notation and Basic Operations

WHILE ALGEBRAIC BASICS can be found in countless texts, I really want to make this book as self contained as reasonably possible. Consequently, in this chapter I introduce some of the basic players of the algebraic drama about the unfold, and the uniform notation I have done my best to adhere to in this book. While chapter 3 is a more formal introduction to linear algebra, in this introductory chapter I also present some of the most basic elements, and permitted manipulations and operations, of linear algebra.

1. *Scalar variables:* Scalars are given in lowercase, slanted, Roman or Greek letters, as in $a, b, x, \alpha, \beta, \theta$.
2. *Stochastic processes and variables:* A stochastic variable is denoted by an italicized uppercase X. A particular value, or realization, of the process X is denoted by x.
3. *Matrix variables:* Matrices are the most fundamental building block of linear algebra. They arise in many, highly diverse situations, which we will get to later. A matrix is a rectangular array of numbers, e.g.,

$$\begin{pmatrix} 1 & 1 & -4 \\ 0 & 3 & 2 \\ 5 & 11 & 24 \\ -1 & 31 & 4 \end{pmatrix}. \tag{2.1}$$

   A matrix is said to be $M \times N$ (M by N) when it comprises M rows and N columns. A vector is a special case of matrix for which either M or N equals 1. By convention, unless otherwise stated, we will treat vectors as column vectors.
4. *Fields:* Fields are sets of elements satisfying the addition and multiplication field axioms (associativity, commutativity, distributivity, identity, and inverses), which can be found in most advanced calculus or abstract algebra texts. In this book, the single most important field is the real line, the set of real numbers, denoted by $\mathbb{R}$. Higher-dimensional spaces over $\mathbb{R}$ are denoted by $\mathbb{R}^N$.
5. *Vector variables:* Vectors are denoted by lowercase, boldfaced, Roman letter, as in $\mathbf{a}$, $\mathbf{b}$, $\mathbf{x}$. When there is risk of ambiguity, and only then, I adhere to normal physics notation, and adorn the vector with an

overhead arrow, as in $\vec{\mathbf{a}}$, $\vec{\mathbf{b}}$, $\vec{\mathbf{x}}$. Unless specifically stated otherwise, all vectors are assumed to be column vectors,

$$\vec{\mathbf{a}} \equiv \mathbf{a} = \begin{pmatrix} a_1 \\ a_2 \\ \vdots \\ a_M \end{pmatrix} \in \mathbb{R}^M, \tag{2.2}$$

where $\mathbf{a}$ is said to be an M-vector (a vector with M elements); "$\equiv$" means "equivalent to"; a_i is $\mathbf{a}$'s ith element ($1 \le i \le M$); "$\in$" means "an element of," so that the object to its left is an element of the object to its right (typically a set); and $\mathbb{R}^M$ is the set (denoted by $\{\cdot\}$) of real M-vectors

$$\mathbb{R}^M = \left\{ \begin{pmatrix} a_1 \\ a_2 \\ \vdots \\ a_M \end{pmatrix} \right\} \quad a_i \in \mathbb{R} \quad \forall i, \tag{2.3}$$

$\mathbb{R}^M$ is the set of all M-vectors $\mathbf{a}$ of which element i, a_i, is real for all i (this is the meaning of $\forall i$). Sometimes, within the text, I use $\mathbf{a} = (a_1\ a_2 \cdots a_M)^T$ (see below).

6. *Vector transpose:* For

$$\mathbf{a} = \begin{pmatrix} a_1 \\ a_2 \\ \vdots \\ a_N \end{pmatrix} \in \mathbb{R}^{N \times 1}, \tag{2.4}$$

$$\mathbf{a}^T = \begin{pmatrix} a_1 \\ a_2 \\ \vdots \\ a_N \end{pmatrix}^T = \begin{pmatrix} a_1 & a_2 & \cdots & a_N \end{pmatrix} \in \mathbb{R}^{1 \times N}, \tag{2.5}$$

where $\mathbf{a}^T$ is pronounced "$\mathbf{a}$ transpose."

7. *Vector addition:* If two vectors share the same dimension N (i.e., $\mathbf{a} \in \mathbb{R}^N$ and $\mathbf{b} \in \mathbb{R}^N$), then their sum or difference $\mathbf{c}$ is defined by

$$\mathbf{a} \pm \mathbf{b} = \mathbf{c} \in \mathbb{R}^N, \qquad c_i = a_i \pm b_i, \quad 1 \le i \le N. \tag{2.6}$$

8. *Linear independence:* Two vectors $\mathbf{a}$ and $\mathbf{b}$ are said to be linearly dependent if there exists a scalar α such that $\mathbf{a} = \alpha\mathbf{b}$. For this to hold, $\mathbf{a}$ and $\mathbf{b}$ must be parallel. If no such α exists, $\mathbf{a}$ and $\mathbf{b}$ are linearly independent.

   In higher dimensions, the situation is naturally a bit murkier. The elements of a set of K $\mathbb{R}^N$ vectors, $\{\mathbf{v}_i\}_{i=1}^K$, are linearly dependent if there exists a set of scalars $\{\alpha_i\}_{i=1}^K$, not all zero, which jointly satisfy

$$\sum_{i=1}^{K} \alpha_i \mathbf{v}_i = \mathbf{0} \in \mathbb{R}^N, \tag{2.7}$$

where the right-hand side is the $\mathbb{R}^N$ zero vector. If the above is only satisfied for $\alpha_i = 0\ \forall i$ (i.e., if the above only holds if all αs vanish), the elements of the set $\{\mathbf{v}_i\}$ are mutually linearly independent.

9. *Inner product of two vectors:* For all practical data analysis purposes, if two vectors share the same dimension N as before, their dot, or inner, product, exists and is the scalar

$$p = \mathbf{a}^T\mathbf{b} = \mathbf{b}^T\mathbf{a} = \sum_{i=1}^{N} a_i b_i \in \mathbb{R}^1 \tag{2.8}$$

(where $\mathbb{R}^1$ is often abbreviated as $\mathbb{R}$).

10. *Projection:* The inner product gives rise to the notion of the projection of one vector on another, explained in fig. 2.1.
11. *Orthogonality:* Two vectors $\mathbf{u}$ and $\mathbf{v}$ are mutually orthogonal, denoted $\mathbf{u} \perp \mathbf{v}$, if $\mathbf{u}^T\mathbf{v} = \mathbf{v}^T\mathbf{u} = 0$. If, in addition to $\mathbf{u}^T\mathbf{v} = \mathbf{v}^T\mathbf{u} = 0$, $\mathbf{u}^T\mathbf{u} = \mathbf{v}^T\mathbf{v} = 1$, $\mathbf{u}$ and $\mathbf{v}$ are mutually orthonormal.
12. *The norm of a vector:* For any $p \in \mathbb{R}$, the p-norm of the vector $\mathbf{a} \in \mathbb{R}^N$ is

$$\|\mathbf{a}\|_p := \sqrt[p]{\sum_{i=1}^{N} |a_i|^p}, \tag{2.9}$$

where the real scalar $|a_i|$ is the absolute value of $\mathbf{a}$'s ith element.

Most often, the definition above is narrowed by setting $p \in \mathbb{N}_1$, where $\mathbb{N}_1$ is the set of positive natural numbers, $\mathbb{N}_1 = \{1, 2, 3, \dots\}$.

A particular norm frequently used in data analysis is the L^2 (also denoted L_2), often used interchangeably with the Euclidean norm,

$$\|\mathbf{a}\| \equiv \|\mathbf{a}\|_2 = \sqrt[2]{\sum_{i=1}^{N} a_i^2} = \sqrt[2]{\mathbf{a}^T\mathbf{a}}, \tag{2.10}$$

where above I use the common convention of omitting the p when $p = 2$, i.e., using "$\|\cdot\|$" as a shorthand for "$\|\cdot\|_2$." The term "Euclidean norm" refers to the fact that in a Euclidean space, a vector's L^2-norm is its length. For example, consider $\mathbf{r} = (\ 1\ 2\ )^T$ shown in fig. 2.2 in its natural habitat, $\mathbb{R}^2$, the geometrical two-dimensional plane intuitively familiar from daily life. The vector $\mathbf{r}$ connects the origin, (0, 0), and the point, (1, 2); how long is it?! Denoting that length by r and invoking the Pythagorean theorem (appropriate here because $\mathbf{x} \perp \mathbf{y}$ in Euclidean spaces),

$$r^2 = 1^2 + 2^2 \quad \text{or} \quad r = \sqrt{1^2 + 2^2} = \sqrt{5}, \tag{2.11}$$

which is exactly

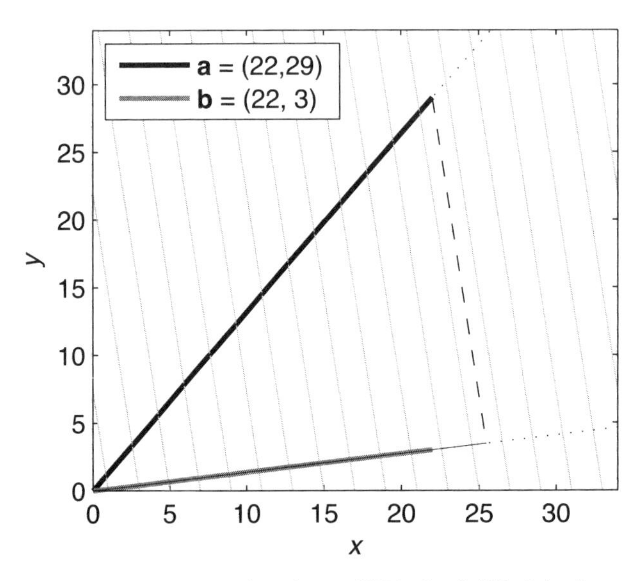

Figure 2.1. Projection of $\mathbf{a} = (\ 22\ 29\ )^T$ (thick solid black line) onto $\mathbf{b} = (\ 22\ 3\ )^T$ (thick solid gray line), shown by the thin black line parallel to $\mathbf{b}$, $\mathbf{p} \equiv [(\mathbf{a}^T\mathbf{b})/(\mathbf{b}^T\mathbf{b})]\mathbf{b} = (\mathbf{a}^T\,\hat{\mathbf{b}})\,\hat{\mathbf{b}}$. The projection is best visualized as the shadow cast by **a** on the **b** direction in the presence of a uniform lighting source shining from upper left to lower right along the thin gray lines, i.e., perpendicular to **b**. The dashed line is the residual of **a**, $\mathbf{r} = \mathbf{a} - \mathbf{p}$, which is normal to $\mathbf{p}$, $(\mathbf{a} - \mathbf{p})^T\mathbf{p} = 0$. Thus, $\mathbf{p} = \mathbf{a}_{\hat{b}}$ (**a**'s part in the direction of **b**) and $\mathbf{r} = \ \mathbf{a}_{\perp\hat{b}}$ (**a**'s part perpendicular to **b**), so **p** and **r** form an orthogonal split of **a**.

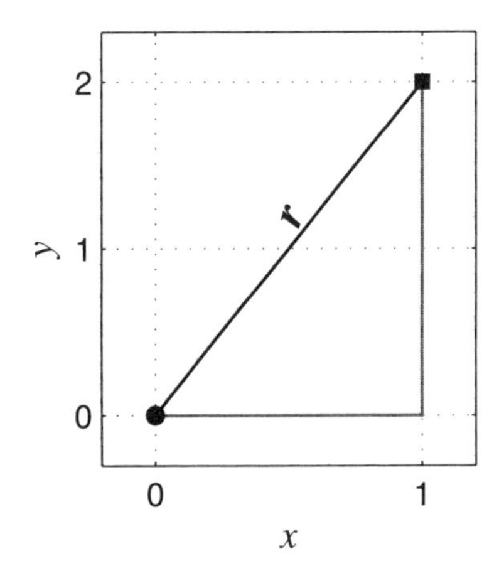

Figure 2.2. A schematic representation of the Euclidean norm as the length of a vector in $\mathbb{R}^2$.

$$\|\mathbf{r}\| = \sqrt{\mathbf{r}^T\mathbf{r}} = \sqrt{\begin{pmatrix}1 & 2\end{pmatrix}\begin{pmatrix}1\\2\end{pmatrix}} = \sqrt{5}, \tag{2.12}$$

demonstrating the "length of a vector" interpretation of the L^2-norm.

13. *Unit vectors:* Vectors of unit length,

$$\hat{\mathbf{a}} := \frac{\mathbf{a}}{\|\mathbf{a}\|}, \tag{2.13}$$

where $\mathbf{a}$, $\hat{\mathbf{a}} \neq \mathbf{0} \in \mathbb{R}^N$, are called unit vectors and are adorned with an overhat.

Note that

$$\|\hat{\mathbf{a}}\| = \left\|\frac{\mathbf{a}}{\|\mathbf{a}\|}\right\| = \sqrt{\sum_{i=1}^{N}\frac{a_i^2}{\|\mathbf{a}\|}} = \frac{1}{\|\mathbf{a}\|}\sqrt{\sum_{i=1}^{N}a_i^2} = \frac{\|\mathbf{a}\|}{\|\mathbf{a}\|} = 1 \tag{2.14}$$

by construction.

14. *Matrix variables:* Matrices are denoted by uppercase, boldfaced, Roman letters, as in **A**, **B**, **M**. When there is any risk of ambiguity, and only then, I adorn matrix variables with two underlines, as in

$$\mathbf{A} \equiv \underline{\underline{\mathbf{A}}}. \tag{2.15}$$

Unless otherwise explicitly stated due to potential ambiguity, matrices are considered to be $M \times N$ (to have dimensions M by N), i.e., to have M rows and N columns,

$$\mathbf{A} = \begin{pmatrix} a_{11} & a_{12} & \cdots & a_{1N} \\ a_{21} & a_{22} & \cdots & a_{2N} \\ \vdots & \vdots & & \vdots \\ a_{M1} & a_{M2} & \cdots & a_{MN} \end{pmatrix} \in \mathbb{R}^{M\times N}, \tag{2.16}$$

where a_{ij} is **A**'s real scalar element in row i and column j.

We sometimes need a column-wise representation of a matrix, for which the notation is

$$\mathbf{A} = \begin{pmatrix}\mathbf{a}_1 & \mathbf{a}_2 & \cdots & \mathbf{a}_N\end{pmatrix} \in \mathbb{R}^{M\times N}, \tag{2.17}$$

where the ith column is $\mathbf{a}_i \in \mathbb{R}^{M\times 1}$ or $\mathbf{a}_i \in \mathbb{R}^M$, and $1 \leq i \leq N$.

15. *Matrix addition:* For $\mathbf{C} = \mathbf{A} \pm \mathbf{B}$ to be defined, **A** and **B** must have the same dimensions. Then, **C** "inherits" these dimensions, and its elements are $c_{ij} = a_{ij} \pm b_{ij}$.
16. *Transpose of a matrix:* The transpose of

$$\mathbf{A} = \begin{pmatrix} a_{11} & a_{12} & \cdots & a_{1N} \\ a_{21} & a_{22} & \cdots & a_{2N} \\ \vdots & \vdots & & \vdots \\ a_{M1} & a_{M2} & \cdots & a_{MN} \end{pmatrix} \tag{2.18}$$

$$= \begin{pmatrix} \mathbf{a}_1 & \mathbf{a}_2 & \cdots & \mathbf{a}_N \end{pmatrix} \in \mathbb{R}^{M \times N}, \tag{2.19}$$

where $\mathbf{a}_i \in \mathbb{R}^M$, is

$$\mathbf{A}^T = \begin{pmatrix} a_{11} & a_{21} & \cdots & a_{M1} \\ a_{12} & a_{22} & \cdots & a_{M2} \\ \vdots & \vdots & & \vdots \\ a_{1N} & a_{2N} & \cdots & a_{MN} \end{pmatrix} = \begin{pmatrix} \mathbf{a}_1^T \\ \mathbf{a}_2^T \\ \vdots \\ \mathbf{a}_N^T \end{pmatrix} \in \mathbb{R}^{N \times M}, \tag{2.20}$$

so that $\mathbf{A}$'s element ij is equal to $\mathbf{A}^T$'s element ji.

17. *Some special matrices:*

- Square diagonal ($M = N$):

$$\mathbf{A} = \begin{pmatrix} a_{11} & 0 & \cdots & 0 \\ 0 & a_{22} & \cdots & 0 \\ \vdots & \vdots & \ddots & \vdots \\ 0 & 0 & \cdots & a_{MM} \end{pmatrix} \in \mathbb{R}^{M \times M}, \tag{2.21}$$

$$a_{ij} = \begin{cases} a_{ii} & i = j \\ 0 & i \neq j \end{cases}. \tag{2.22}$$

- Rectangular diagonal, $M > N$:

$$\mathbf{A} = \begin{pmatrix} a_{11} & 0 & \cdots & 0 \\ 0 & a_{22} & \cdots & 0 \\ \vdots & \vdots & & \vdots \\ 0 & 0 & \cdots & a_{NN} \\ 0 & 0 & \cdots & 0 \\ \vdots & \vdots & & \vdots \\ 0 & 0 & \cdots & 0 \end{pmatrix} \in \mathbb{R}^{M \times N}, \tag{2.23}$$

i.e.,

$$a_{ij} = \begin{cases} a_{ii} & i = j \leq N \\ 0 & i \neq j \end{cases}. \tag{2.24}$$

- Rectangular diagonal, $M < N$:

$$\mathbf{A} = \begin{pmatrix} a_{11} & 0 & \cdots & 0 & 0 & \cdots & 0 \\ 0 & a_{22} & \cdots & 0 & 0 & \cdots & 0 \\ \vdots & \vdots & & \vdots & \vdots & & \vdots \\ 0 & 0 & \cdots & a_{MM} & 0 & \cdots & 0 \end{pmatrix} \in \mathbb{R}^{M \times N}, \tag{2.25}$$

i.e.,

$$a_{ij} = \begin{cases} a_{ii} & i = j \leq M \\ 0 & i \neq j \end{cases}. \tag{2.26}$$

- Square symmetric, $M = N$:

$$\mathbf{A} = \mathbf{A}^T = \begin{pmatrix} a_{11} & a_{12} & \cdots & a_{1M} \\ a_{21} = a_{12} & a_{22} & \cdots & a_{2M} \\ \vdots & \vdots & \vdots & \vdots \\ a_{M1} = a_{1M} & a_{M2} = a_{2M} & \cdots & a_{MM} \end{pmatrix}, \tag{2.27}$$

i.e., $a_{ij} = a_{ji}$ with $\mathbf{A} = \mathbf{A}^T \in \mathbb{R}^{M \times M}$.

18. *Matrix product:* **AB** is possible only if **A** and **B** share their second and first dimensions, respectively. That is, for **AB** to exist $\mathbf{A} \in \mathbb{R}^{M \times N}$, $\mathbf{B} \in \mathbb{R}^{N \times K}$, where M and K are positive integers, must hold. When the matrix multiplication is permitted,

$$\mathbf{AB} = \begin{pmatrix} a_{11} & a_{12} & \cdots & a_{1N} \\ a_{21} & a_{22} & \cdots & a_{2N} \\ \vdots & \vdots & & \vdots \\ a_{M1} & a_{M2} & \cdots & a_{MN} \end{pmatrix} \begin{pmatrix} b_{11} & b_{12} & \cdots & b_{1K} \\ b_{21} & b_{22} & \cdots & b_{2K} \\ \vdots & \vdots & & \vdots \\ b_{N1} & b_{N2} & \cdots & b_{NK} \end{pmatrix}$$

$$= \begin{pmatrix} \sum a_{1i} b_{i1} & \sum a_{1i} b_{i2} & \cdots & \sum a_{1i} b_{iK} \\ \sum a_{2i} b_{i1} & \sum a_{2i} b_{i2} & \cdots & \sum a_{2i} b_{iK} \\ \vdots & \vdots & & \vdots \\ \sum a_{Mi} b_{i1} & \sum a_{Mi} b_{i2} & \cdots & \sum a_{Mi} b_{iK} \end{pmatrix}, \tag{2.28}$$

where $\mathbf{AB} \in \mathbb{R}^{M \times K}$, and all sums run over $[1, N]$, i.e., $\sum$ is shorthand for $\sum_{i=1}^{N}$.

If we denote **A**'s ith row by $\mathbf{a}_i^T$ and **B**'s jth column by $\mathbf{b}_j$ and take advantage of the summation implied by the inner product definition, **AB** can also be written more succinctly as

$$\mathbf{AB} = \begin{pmatrix} \mathbf{a}_1^T \mathbf{b}_1 & \mathbf{a}_1^T \mathbf{b}_2 & \cdots & \mathbf{a}_1^T \mathbf{b}_K \\ \mathbf{a}_2^T \mathbf{b}_1 & \mathbf{a}_2^T \mathbf{b}_2 & \cdots & \mathbf{a}_2^T \mathbf{b}_K \\ \vdots & \vdots & & \vdots \\ \mathbf{a}_M^T \mathbf{b}_1 & \mathbf{a}_M^T \mathbf{b}_2 & \cdots & \mathbf{a}_M^T \mathbf{b}_K \end{pmatrix}. \tag{2.29}$$

To check whether a given matrix product is possible, multiply the dimensions: if **AB** is possible, its dimensions will be $(M \times N)(N \times K) \sim (M \times \not{N})(\not{N} \times K) \sim M \times K$, where "$\sim$" means loosely "goes dimensionally as," and the crossing means that the matching inner dimension (N in this case) is annihilated by the permitted multiplication (or, put differently, N is the number of terms summed when evaluating the inner product of **A**'s ith row and **B**'s jth column to obtain **AB**'s element ij). When there is no cancellation, as in $\mathbf{CD} \sim (M \times N)(J \times K)$, $J \neq N$, the operation is not permitted and **CD** does not exist.

In general, matrix products do not commute; $\mathbf{AB} \neq \mathbf{BA}$. One or both of these may not even be permitted because of failure to meet the requirement for a common inner dimension. For this reason, we must distinguish post- from premultiplication: in $\mathbf{AB}$, $\mathbf{A}$ premultiplies $\mathbf{B}$ and $\mathbf{B}$ postmultiplies $\mathbf{A}$.

19. *Outer product:* A vector pair $\{\mathbf{a} \in \mathbb{R}^M, \mathbf{b} \in \mathbb{R}^N\}$ can generate

$$\mathbf{C} = \mathbf{ab}^T = \begin{pmatrix} a_1 \\ a_2 \\ \vdots \\ a_M \end{pmatrix} \begin{pmatrix} b_1 & b_2 & \cdots & b_N \end{pmatrix} \in \mathbb{R}^{M \times N}, \tag{2.30}$$

where $\mathbf{ab}^T$ is the outer product of $\mathbf{a}$ and $\mathbf{b}$. (A more formal and general notation is $\mathbf{C} = \mathbf{a} \otimes \mathbf{b}$. However, in the context of most practical data analyses, $\mathbf{a} \otimes \mathbf{b}$ and $\mathbf{ab}^T$ are interchangeable.) Expanded, the outer product is

$$\mathbf{C} = \begin{pmatrix} a_1 b_1 & a_1 b_2 & \cdots & a_1 b_N \\ a_2 b_1 & a_2 b_2 & \cdots & a_2 b_N \\ \vdots & \vdots & & \vdots \\ a_M b_1 & a_M b_2 & \cdots & a_M b_N \end{pmatrix} \in \mathbb{R}^{M \times N}, \qquad c_{ij} = a_i b_j \tag{2.31}$$

a degenerate form of eq. 2.28. (The above $\mathbf{C}$ matrix can only be rank 1 because it is the outer product of a single vector pair. More on rank later.)

20. *Matrix outer product:* By extension of the above with $\mathbf{a}_i \in \mathbb{R}^M$ and $\mathbf{b}_i \in \mathbb{R}^N$ denoting the ith columns of $\mathbf{A} \in \mathbb{R}^{M \times J}$ and $\mathbf{B} \in \mathbb{R}^{N \times J}$,

$$\mathbf{C} = \mathbf{AB}^T = \begin{pmatrix} \vdots & \vdots & & \vdots \\ \mathbf{a}_1 & \mathbf{a}_2 & \cdots & \mathbf{a}_J \\ \vdots & \vdots & & \vdots \end{pmatrix} \begin{pmatrix} \cdots & \mathbf{b}_1^T & \cdots \\ \cdots & \mathbf{b}_2^T & \cdots \\ & \vdots & \\ \cdots & \mathbf{b}_J^T & \cdots \end{pmatrix} \tag{2.32}$$

$$= \begin{pmatrix} \sum a_{1j} b_{1j} & \sum a_{1j} b_{2j} & \cdots & \sum a_{1j} b_{Nj} \\ \sum a_{2j} b_{1j} & \sum a_{2j} b_{2j} & \cdots & \sum a_{2j} b_{Nj} \\ \vdots & \vdots & & \vdots \\ \sum a_{Mj} b_{1j} & \sum a_{Mj} b_{2j} & \cdots & \sum a_{Mj} b_{Nj} \end{pmatrix} \in \mathbb{R}^{M \times N},$$

where the summation is carried out along the annihilated inner dimension, i.e., $\sum \equiv \sum_{j=1}^{J}$. Because the same summation is applied to each term, it can be applied to the whole matrix rather than to individual elements. That is, $\mathbf{C}$ can also be expressed as the J element series of $M \times N$ rank 1 matrices

$$\mathbf{C} = \mathbf{A}\mathbf{B}^T = \sum_{j=1}^{J} \begin{pmatrix} a_{1j}b_{1j} & a_{1j}b_{2j} & \cdots & a_{1j}b_{Nj} \\ a_{2j}b_{1j} & a_{2j}b_{2j} & \cdots & a_{2j}b_{Nj} \\ \vdots & \vdots & & \vdots \\ a_{Mj}b_{1j} & a_{Mj}b_{2j} & \cdots & a_{Mj}b_{Nj} \end{pmatrix}. \tag{2.33}$$

It may not be obvious at first, but the jth element of this series is $\mathbf{a}_j\mathbf{b}_j^T$. To show this, recall that

$$\mathbf{a}_j = \begin{pmatrix} a_{1j} \\ a_{2j} \\ \vdots \\ a_{Mj} \end{pmatrix} \in \mathbb{R}^{M\times 1} \tag{2.34}$$

and

$$\mathbf{b}_j^T = \begin{pmatrix} b_{1j} & b_{2j} & \cdots & b_{Nj} \end{pmatrix} \in \mathbb{R}^{1\times N}, \tag{2.35}$$

so that

$$\mathbf{a}_j\mathbf{b}_j^T = \begin{pmatrix} a_{1j}b_{1j} & a_{1j}b_{2j} & \cdots & a_{1j}b_{Nj} \\ a_{2j}b_{1j} & a_{2j}b_{2j} & \cdots & a_{2j}b_{Nj} \\ \vdots & \vdots & & \vdots \\ a_{Mj}b_{1j} & a_{Mj}b_{2j} & \cdots & a_{Mj}b_{Nj} \end{pmatrix} \in \mathbb{R}^{M\times N} \tag{2.36}$$

the jth element of the series in eq. 2.33. That is,

$$\mathbf{C} = \mathbf{A}\mathbf{B}^T = \sum_{j=1}^{N} \mathbf{a}_j\mathbf{b}_j^T. \tag{2.37}$$

Because some terms in this sum can be mutually redundant, **C**'s rank need not be full.

THREE

Matrix Properties, Fundamental Spaces, Orthogonality

3.1 Vector Spaces

3.1.1 Introduction

For our purposes, it is sufficient to think of a vector space as the set of all vectors of a certain type. While the vectors need not be actual vectors (they can also be functions, matrices, etc.), in this book "vectors" are literally column vectors of real number elements, which means we consider vector spaces over $\mathbb{R}$.

The lowest dimensional vector space is $\mathbb{R}^0$, comprising a single point, 0; not too interesting. In $\mathbb{R}$, the real line, one and only one kind of inhabitant is found: 1-vectors (scalars) whose single element is any one of the real numbers from $-\infty$ to ∞. The numerical value of $v \in \mathbb{R}$ ("v which is an element of R-one") is the distance along the real line from the origin (0, not boldfaced because it is a scalar) to v. Note that the rigid distinction between scalars and vectors, while traditional in physics, is not really warranted because $\mathbb{R}$ contains vectors, just like any other $\mathbb{R}^N$, but they all point in a single direction, the one stretching from $-\infty$ to ∞.

Next up is the familiar geometrical plane, or $\mathbb{R}^2$ (fig. 3.1), home to all 2-vectors. Each 2-vector $(\, x \; y \,)^T$ connects the origin $(0, 0)$ and the point (x, y) on the plane. Thus, the two elements are the projections of the vector on the two coordinates (the dashed projections in fig. 3.1). Likewise, $\mathbb{R}^3$, the three-dimensional Euclidean space in which our everyday life unfolds, is home to 3-vectors $\mathbf{v} = (\, v_1 \; v_2 \; v_3)^T$ stretched in three-dimensional space between the origin $(0, 0, 0)$ and (v_1, v_2, v_3). While $\mathbb{R}^{N \geq 4}$ may be harder to visualize, such vector spaces are direct generalizations of the more intuitive $\mathbb{R}^2$ or $\mathbb{R}^3$.

Vector spaces follow a few rules. Multiplication by a scalar and vector addition are defined, yielding vectors in the same space: with $\alpha \in \mathbb{R}$, $\mathbf{u} \in \mathbb{R}^N$ and $\mathbf{v} \in \mathbb{R}^N$, $\alpha\mathbf{u} \in \mathbb{R}^N$ and $(\mathbf{u}+\mathbf{v}) \in \mathbb{R}^N$ are defined. Addition is commutative $(\mathbf{u}+\mathbf{v} = \mathbf{v}+\mathbf{u})$ and associative $(\mathbf{u}+(\mathbf{v}+\mathbf{w}) = \mathbf{w}+(\mathbf{u}+\mathbf{v}) = \mathbf{v}+(\mathbf{u}+\mathbf{w})$ or any other permutation of $\mathbf{u}$, $\mathbf{v}$, and $\mathbf{w}$). There exists a zero-vector $\mathbf{0}$ satisfying $\mathbf{v} + \mathbf{0} = \mathbf{v}$, and vectors and their negative counterparts ("additive inverses"; unlike scalars, vectors do not have multiplicative inverse, so $1/\mathbf{u}$ is meaningless) satisfy $\mathbf{v}+(-\mathbf{v}) = \mathbf{0}$. Multiplication by a scalar is distributive, $\alpha(\mathbf{u} + \mathbf{v}) = \alpha\mathbf{u} + \alpha\mathbf{v}$ and $(\alpha + \beta)\mathbf{u} = \alpha\mathbf{u} + \beta\mathbf{u}$, and satisfies $\alpha(\beta\mathbf{u}) = (\alpha\beta)\mathbf{u} = \alpha\beta\mathbf{u}$. Additional vector space rules and axioms, more general but less germane to data analysis, can be found in most linear algebra texts.

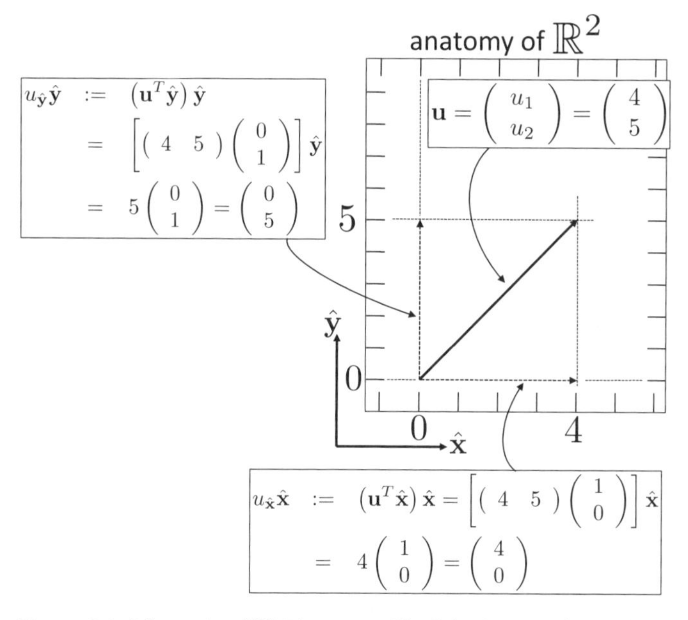

Figure 3.1. Schematic of $\mathbb{R}^2$. The vector (thick line) is an arbitrarily chosen $\mathbf{u} = (\,4\;5\,)^T \in \mathbb{R}^2$. The vector components of $\mathbf{u}$ in the direction of $\hat{\mathbf{x}}$ and $\hat{\mathbf{y}}$, with (scalar) magnitudes given by $\mathbf{u}^T\hat{\mathbf{x}}$ and $\mathbf{u}^T\hat{\mathbf{y}}$, are shown by the dashed horizontal and vertical lines, respectively.

3.1.2 Normed Inner-Product Vector Spaces

Throughout this book we will treat $\mathbb{R}^N$ as a normed inner-product vector space, i.e., one in which both the norm and the inner product, introduced in chapter 2, are well defined.

3.1.3 Vector Space Spanning

An N-dimensional vector space is minimally spanned by a particular (non-unique) choice of N linearly independent $\mathbb{R}^N$ vectors in terms of which each $\mathbb{R}^N$ vector can be uniquely expressed. Once the choice of these N vectors is made, the vectors are collectively referred to as a "basis" for $\mathbb{R}^N$, and each one of them is a basis vector. The term "spanning" refers to the property that because of their linear independence, the basis vectors can express—or span—any arbitrary $\mathbb{R}^N$ vector. Pictorially, spanning is explained in fig. 3.2. Imagine a (semi-transparent gray) curtain suspended from a telescopic rod attached to a wall (left thick vertical black line). When the rod is retracted (left panel), the curtain collapses to a vertical line, and is thus a one-dimensional object. When the rod is extended

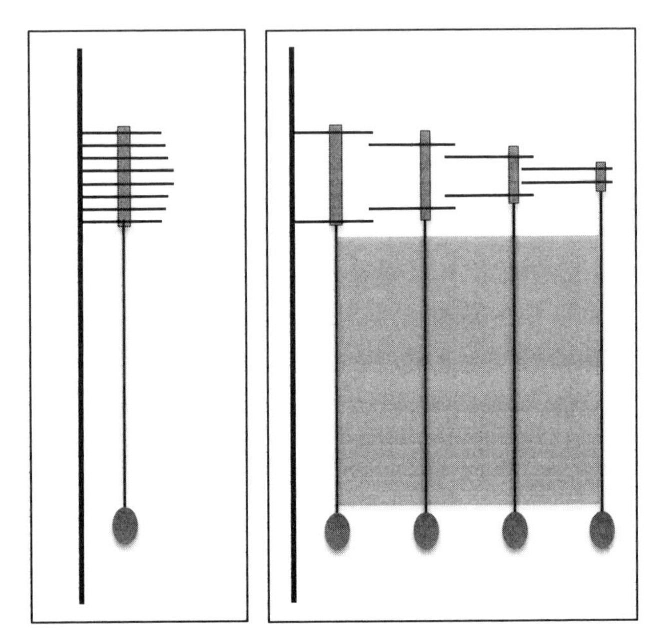

Figure 3.2. Schematic explanation of vector space spanning by the basis set, discussed in the text.

(right panel), it spans the curtain, which therefore becomes two dimensional. In the former (left panel) case, gravity is the spanning force, and—since it operates in the up–down direction—the curtain's only relevant dimension is its height, the length along the direction of gravity. In the extended case (right panel), gravity is joined by the rod, which extends, or spans, the curtain sideways. Now the curtain has two relevant dimensions, along gravity and along the rod. These two thus form a spanning set, a basis, for the two-dimensional curtain.

Let us consider some examples. For spanning $\mathbb{R}^3$, the Cartesian basis set

$$\left\{ \hat{\mathbf{i}} = \begin{pmatrix} 1 \\ 0 \\ 0 \end{pmatrix}, \hat{\mathbf{j}} = \begin{pmatrix} 0 \\ 1 \\ 0 \end{pmatrix}, \hat{\mathbf{k}} = \begin{pmatrix} 0 \\ 0 \\ 1 \end{pmatrix} \right\} \tag{3.1}$$

(sometimes denoted $\{\hat{\mathbf{x}}, \hat{\mathbf{y}}, \hat{\mathbf{z}}\}$) is often chosen. This set is suitable for spanning $\mathbb{R}^3$ because any $\mathbb{R}^3$ vector can be expressed as a linear combination of $\{\hat{\mathbf{i}}, \hat{\mathbf{j}}, \hat{\mathbf{k}}\}$:

$$\begin{aligned} \mathbf{v} = \begin{pmatrix} v_1 \\ v_2 \\ v_3 \end{pmatrix} &= v_1 \begin{pmatrix} 1 \\ 0 \\ 0 \end{pmatrix} + v_2 \begin{pmatrix} 0 \\ 1 \\ 0 \end{pmatrix} + v_3 \begin{pmatrix} 0 \\ 0 \\ 1 \end{pmatrix} \\ &= v_1 \hat{\mathbf{i}} + v_2 \hat{\mathbf{j}} + v_3 \hat{\mathbf{k}}. \end{aligned} \tag{3.2}$$

Note, again, that this is not a unique choice for spanning $\mathbb{R}^3$; there are infinitely many such choices. The only constraint on the choice, again, is that to span $\mathbb{R}^3$, the 3 vectors must be linearly independent, that is, that no nontrivial $\{\alpha, \beta, \gamma\}$ satisfying $\alpha\hat{\mathbf{i}} + \beta\hat{\mathbf{j}} + \gamma\hat{\mathbf{k}} = \mathbf{0} \in \mathbb{R}^3$ can be found.

The requirement for mutual linear independence of the basis vectors follows from the fact that a 3-vector has 3 independent pieces of information, v_1, v_2, and v_3. Given these 3 degrees of freedom (three independent choices in making up $\mathbf{v}$; much more on that later in the book), we must have 3 corresponding basis vectors with which to work. If one of the basis vectors is a linear combination of other ones, e.g., if $\hat{\mathbf{j}} = \alpha\hat{\mathbf{k}}$ say, then $\hat{\mathbf{j}}$ and $\hat{\mathbf{k}}$ no longer represent two directions in $\mathbb{R}^3$, but just one. To show how this happens, consider the choice

$$\left\{\hat{\mathbf{i}} = \begin{pmatrix} 1 \\ 0 \\ 0 \end{pmatrix}, \hat{\mathbf{j}} = \begin{pmatrix} 0 \\ 1 \\ 0 \end{pmatrix}, \mathbf{k} = \begin{pmatrix} 2 \\ 3 \\ 0 \end{pmatrix} = 2\hat{\mathbf{i}} + 3\hat{\mathbf{j}}\right\}, \tag{3.3}$$

which cannot represent $(\ 0\ 0\ v_3\ )^T$. Thus, this choice of a basis doesn't span $\mathbb{R}^3$ (while it does span $\mathbb{R}^2 \subset \mathbb{R}^3$, just as well as $\{\hat{\mathbf{i}}, \hat{\mathbf{j}}\}$ alone, for 3 vectors to span $\mathbb{R}^2$, a two-dimensional subspace of $\mathbb{R}^3$, is not very impressive). To add a third basis vector that will complete the spanning of $\mathbb{R}^3$, we need a vector not contained in any z = constant plane. Fully contained within the $z = 0$ plane already successfully spanned by the previous two basis vectors, $(\ 2\ 3\ 0\ )^T$ doesn't help.

Note that the above failure to span $\mathbb{R}^3$ is *not* because none of our basis vectors has a nonzero third element; try finding $\{\alpha, \beta, \gamma\}$ satisfying

$$\begin{pmatrix} v_1 \\ v_2 \\ v_3 \end{pmatrix} = \alpha \begin{pmatrix} 1 \\ 1 \\ 1 \end{pmatrix} + \beta \begin{pmatrix} 1 \\ 0 \\ -1 \end{pmatrix} + \gamma \begin{pmatrix} 2 \\ 1 \\ 0 \end{pmatrix} \tag{3.4}$$

(i.e., consider the $\mathbb{R}^3$ spanning potential of the above three $\mathbb{R}^3$ vectors). The second and third rows give

$$v_2 = \alpha + \gamma \ \Rightarrow\ \gamma = v_2 - \alpha \quad \text{and} \quad v_3 = \alpha - \beta \ \Rightarrow\ \beta = \alpha - v_3,$$

so the first row becomes

$$v_1 = \alpha + \beta + 2\gamma = \alpha + \alpha - v_3 + 2v_2 - 2\alpha = 2v_2 - v_3.$$

Thus, the considered set can span the subset of $\mathbb{R}^3$ vectors of the general form $(\ 2v_2 - v_3\ v_2\ v_3\ )^T$, but not arbitrary ones (for which $v_1 \neq 2v_2 - v_3$). This is because

$$\begin{pmatrix} 2 \\ 1 \\ 0 \end{pmatrix} = \begin{pmatrix} 1 \\ 1 \\ 1 \end{pmatrix} + \begin{pmatrix} 1 \\ 0 \\ -1 \end{pmatrix}, \tag{3.5}$$

i.e., the third spanning vector in this deficient spanning set, the sum of the earlier two, fails to add a third dimension required for fully spanning $\mathbb{R}^3$.

To better understand the need for linear independence of basis vectors, it is useful to visualize the geometry of the problem. Consider

$$\left\{ \mathbf{i} = \begin{pmatrix} 1 \\ 1 \\ 1 \end{pmatrix}, \mathbf{j} = \begin{pmatrix} 1 \\ 0 \\ -1 \end{pmatrix}, \mathbf{k} = \begin{pmatrix} 2 \\ 1 \\ 0 \end{pmatrix} = \mathbf{i} + \mathbf{j} \right\}, \tag{3.6}$$

which fail to span $\mathbb{R}^3$, because $\mathbf{k}$ is linearly dependent on $\mathbf{i}$ and $\mathbf{j}$. What does this failure look like? While this more interesting and general situation is not obvious to visualize—the redundancy occurs in a plane parallel to neither of $(1\ 0\ 0)^T$, $(0\ 1\ 0)^T$, or $(0\ 0\ 1)^T$ but inclined with respect to all of them—visualization may be facilitated by fig. 3.3. (We will learn later how to transform the coordinates so that the redundant plane becomes a fixed value of one coordinate, which we can then eliminate from the problem, thus reducing the apparent dimensionality, 3, to the actual dimensionality, 2.)

Now let's go back to the easier to visualize vectors $\hat{\mathbf{i}} = (1\,0\,0)^T$, $\hat{\mathbf{j}} = (0\,1\,0)^T$. We have realized above that to assist in spanning $\mathbb{R}^3$, the additional basis vector $\hat{\mathbf{k}} = (\hat{k}_1\ \hat{k}_2\ \hat{k}_3)^T$ must not be fully contained within any z = constant plane. To meet this criterion,

$$\hat{\mathbf{k}} - (\hat{\mathbf{k}}^T\hat{\mathbf{i}})\hat{\mathbf{i}} - (\hat{\mathbf{k}}^T\hat{\mathbf{j}})\hat{\mathbf{j}} \neq 0 \in \mathbb{R}^3, \tag{3.7}$$

i.e., $\hat{\mathbf{k}}$ must have a nonzero remainder after subtracting its projections on $\hat{\mathbf{i}}$ and $\hat{\mathbf{j}}$. Because $\hat{\mathbf{k}}^T\hat{\mathbf{i}} = \hat{k}_1$ and $\hat{\mathbf{k}}^T\hat{\mathbf{j}} = \hat{k}_2$, this requirement reduces to

$$\hat{\mathbf{k}} - \hat{k}_1\hat{\mathbf{i}} - \hat{k}_2\hat{\mathbf{j}} = \begin{pmatrix} \hat{k}_1 \\ \hat{k}_2 \\ \hat{k}_3 \end{pmatrix} - \begin{pmatrix} \hat{k}_1 \\ 0 \\ 0 \end{pmatrix} - \begin{pmatrix} 0 \\ \hat{k}_2 \\ 0 \end{pmatrix} = \begin{pmatrix} 0 \\ 0 \\ \hat{k}_3 \end{pmatrix}, \tag{3.8}$$

which can vanish only when $\hat{k}_3 = 0$. Thus, any $\hat{\mathbf{k}}$ with nonzero $\hat{k}_3$ will complement $(1\,0\,0)^T$ and $(0\,1\,0)^T$ in spanning $\mathbb{R}^3$.

However, we are still left with a choice of exactly which $\mathbf{k}$ among all those satisfying $\hat{k}_3 \neq 0$ we choose; we can equally well add $(1\,1\,1)^T$, $(0\,0\,1)^T$, $(1\,1\,-4)^T$, etc. Given this indeterminacy, the choice is ours; any one of these vectors will do just fine. It is often useful, but not algebraically essential, to choose mutually orthogonal basis vectors so that the information contained in one is entirely absent from the others. With $(1\,0\,0)^T$ and $(0\,1\,0)^T$ already chosen, the vector orthogonal to both must satisfy

$$\begin{pmatrix} 1 & 0 & 0 \end{pmatrix} \begin{pmatrix} \hat{k}_1 \\ \hat{k}_2 \\ \hat{k}_3 \end{pmatrix} = \begin{pmatrix} 0 & 1 & 0 \end{pmatrix} \begin{pmatrix} \hat{k}_1 \\ \hat{k}_2 \\ \hat{k}_3 \end{pmatrix} = 0,$$

which can hold only if $\hat{k}_1 = \hat{k}_2 = 0$. When these conditions are met, *any* $\hat{k}_3$ will satisfy the orthogonality conditions.

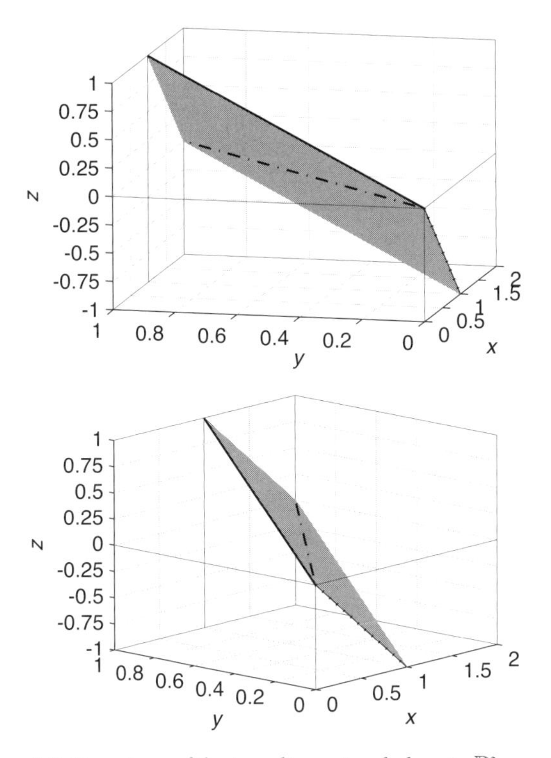

Figure 3.3. Two views of the two-dimensional plane in $\mathbb{R}^3$ spanned by the spanning set of eq. 3.6. Solid thick and thin solid-dotted lines show $(\,1\ 1\ 1\,)^T$ and $(\,1\ 0\ -1\,)^T$, while their linear combination, $(\,2\ 1\ 0\,)^T$, given by the dash-dotted line, is clearly contained in the plane they span.

The arbitrariness of $\hat{k}_3$ can be alleviated by the (customary but not essential) custom of choosing unit norm basis vectors. Employing the L^2 norm, $\hat{k}_1 = \hat{k}_2 = 0$ and $\mathbf{k} = \hat{\mathbf{k}}$ can only mean

$$\begin{pmatrix} 0 & 0 & \hat{k}_3 \end{pmatrix} \begin{pmatrix} 0 \\ 0 \\ \hat{k}_3 \end{pmatrix} = 1 \;\Rightarrow\; \hat{k}_3 = 1 \;\Rightarrow\; \hat{\mathbf{k}} = \begin{pmatrix} 0 & 0 & 1 \end{pmatrix}^T.$$

3.1.4 Subspaces

Vector spaces have subspaces. If $\mathcal{U}$ is a real, finite-dimensional vector space (such as, but not restricted to, the class defined in section 3.1.2), then $\mathcal{V}$ is a subspace of $\mathcal{U}$ if $\mathcal{V} \subset \mathcal{U}$ ($\mathcal{V}$ is a subset of $\mathcal{U}$), for any two $\mathbf{v}_1, \mathbf{v}_2 \in \mathcal{V}$, $\mathbf{v}_1 + \mathbf{v}_2 \in \mathcal{V}$ (the sum of any two vectors from $\mathcal{V}$ is still in $\mathcal{V}$), and for any $\alpha \in \mathbb{R}$, $\alpha\mathbf{v}_i \in \mathcal{V}$ for any i (the product of any vector from $\mathcal{V}$ and a real scalar is still in $\mathcal{V}$).

3.2 Matrix Rank

There are various ways to define the matrix rank $q_A := \text{rank}(\mathbf{A})$ (also denoted q when there is no risk of ambiguity). The simplest is the number of independent columns in $\mathbf{A} \in \mathbb{R}^{M \times N}$, $q \le \min(M, N)$. A related intuitive geometrical interpretation of the rank is as follows. When an $\mathbf{A} \in \mathbb{R}^{M \times N}$ premultiplies a vector $\mathbf{x} \in \mathbb{R}^N$ to generate a $\mathbf{b} \in \mathbb{R}^M$, $\mathbf{A}$'s rank is the highest number of independent $\mathbb{R}^M$ directions along which $\mathbf{b}$ can lie. This may be initially confusing to some—if $\mathbf{b}$ has M elements, can it not lie along any direction in $\mathbb{R}^M$? Not in this context, because here $\mathbf{b}$ linearly combines $\mathbf{A}$'s columns,

$$\mathbf{b} = x_1 \mathbf{a}_1 + x_2 \mathbf{a}_2 + \cdots$$

(where x_i is $\mathbf{x}$'s ith element and $\mathbf{a}_i$ is $\mathbf{A}$'s ith column), so at most there are N independent $\mathbb{R}^M$ directions along which $\mathbf{b}$ can lie, and it is entirely possible that $N < M$. Because one or several of $\mathbf{A}$'s columns may depend on other columns, there may be less than N such dimensions; in those cases $q < N$, and the dimension of the $\mathbb{R}^M$ subspace $\mathbf{b}$ can occupy is $q < M$.

There are several ways of obtaining q. Arguably, the algebraically simplest is to seek an $\mathbf{x}$ that satisfies $\mathbf{A}\mathbf{x} = \mathbf{0} \in \mathbb{R}^M$. Let's clarify this with some examples that are only subtly different yet give rise to rather different behaviors.

3.2.1 *Example I*

Consider

$$\mathbf{A} = \begin{pmatrix} 1 & 1 & 5 \\ 1 & 2 & 3 \\ 1 & 3 & 2 \\ 1 & 4 & 1 \end{pmatrix}. \tag{3.9}$$

Because $\mathbf{A}$ has only 3 columns, its rank is at most 3. To find q, recall that for $q = 3$, $\mathbf{A}$'s 3 columns must be linearly independent, which requires that no nontrivial $\mathbf{x} = (\,\alpha\;\beta\;\gamma\,)^T$ satisfying

$$\begin{pmatrix} 1 & 1 & 5 \\ 1 & 2 & 3 \\ 1 & 3 & 2 \\ 1 & 4 & 1 \end{pmatrix} \begin{pmatrix} \alpha \\ \beta \\ \gamma \end{pmatrix} = \begin{pmatrix} 0 \\ 0 \\ 0 \\ 0 \end{pmatrix} = \mathbf{0} \tag{3.10}$$

can be found. A reasonable way to proceed is Gaussian elimination. If this is your first encounter with this procedure, you will probably want to consult a more thorough treatment than the following in most any linear algebra text. The essence of Gaussian elimination, however, is as as follows. We operate on $\mathbf{A}$ with a sequence of so-called elementary operations (adding to various rows multiples of other rows), to reduce $\mathbf{A}$'s elements below the main diagonal to zero. While not essential in this context, it is useful to carry out the elementary operations by constructing matrices $\mathbf{E}_i$ that execute the operations upon premultiplying $\mathbf{A}$ or, at later stages, products of earlier $\mathbf{E}_i$s and $\mathbf{A}$.

To reduce the above $\mathbf{A}$ to upper diagonal form, we subtract the first row from rows 2–4, so that their left-most elements will vanish,

$$\mathbf{E}_1\mathbf{A} = \begin{pmatrix} 1 & 0 & 0 & 0 \\ -1 & 1 & 0 & 0 \\ -1 & 0 & 1 & 0 \\ -1 & 0 & 0 & 1 \end{pmatrix} \begin{pmatrix} 1 & 1 & 5 \\ 1 & 2 & 3 \\ 1 & 3 & 2 \\ 1 & 4 & 1 \end{pmatrix} = \begin{pmatrix} 1 & 1 & 5 \\ 0 & 1 & -2 \\ 0 & 2 & -3 \\ 0 & 3 & -4 \end{pmatrix}. \tag{3.11}$$

Next, we subtract twice and three times row 2 from rows 3 and 4, respectively,

$$\mathbf{E}_2\mathbf{E}_1\mathbf{A} = \begin{pmatrix} 1 & 0 & 0 & 0 \\ 0 & 1 & 0 & 0 \\ 0 & -2 & 1 & 0 \\ 0 & -3 & 0 & 1 \end{pmatrix} \begin{pmatrix} 1 & 1 & 5 \\ 0 & 1 & -2 \\ 0 & 2 & -3 \\ 0 & 3 & -4 \end{pmatrix} = \begin{pmatrix} 1 & 1 & 5 \\ 0 & 1 & -2 \\ 0 & 0 & 1 \\ 0 & 0 & 2 \end{pmatrix}. \tag{3.12}$$

Finally, we subtract twice row 3 from row 4,

$$\begin{aligned} \mathbf{E}_3\mathbf{E}_2\mathbf{E}_1\mathbf{A} &= \begin{pmatrix} 1 & 0 & 0 & 0 \\ 0 & 1 & 0 & 0 \\ 0 & 0 & 1 & 0 \\ 0 & 0 & -2 & 1 \end{pmatrix} \begin{pmatrix} 1 & 1 & 5 \\ 0 & 1 & -2 \\ 0 & 0 & 1 \\ 0 & 0 & 2 \end{pmatrix} \\ &= \begin{pmatrix} \boxed{1} & 1 & 5 \\ 0 & \boxed{1} & -2 \\ 0 & 0 & \boxed{1} \\ 0 & 0 & 0 \end{pmatrix} \equiv \mathbf{U}, \end{aligned} \tag{3.13}$$

where $\mathbf{U}$ is $\mathbf{A}$'s upper diagonal counterpart, whose emergence signals the conclusion of the Gaussian elimination.

The nonzero diagonal elements of $\mathbf{U}$, boxed in eq. 3.13, are called pivots, of which we have three in this case. The number of nonzero pivots in $\mathbf{U}$ is the rank of $\mathbf{A}$ from which $\mathbf{U}$ was derived. Thus, in this case, $q = 3 = \min(M, N)$; $\mathbf{A}$ is full rank.

While we achieved our original objective, obtaining q, there is more to learn about the rank from this example. So let's continue our exploration, recalling that our overall goal is to find an $\mathbf{x}$ satisfying $\mathbf{Ax} = \mathbf{0}$. Note that whereas premultiplying by $\mathbf{E}_3\mathbf{E}_2\mathbf{E}_1$ transforms $\mathbf{A}$ to $\mathbf{U}$ on the left-hand side, it does nothing to the right-hand side of the equation, because any $M \times N$ matrix premultiplying the zero N-vector will yield the zero M-vector. Thus, $\mathbf{Ax} = \mathbf{0}$ is solved by solving $\mathbf{Ux} = \mathbf{0}$,

$$\begin{pmatrix} 1 & 1 & 5 \\ 0 & 1 & -2 \\ 0 & 0 & 1 \\ 0 & 0 & 0 \end{pmatrix} \begin{pmatrix} \alpha \\ \beta \\ \gamma \end{pmatrix} = \begin{pmatrix} 0 \\ 0 \\ 0 \\ 0 \end{pmatrix}, \tag{3.14}$$

which is the point and great utility of the Gaussian elimination procedure. This is solved by back-substitution, starting from **U**'s lowermost nonzero row, the third, which reads $\gamma = 0$. The next row up, the second, states that $\beta - 2\gamma = 0$ or $\beta = 2\gamma = 0$. Finally, with $\beta = \gamma = 0$, the first row reads $\alpha = 0$.

Thus, the only **x** that satisfies $\mathbf{Ax} = \mathbf{0}$ is the trivial one, $\alpha = \beta = \gamma = 0$. This indicates that **A**'s columns are linearly independent, and, since there are 3 of them, that $q = 3$.

3.2.2 Example II

Next, consider

$$\mathbf{A} = \begin{pmatrix} 1 & 1 & 4 \\ 1 & 2 & 3 \\ 1 & 3 & 2 \\ 1 & 4 & 1 \end{pmatrix}, \tag{3.15}$$

only a slight change from example I. To find this **A**'s rank q, we again seek an $\mathbf{x} = (\ \alpha\ \beta\ \gamma\ )^T$ satisfying

$$\begin{pmatrix} 1 & 1 & 4 \\ 1 & 2 & 3 \\ 1 & 3 & 2 \\ 1 & 4 & 1 \end{pmatrix} \begin{pmatrix} \alpha \\ \beta \\ \gamma \end{pmatrix} = \begin{pmatrix} 0 \\ 0 \\ 0 \\ 0 \end{pmatrix} = \mathbf{0} \tag{3.16}$$

employing Gaussian elimination.

The first step is as before,

$$\mathbf{E}_1\mathbf{A} = \begin{pmatrix} 1 & 0 & 0 & 0 \\ -1 & 1 & 0 & 0 \\ -1 & 0 & 1 & 0 \\ -1 & 0 & 0 & 1 \end{pmatrix} \begin{pmatrix} 1 & 1 & 4 \\ 1 & 2 & 3 \\ 1 & 3 & 2 \\ 1 & 4 & 1 \end{pmatrix} = \begin{pmatrix} 1 & 1 & 4 \\ 0 & 1 & -1 \\ 0 & 2 & -2 \\ 0 & 3 & -3 \end{pmatrix}, \tag{3.17}$$

as is the next,

$$\mathbf{E}_2 = \begin{pmatrix} 1 & 0 & 0 & 0 \\ 0 & 1 & 0 & 0 \\ 0 & -2 & 1 & 0 \\ 0 & -3 & 0 & 1 \end{pmatrix}, \quad \mathbf{E}_2\mathbf{E}_1\mathbf{A} = \begin{pmatrix} \boxed{1} & 1 & 4 \\ 0 & \boxed{1} & -1 \\ 0 & 0 & 0 \\ 0 & 0 & 0 \end{pmatrix}, \tag{3.18}$$

but now this concludes **A**'s reduction to upper diagonal form, with only two nonzero pivots. Thus, this **A**'s rank is $q = 2 < \min(M, N)$; it is rank deficient.

To use this reduction to solve $\mathbf{Ax} = \mathbf{0}$, recall that $\mathbf{E}_2\mathbf{E}_1\mathbf{0} = \mathbf{0}$, so the question becomes

$$\mathbf{Ax} = \mathbf{E}_2\mathbf{E}_1\mathbf{Ax} = \mathbf{0} \tag{3.19}$$

or

$$\begin{pmatrix} 1 & 1 & 4 \\ 0 & 1 & -1 \\ 0 & 0 & 0 \\ 0 & 0 & 0 \end{pmatrix} \begin{pmatrix} \alpha \\ \beta \\ \gamma \end{pmatrix} = \begin{pmatrix} 0 \\ 0 \\ 0 \\ 0 \end{pmatrix}, \tag{3.20}$$

which we again solve using back-substitution. Because the two lowest rows are satisfied identically for any $\mathbf{x}$, back-substitution starts in row 2, $\beta = \gamma$. Using this in the next row up, row 1, yields $\alpha + 5\beta = 0$ or $\beta = -\alpha/5$. Thus, the final form of the solution to $\mathbf{Ax} = \mathbf{0}$ is $\mathbf{x} = (\ \alpha \ -\alpha/5 \ -\alpha/5\ )^T = (-\alpha/5)(\ -5\ 1\ 1\ )^T$. Importantly, α is entirely unconstrained, so

$$\mathbf{x} = \xi \begin{pmatrix} -5 \\ 1 \\ 1 \end{pmatrix}, \tag{3.21}$$

with any ξ, will satisfy $\mathbf{Ax} = \mathbf{0}$, because

$$\mathbf{Ax} = \begin{pmatrix} 1 & 1 & 4 \\ 1 & 2 & 3 \\ 1 & 3 & 2 \\ 1 & 4 & 1 \end{pmatrix} \left[\xi \begin{pmatrix} -5 \\ 1 \\ 1 \end{pmatrix} \right] \tag{3.22}$$

$$= \xi \begin{pmatrix} 1 & 1 & 4 \\ 1 & 2 & 3 \\ 1 & 3 & 2 \\ 1 & 4 & 1 \end{pmatrix} \begin{pmatrix} -5 \\ 1 \\ 1 \end{pmatrix} = \xi \begin{pmatrix} 0 \\ 0 \\ 0 \\ 0 \end{pmatrix} = \begin{pmatrix} 0 \\ 0 \\ 0 \\ 0 \end{pmatrix}.$$

There is one direction in $\mathbb{R}^3$ along which any $\mathbf{x} = \xi(\ -5\ 1\ 1\ )^T$ yields, upon premultiplication by $\mathbf{Ax}$, $(\ 0\ 0\ 0\ 0\ )^T$. This means that nontrivial $\mathbf{x}$ can only exist in a residual two-dimensional subspace (see section 3.1.4) of $\mathbb{R}^3$ (spanned by any two linearly independent $\mathbb{R}^3$ vectors $\{\hat{\mathbf{a}}, \hat{\mathbf{b}}\}$ orthogonal to $(\ -5\ 1\ 1\ )^T$), which is another way to view this $\mathbf{A}$'s rank deficiency, $q = \min(M, N) - 1 = 3 - 1 = 2$ (this is a specific example of the *rank–nullity theorem* introduced in section 3.3.1).

Finally, note that the rank deficiency of this $\mathbf{A}$, the fact that its columns are not all linearly independent, stems from its first column being one-fifth the sum of the other two columns:

$$\frac{1}{5}\left[\begin{pmatrix} 4 \\ 3 \\ 2 \\ 1 \end{pmatrix} + \begin{pmatrix} 1 \\ 2 \\ 3 \\ 4 \end{pmatrix} \right] = \begin{pmatrix} 1 \\ 1 \\ 1 \\ 1 \end{pmatrix},$$

which is why

$$\mathbf{A}\begin{pmatrix} -5 \\ 1 \\ 1 \end{pmatrix} = \begin{pmatrix} 1 & 1 & 4 \\ 1 & 2 & 3 \\ 1 & 3 & 2 \\ 1 & 4 & 1 \end{pmatrix}\begin{pmatrix} -5 \\ 1 \\ 1 \end{pmatrix} = \begin{pmatrix} 0 \\ 0 \\ 0 \\ 0 \end{pmatrix} = \mathbf{0}.$$

3.2.3 Example III

As a final example, consider

$$\mathbf{A} = \begin{pmatrix} 1 & -2 & 4 \\ 1 & -2 & 4 \\ 1 & -2 & 4 \\ 1 & -2 & 4 \end{pmatrix}. \tag{3.23}$$

With

$$\mathbf{E}_1 = \begin{pmatrix} 1 & 0 & 0 & 0 \\ -1 & 1 & 0 & 0 \\ -1 & 0 & 1 & 0 \\ -1 & 0 & 0 & 1 \end{pmatrix}, \quad \mathbf{E}_1\mathbf{A} = \begin{pmatrix} \boxed{1} & -2 & 4 \\ 0 & 0 & 0 \\ 0 & 0 & 0 \\ 0 & 0 & 0 \end{pmatrix}, \tag{3.24}$$

i.e., this **A**'s **U** has only a single nonzero pivot, so its rank is $q = 1$. Back-substitution is equally simple, with the single (top) nontrivial row stating that $\gamma = \beta/2 - \alpha/4$, so that any **x** of the general form

$$\mathbf{x} = \begin{pmatrix} \alpha \\ \beta \\ \frac{\beta}{2} - \frac{\alpha}{4} \end{pmatrix} = \xi\begin{pmatrix} 4 \\ 0 \\ -1 \end{pmatrix} + \phi\begin{pmatrix} 0 \\ 2 \\ 1 \end{pmatrix}, \tag{3.25}$$

for any ξ and ϕ, will produce the required zero vector.

Consistent with the single nonzero pivot in **U** of this problem, there is now a plane in (a two-dimensional subspace of) $\mathbb{R}^3$ from which any **x** yields a zero right-hand side. This leaves a single direction along which $\mathbb{R}^3$ vectors producing nontrivial right-hand sides can reside, which is another way to view the current $q = 1$.

While not essential here, note that to identify this direction, we use the fact that this nontrivial direction, $\mathbf{x}_n = (\, p \; q \; r \,)^T$, must be orthogonal to the trivial plane's spanning vectors. Orthogonality with the first entails $4p = r$, and with the second $2q = -r$, so that $\mathbf{x}_n = p(\, 1 \; -2 \; 4 \,)^T$, with any p, yields a nontrivial right-hand side.

In summary, to find the rank of **A**, use elementary operations to reduce **A** to **U**; the number of nonzero pivots (diagonal elements of **U**) is **A**'s rank.

3.2.4 Row Swap

There are a few practical considerations worth mentioning briefly here that may be present themselves when reducing large numerical matrices to their **U** form. First, a row swap may be needed at some intermediate step. For example, suppose that after obtaining two elementary matrices while reducing an **A** to its **U** form, the matrix

$$\mathbf{E}_2\mathbf{E}_1\mathbf{A} = \begin{pmatrix} 1 & 0 & -1 & 2 \\ 0 & 0 & 0 & 0 \\ 0 & 1 & 2 & 0 \end{pmatrix} \tag{3.26}$$

presents itself. For an automated procedure, it would clearly be better to introduce at this point a

$$\mathbf{P} = \begin{pmatrix} 1 & 0 & 0 \\ 0 & 0 & 1 \\ 0 & 1 & 0 \end{pmatrix} \tag{3.27}$$

with which

$$\mathbf{P}\mathbf{E}_2\mathbf{E}_1\mathbf{A} = \begin{pmatrix} 1 & 0 & -1 & 2 \\ 0 & 1 & 2 & 0 \\ 0 & 0 & 0 & 0 \end{pmatrix}. \tag{3.28}$$

Yet **P** does not naturally belong to the $\mathbf{E}_i$ family because it has a nonzero element above the main diagonal. When such a situation arises in the context of a nonzero right-hand side, the needed **P**s and their location in the sequence must be stored.

In addition, for large rank deficient matrices that produce **U**s with multiple trivial (all zero) bottom rows, back-substitution will be of course more efficient if those rows are removed, but a record must be kept of the removed rows if the need to address a nontrivial right-hand side vector is envisioned.

3.3 Fundamental Spaces Associated with $\mathbf{A} \in \mathbb{R}^{M \times N}$

To expand our understanding of the rank, its meaning, and why it is important, let us introduce a new view of matrices, that of a matrix as a map.

When an $M \times N$ matrix $\mathbf{A} \in \mathbb{R}^{M \times N}$ operates on (premultiplies) a vector (as in $\mathbf{b} = \mathbf{A}\mathbf{x}$, where **A** operates on **x** to yield **b**), **x** must be a column vector of the same length as **A**'s rows; $\mathbf{x} \in \mathbb{R}^N$. But, in general, $M \neq N$, and M is almost never equal to N in data matrices. Consequently, $\mathbf{b} \in \mathbb{R}^M$; the outcome vector **b** is from $\mathbb{R}^M$, not $\mathbb{R}^N$. Thus, we say that **A** maps $\mathbb{R}^N$ to $\mathbb{R}^M$; $\mathbf{x} \in \mathbb{R}^N \overset{\mathbf{A}}{\mapsto} \mathbf{b} \in \mathbb{R}^M$ (read "**x** from R-N is mapped by **A** to **b** from R-M").

The M and N spaces have an important difference; while **A** can operate on any $\mathbb{R}^N$ vector (the set of all possible **x** vectors is always $\mathbb{R}^N$ in its entirety), the set $\{\mathbf{b}:\mathbf{b}=\mathbf{Ax}\}$ (which reads "the set of all vectors **b** such that $\mathbf{b}=\mathbf{Ax}$"; all $\mathbb{R}^M$ vectors that are linear combinations of **A**'s columns) can be $\mathbb{R}^M$, but it can also be only a subspace of $\mathbb{R}^M$ (see section 3.1.4). For example, let's seek an $\mathbf{x}=(\,x_1\;x_2\,)^T$ satisfying $\mathbf{Ax}=\mathbf{b}$ when

$$\mathbf{A}=\begin{pmatrix}1&0\\2&1\\-1&1\end{pmatrix}\quad\text{and}\quad\mathbf{b}=\begin{pmatrix}3\\-1\\1\end{pmatrix}.\tag{3.29}$$

The top row indicates that $x_1=3$, so the middle row is $6+x_2=-1$ or $x_2=-7$. The third row then becomes the impossibility $-3-7=1$, which demonstrates that no **x** can combine **A**'s columns to yield the chosen **b**. Conversely, when $\mathbf{b}=(\,1\;4\;1\,)^T$, the first row yields $x_1=1$, the second row $x_2=2$, and the third has nothing to say other than the obviously true $1=1$, i.e., and unlike the previous example, the third row does not contradict the above two, so that $\mathbf{x}=(\,1\;2\,)^T$ solves the problem, demonstrating that this **b** *is* a linear combination of **A**'s columns.

3.3.1 The Fundamental Theorem of Linear Algebra

The two distinct possibilities above define by example the two complementary $\mathbb{R}^M$ subspaces associated with an $\mathbf{A}\in\mathbb{R}^{M\times N}$:

1. **A***'s column space*, $\mathcal{R}(\mathbf{A})=\{\mathbf{c}\in\mathbb{R}^M:\mathbf{c}=\mathbf{Ax},\ \mathbf{x}\in\mathbb{R}^N\}$, the set of all $\mathbb{R}^M$ vectors that are linear combinations of **A**'s columns (with **x**'s ith element, x_i, being the weight of **A**'s ith column). The dimension of $\mathcal{R}(\mathbf{A})$—also known as **A**'s image, im(**A**), or range, range(**A**)—is the number of independent columns in **A**, **A**'s rank q.
2. **A***'s left null space*, $\mathcal{N}(\mathbf{A}^T)=\{\mathbf{n}_l\in\mathbb{R}^M:\mathbf{A}^T\mathbf{n}_l=\mathbf{0}\in\mathbb{R}^N\}$, the set of all $\mathbb{R}^M$ vectors that $\mathbf{A}^T$ annihilates into $\mathbf{0}\in\mathbb{R}^N$. The left null space is reasonably viewed as $\mathbf{A}^T$'s null space (see below). The dimension of $\mathcal{N}(\mathbf{A}^T)$—also known as **A**'s cokernel, ker($\mathbf{A}^T$), and sometimes denoted null($\mathbf{A}^T$)—is whatever is left of $\mathbb{R}^M$, $M-q$.

By simple analogy and the definition of the transpose, $\mathbb{R}^N$ too can be split into two complementary subspaces:

3. **A***'s row space*, $\mathcal{R}(\mathbf{A}^T)=\{\mathbf{c}_l\in\mathbb{R}^N:\mathbf{c}_l=\mathbf{A}^T\mathbf{y},\ \mathbf{y}\in\mathbb{R}^M\}$, the set of all $\mathbb{R}^N$ vectors that are linear combinations of **A**'s rows. The row space of **A** is $\mathbf{A}^T$'s column space. The dimension of $\mathcal{R}(\mathbf{A}^T)$—also known as the coimage of **A**, im($\mathbf{A}^T$), or range($\mathbf{A}^T$)—is the number of independent rows in **A**, also $q=\text{rank}(\mathbf{A})$.
4. **A***'s null space*, $\mathcal{N}(\mathbf{A})=\{\mathbf{n}\in\mathbb{R}^N:\mathbf{An}=\mathbf{0}\in\mathbb{R}^M\}$, the set of all $\mathbb{R}^N$ vectors that **A** annihilates into $\mathbf{0}\in\mathbb{R}^M$, or the set of all $\mathbb{R}^N$ vectors whose elements provide a recipe for combining **A**'s columns, $\mathbf{a}_i$, to produce—if

possible—zero; $\{\mathbf{n} \in \mathcal{N}(\mathbf{A}) : \sum_i n_i \mathbf{a}_i = \mathbf{0} \in \mathbb{R}^M\}$. (The "if possible" qualifier addresses the possibility that the only combination of **A**'s columns that yields $\mathbf{0} \in \mathbb{R}^M$ is $\mathbf{x} = (\, 0 \; \cdots \; 0 \,)^T$, the *trivial* zero-dimensional, single point $\mathcal{N}(\mathbf{A})$ *any* **A** has.) The dimension of $\mathcal{N}(\mathbf{A})$—sometimes denoted null(**A**)—is whatever is left of $\mathbb{R}^N$, $N - q$. This dimension is said to be **A**'s *nullity*.

The existence and dimensions of the four fundamental spaces constitute the first part of the *fundamental theorem of linear algebra*.[1] The dimensional statement follows from the *rank–nullity theorem*,[2] which, applied narrowly to matrices, guarantees that for any finite-dimensional linear map from R-N to R-M, $\mathbf{A} \in \mathbb{R}^{M \times N} : \mathbb{R}^N \mapsto \mathbb{R}^M$, the rank and nullity sum to N, the number of **A**'s columns,

$$\begin{aligned} q + \dim\left[\mathcal{N}(\mathbf{A})\right] &= \dim\left[\mathcal{R}(\mathbf{A})\right] + \dim\left[\mathcal{N}(\mathbf{A})\right] \\ &= \dim\left[\mathcal{R}(\mathbf{A}^T)\right] + \dim\left[\mathcal{N}(\mathbf{A})\right] = N. \end{aligned} \tag{3.30}$$

Because $\dim[\mathcal{R}(\mathbf{A})] = \dim[\mathcal{R}(\mathbf{A}^T)] = q$, by definition of the transpose (3.30) also implies that

$$\dim[\mathcal{R}(\mathbf{A})] + \dim[\mathcal{N}(\mathbf{A}^T)] = M. \tag{3.31}$$

The fundamental spaces form mutually orthogonal pairs. In $\mathbb{R}^N$, $\mathbf{c}_l^T \mathbf{n} = (\mathbf{A}^T\mathbf{y})^T\mathbf{n}$ for any $\mathbf{y} \in \mathbb{R}^M$, which expands into $\mathbf{y}^T\mathbf{A}\mathbf{n}$. But by the definition of the null space, $\mathbf{A}\mathbf{n} = \mathbf{0} \in \mathbb{R}^M$, so $\mathbf{y}^T\mathbf{A}\mathbf{n} = 0$ as well; $\mathbf{c}_l$ and $\mathbf{n}$ are mutually orthogonal. Since we imposed no restrictions on $\mathbf{c}_l$ and $\mathbf{n}$ other than $\mathbf{c}_l \in \mathcal{R}(\mathbf{A}^T)$ and $\mathbf{n} \in \mathcal{N}(\mathbf{A})$, it follows that $\mathcal{R}(\mathbf{A}^T) \perp \mathcal{N}(\mathbf{A})$, the row and null spaces are mutually orthogonal, with any vector from one being orthogonal to any vector from the other. With $\mathcal{R}(\mathbf{A}^T) \perp \mathcal{N}(\mathbf{A})$ and the dimensions of the two $\mathbb{R}^N$ subspaces satisfying eq. 3.30, $\mathcal{R}(\mathbf{A}^T) + \mathcal{N}(\mathbf{A}) = \mathbb{R}^N$ must follow.

By analogy, in $\mathbb{R}^M$, $\mathbf{c}^T\mathbf{n}_l = (\mathbf{A}\mathbf{x})^T\mathbf{n}_l$ for any $\mathbf{x} \in \mathbb{R}^N$, or $\mathbf{x}^T\mathbf{A}^T\mathbf{n}_l$. By the left null space's definition, $\mathbf{A}^T\mathbf{n}_l = \mathbf{0} \in \mathbb{R}^N$, so $\mathbf{x}^T\mathbf{A}^T\mathbf{n}_l = 0$ as well; **c** and $\mathbf{n}_l$ are mutually orthogonal. Since the only requirements of **c** and $\mathbf{n}_l$ are $\mathbf{c} \in \mathcal{R}(\mathbf{A})$ and $\mathbf{n}_l \in \mathcal{N}(\mathbf{A}^T)$, $\mathcal{R}(\mathbf{A}) \perp \mathcal{N}(\mathbf{A}^T)$, the column and left null spaces are mutually orthogonal. With $\mathcal{R}(\mathbf{A}) \perp \mathcal{N}(\mathbf{A}^T)$ and the dimensions of the two $\mathbb{R}^M$ subspaces satisfying eq. 3.31, $\mathcal{R}(\mathbf{A}) + \mathcal{N}(\mathbf{A}^T) = \mathbb{R}^M$ must follow.

The second part of the *fundamental theorem of linear algebra* is a summary of the above:

[1] For a more complete, beautiful yet heuristic treatment of the theorem, see Strang, G. (1993) The fundamental theorem of linear algebra, *The American Mathematical Monthly*, **100**(9), 848–855.

[2] For description and proof of this theorem, see, e.g., Cantrell, C. D. (2000) *Modern Mathematical Methods for Physicists and Engineers*, Cambridge University Press, New York, 763 pp., ISBN-13: 978-0521598279.

$$\mathcal{R}(\mathbf{A}^T) \perp \mathcal{N}(\mathbf{A}), \quad \mathcal{R}(\mathbf{A}^T) + \mathcal{N}(\mathbf{A}) = \mathbb{R}^N$$
$$\mathcal{R}(\mathbf{A}) \perp \mathcal{N}(\mathbf{A}^T), \quad \mathcal{R}(\mathbf{A}) + \mathcal{N}(\mathbf{A}^T) = \mathbb{R}^M.$$

3.3.2 Further Examples

There are several ways to find a basis for $\mathcal{R}(\mathbf{A})$. A reasonable one is to

- reduce **A** to **U** using elementary operations
- take the columns, back in **A**, that correspond to nonzero pivots in **U**
- (optional but useful) normalize those vectors

This procedure will definitely identify a choice for the needed basis vectors. However, it will, in general, produce a linearly independent but not orthogonal basis. Later on in this chapter, we will learn how to achieve orthonormality of the basis (which, for a basis set $\{\mathbf{b}_i\}$, means that

$$\mathbf{b}_i^T \mathbf{b}_j = \begin{cases} 1 & i = j \\ 0 & i \neq j \end{cases}, \tag{3.32}$$

i.e., the basis set is actually $\{\hat{\mathbf{b}}_i\}$). For now, we will settle for linear independence and simply produce a legitimate basis.

Let's consider the example

$$\mathbf{A} = \begin{pmatrix} 1 & 0 & 1 \\ 2 & 1 & 1 \\ -1 & 1 & 0 \end{pmatrix}. \tag{3.33}$$

With

$$\mathbf{E}_1 = \begin{pmatrix} 1 & 0 & 0 \\ -2 & 1 & 0 \\ 1 & 0 & 1 \end{pmatrix} \text{ and } \mathbf{E}_2 = \begin{pmatrix} 1 & 0 & 0 \\ 0 & 1 & 0 \\ 0 & -1 & 1 \end{pmatrix}, \tag{3.34}$$

$$\mathbf{E}_2\mathbf{E}_1\mathbf{A} = \begin{pmatrix} \boxed{1} & 0 & 1 \\ 0 & \boxed{1} & -1 \\ 0 & 0 & \boxed{2} \end{pmatrix}. \tag{3.35}$$

Thus, this **A**'s columns are all linearly independent. Going back next to **A**'s columns, a normalized but not orthogonal basis for $\mathcal{R}(\mathbf{A})$ is

$$\left\{ \frac{\mathbf{a}_1}{\|\mathbf{a}_1\|}, \frac{\mathbf{a}_2}{\|\mathbf{a}_2\|}, \frac{\mathbf{a}_3}{\|\mathbf{a}_3\|} \right\}, \tag{3.36}$$

where $\mathbf{a}_i$ is **A**'s ith column. If we use the same $\mathbf{E}_1$ and $\mathbf{E}_2$ but choose, by contrast, the only slightly different

$$\mathbf{A} = \begin{pmatrix} 1 & 0 & 1 \\ 2 & 1 & 3 \\ -1 & 1 & 0 \end{pmatrix}, \quad \mathbf{E}_2\mathbf{E}_1\mathbf{A} = \begin{pmatrix} \boxed{1} & 0 & 1 \\ 0 & \boxed{1} & 1 \\ 0 & 0 & 0 \end{pmatrix}, \tag{3.37}$$

a normalized but not orthogonal basis for $\mathcal{R}(\mathbf{A})$ is in this case

$$\left\{ \frac{\mathbf{a}_1}{\|\mathbf{a}_1\|}, \frac{\mathbf{a}_2}{\|\mathbf{a}_2\|} \right\}. \tag{3.38}$$

Let's consider next the richer example

$$\mathbf{A} = \begin{pmatrix} 1 & 0 & 1 & 1 & 2 \\ 2 & 1 & 3 & 1 & 1 \\ -1 & 1 & 0 & -2 & 3 \end{pmatrix}. \tag{3.39}$$

Setting

$$\mathbf{E}_1 = \begin{pmatrix} 1 & 0 & 0 \\ -2 & 1 & 0 \\ 1 & 0 & 1 \end{pmatrix} \text{ and } \mathbf{E}_2 = \begin{pmatrix} 1 & 0 & 0 \\ 0 & 1 & 0 \\ 0 & -1 & 1 \end{pmatrix}, \tag{3.40}$$

$$\mathbf{E}_2\mathbf{E}_1\mathbf{A} = \begin{pmatrix} \boxed{1} & 0 & 1 & 1 & 2 \\ 0 & \boxed{1} & 1 & -1 & -3 \\ 0 & 0 & 0 & 0 & \boxed{8} \end{pmatrix}, \tag{3.41}$$

so a normalized but not orthogonal basis for $\mathcal{R}(\mathbf{A})$ is

$$\left\{ \frac{\mathbf{a}_1}{\|\mathbf{a}_1\|}, \frac{\mathbf{a}_2}{\|\mathbf{a}_2\|}, \frac{\mathbf{a}_5}{\|\mathbf{a}_5\|} \right\}, \tag{3.42}$$

Similarly, there are various ways to obtain a suitable basis for $\mathcal{N}(\mathbf{A}^T)$. The most straightforward is simply to solve $\mathbf{A}^T\mathbf{n}_l = \mathbf{0} \in \mathbb{R}^N$. For example, with

$$\mathbf{A} = \begin{pmatrix} 1 & 0 & 1 \\ 2 & 1 & 3 \\ -1 & 1 & 0 \end{pmatrix}, \tag{3.43}$$

$$\mathbf{E}_1 = \begin{pmatrix} 1 & 0 & 0 \\ 0 & 1 & 0 \\ 1 & 0 & -1 \end{pmatrix} \text{ and } \mathbf{E}_2 = \begin{pmatrix} 1 & 0 & 0 \\ 0 & 1 & 0 \\ 0 & 1 & 1 \end{pmatrix}, \tag{3.44}$$

$$\mathbf{E}_2\mathbf{E}_1\mathbf{A}^T = \begin{pmatrix} \boxed{1} & 2 & -1 \\ 0 & \boxed{1} & 1 \\ 0 & 0 & 0 \end{pmatrix} \equiv \mathbf{U}. \tag{3.45}$$

With this $\mathbf{U}$, we can readily employ back-substitution to obtain $\mathbf{n}_l$ satisfying $\mathbf{A}^T\mathbf{n}_l = \mathbf{0} \in \mathbb{R}^N$. Because premultiplying the zero vector rhs (right-hand side) by

$\mathbf{E}_2\mathbf{E}_1$ leaves it intact, our problem is $\mathbf{U}\mathbf{n}_l = \mathbf{0} \in \mathbb{R}^N$. Back-substituting our way up, starting from the lowermost nonvanishing row, row 2 is $\gamma = -\beta$, with which the top row becomes $\alpha = -3\beta$, so

$$\mathbf{n}_l = -\beta \begin{pmatrix} 3 \\ -1 \\ 1 \end{pmatrix} \quad \text{and} \quad \hat{\mathbf{n}}_l = \frac{1}{\sqrt{9+1+1}} \begin{pmatrix} 3 \\ -1 \\ 1 \end{pmatrix}. \tag{3.46}$$

This can be readily verified using

$$\mathbf{A}^T\mathbf{n}_l = \begin{pmatrix} 1 & 2 & -1 \\ 0 & 1 & 1 \\ 1 & 3 & 0 \end{pmatrix} \begin{pmatrix} 3 \\ -1 \\ 1 \end{pmatrix} = \begin{pmatrix} 3-2-1 \\ -1+1 \\ 3-3 \end{pmatrix} = \begin{pmatrix} 0 \\ 0 \\ 0 \end{pmatrix} \tag{3.47}$$

as required.

As a second example of calculating a spanning set for $\mathcal{N}(\mathbf{A}^T)$, let's consider the transpose of an earlier example,

$$\mathbf{A} = \begin{pmatrix} 1 & 2 & -1 \\ 0 & 1 & 1 \\ 1 & 3 & 0 \\ 1 & 1 & -2 \\ 2 & 1 & 3 \end{pmatrix}, \tag{3.48}$$

which reduces to

$$\begin{aligned} \mathbf{E}_2\mathbf{E}_1\mathbf{A}^T &= \begin{pmatrix} 1 & 0 & 0 \\ 0 & 1 & 0 \\ 0 & -1 & 1 \end{pmatrix} \begin{pmatrix} 1 & 0 & 0 \\ -2 & 1 & 0 \\ 1 & 0 & 1 \end{pmatrix} \mathbf{A}^T \\ &= \begin{pmatrix} \boxed{1} & 0 & 1 & 1 & 2 \\ 0 & \boxed{1} & 1 & -1 & -3 \\ 0 & 0 & 0 & 0 & \boxed{8} \end{pmatrix} \equiv \mathbf{U}. \end{aligned} \tag{3.49}$$

Because here $\mathbf{n}_l \in \mathbb{R}^5$ while $q = 3$ (corresponding to the three boxed nontrivial pivots in $\mathbf{U}$), it is clear that $\mathcal{N}(\mathbf{A}^T)$ must be a plane. Recalling that the rhs is invariant under premultiplication by $\mathbf{E}_i$, to span this $\mathbf{A}$'s $\mathcal{N}(\mathbf{A}^T)$, we need to solve $\mathbf{U}\mathbf{n}_l = \mathbf{0} \in \mathbb{R}^3$ with the above $\mathbf{U}$ for $\mathbf{n}_l = (\,\alpha\;\beta\;\gamma\;\delta\;\eta\,)^T$. Row 3 yields $\eta = 0$, so row 2, $\beta + \gamma - \delta = 3\eta$, becomes $\delta = \beta + \gamma$. The top row, $\alpha = -\gamma - \delta - 2\eta$, then becomes $\alpha = -\beta - 2\gamma$, and

$$\mathbf{n}_l = \begin{pmatrix} -\beta - 2\gamma \\ \beta \\ \gamma \\ \beta + \gamma \\ 0 \end{pmatrix} = \beta \begin{pmatrix} -1 \\ 1 \\ 0 \\ 1 \\ 0 \end{pmatrix} + \gamma \begin{pmatrix} -2 \\ 0 \\ 1 \\ 1 \\ 0 \end{pmatrix}. \tag{3.50}$$

A suitable, neither normalized nor necessarily orthogonal, basis for $\mathcal{N}(\mathbf{A}^T)$ is therefore

$$\left\{ \begin{pmatrix} -1 \\ 1 \\ 0 \\ 1 \\ 0 \end{pmatrix}, \begin{pmatrix} -2 \\ 0 \\ 1 \\ 1 \\ 0 \end{pmatrix} \right\}, \tag{3.51}$$

which is readily verified by premultiplying the above two spanning vectors by $\mathbf{A}^T$; they vanish. This $\mathbf{A}$'s $\mathcal{N}(\mathbf{A}^T)$ is thus the plane in $\mathbb{R}^5$ spanned by the above 2 vectors.

Finding bases for $\mathcal{R}(\mathbf{A}^T)$ and $\mathcal{N}(\mathbf{A})$, the $\mathbb{R}^N$ counterparts of the above fundamental spaces, follows straightforwardly the same procedure, with transposition where appropriate. Let's analyze one additional example,

$$\mathbf{A} = \begin{pmatrix} 1 & 0 & 1 & 1 & 2 \\ 2 & 1 & 3 & 1 & 1 \\ -1 & 1 & 0 & -2 & 3 \\ 3 & 1 & 4 & 2 & 3 \end{pmatrix}, \tag{3.52}$$

to address both simultaneously. With

$$\mathbf{E}_3\mathbf{E}_2\mathbf{E}_1 = \begin{pmatrix} 1 & 0 & 0 & 0 & 0 \\ 0 & 1 & 0 & 0 & 0 \\ 0 & 0 & 0 & 0 & 1 \\ 0 & 0 & 0 & 1 & 0 \\ 0 & 0 & 1 & 0 & 0 \end{pmatrix} \begin{pmatrix} 1 & 0 & 0 & 0 & 0 \\ 0 & 1 & 0 & 0 & 0 \\ 0 & -1 & 1 & 0 & 0 \\ 0 & 1 & 0 & 1 & 0 \\ 0 & -3 & 0 & 0 & 1 \end{pmatrix} \begin{pmatrix} 1 & 0 & 0 & 0 & 0 \\ 0 & 1 & 0 & 0 & 0 \\ -1 & 0 & 1 & 0 & 0 \\ -1 & 0 & 0 & 1 & 0 \\ -2 & 0 & 0 & 0 & 1 \end{pmatrix}, \tag{3.53}$$

$$\mathbf{E}_3\mathbf{E}_2\mathbf{E}_1\mathbf{A}^T = \begin{pmatrix} \boxed{1} & 2 & -1 & 3 \\ 0 & \boxed{1} & 1 & 1 \\ 0 & 0 & \boxed{8} & 0 \\ 0 & 0 & 0 & 0 \\ 0 & 0 & 0 & 0 \end{pmatrix} \equiv \mathbf{U}. \tag{3.54}$$

A raw nonorthogonal basis for $\mathcal{R}(\mathbf{A}^T)$ is therefore the set of $\mathbf{A}$'s rows 1–3,

$$\left\{ \begin{pmatrix} 1 \\ 0 \\ 1 \\ 1 \\ 2 \end{pmatrix}, \begin{pmatrix} 2 \\ 1 \\ 3 \\ 1 \\ 1 \end{pmatrix}, \begin{pmatrix} -1 \\ 1 \\ 0 \\ -2 \\ 3 \end{pmatrix} \right\}. \tag{3.55}$$

To find a basis for $\mathcal{N}(\mathbf{A})$, we solve $\mathbf{An} = \mathbf{Un} = \mathbf{0} \in \mathbb{R}^4$ for $\mathbf{n} = (\,\alpha\, \beta\, \gamma\, \delta\, \eta\,)^T$. With

$$\mathbf{E}_2\mathbf{E}_1 = \begin{pmatrix} 1 & 0 & 0 & 0 \\ 0 & 1 & 0 & 0 \\ 0 & -1 & 1 & 0 \\ 0 & -1 & 0 & 1 \end{pmatrix} \begin{pmatrix} 1 & 0 & 0 & 0 \\ -2 & 1 & 0 & 0 \\ 1 & 0 & 1 & 0 \\ -3 & 0 & 0 & 1 \end{pmatrix}, \tag{3.56}$$

$$\mathbf{E}_2\mathbf{E}_1\mathbf{A} = \begin{pmatrix} \boxed{1} & 0 & 1 & 1 & 2 \\ 0 & \boxed{1} & 1 & -1 & -3 \\ 0 & 0 & 0 & 0 & \boxed{8} \\ 0 & 0 & 0 & 0 & 0 \end{pmatrix} \equiv \mathbf{U}. \tag{3.57}$$

The third row dictates that $\eta = 0$. With this, the second row, $\beta + \gamma - \delta = 3\eta$, becomes $\beta = \delta - \gamma$. The top row, $\alpha + \gamma + \delta = -2\eta$ becomes $\alpha = -\gamma - \delta$, so

$$\mathbf{n} = \begin{pmatrix} -\gamma - \delta \\ \delta - \gamma \\ \gamma \\ \delta \\ 0 \end{pmatrix} = \xi \begin{pmatrix} 1 \\ 1 \\ -1 \\ 0 \\ 0 \end{pmatrix} + \nu \begin{pmatrix} -1 \\ 1 \\ 0 \\ 1 \\ 0 \end{pmatrix}, \tag{3.58}$$

and

$$\text{basis}\left[\mathcal{N}(\mathbf{A})\right] = \left\{ \begin{pmatrix} 1 \\ 1 \\ -1 \\ 0 \\ 0 \end{pmatrix}, \begin{pmatrix} -1 \\ 1 \\ 0 \\ 1 \\ 0 \end{pmatrix} \right\}, \tag{3.59}$$

which can be readily tested by premultiplying each of these vectors by $\mathbf{A}$ and ascertaining that the products are $\mathbf{0} \in \mathbb{R}^4$; they are. Finally, for completeness, note that the $\mathbf{U}$ we obtained from $\mathbf{A}$, with nonzero pivots in columns 1, 2, and 5, can also be used to obtain

$$\text{basis}\left[\mathcal{R}(\mathbf{A})\right] = \left\{ \frac{\mathbf{a}_1}{\|\mathbf{a}_1\|}, \frac{\mathbf{a}_2}{\|\mathbf{a}_2\|}, \frac{\mathbf{a}_5}{\|\mathbf{a}_5\|} \right\}. \tag{3.60}$$

3.3.3 *Summary: Full Example of* A*'s Fundamental Spaces*

While possibly unnecessary for some, many readers will benefit from a single, exhaustively analyzed, example. Let's therefore consider

$$\mathbf{A} = \begin{pmatrix} 1 & 2 & 3 \\ -1 & 1 & 4 \end{pmatrix} \tag{3.61}$$

$(M = 2, N = 3)$ and analyze it fully.

3.3.3.1 RANK

Choosing

$$\mathbf{E} = \begin{pmatrix} 1 & 0 \\ 1 & 1 \end{pmatrix}, \tag{3.62}$$

$$\mathbf{EA} = \begin{pmatrix} 1 & 0 \\ 1 & 1 \end{pmatrix} \begin{pmatrix} 1 & 2 & 3 \\ -1 & 1 & 4 \end{pmatrix} = \begin{pmatrix} \boxed{1} & 2 & 3 \\ 0 & \boxed{3} & 7 \end{pmatrix} \equiv \mathbf{U}. \tag{3.63}$$

Because there are two nonzero (boxed) pivots, $\mathbf{A}$ is full rank, $q = 2 = \min(2, 3)$.

3.3.3.2 COLUMN SPACE $\mathcal{R}(\mathbf{A}) \subseteq \mathbb{R}^2$

The two leftmost columns of $\mathbf{A}$ correspond to nonzero pivots in $\mathbf{U}$. Normalized, they thus form a basis for $\mathcal{R}(\mathbf{A})$,

$$\text{basis}\left[\mathcal{R}(\mathbf{A})\right] = \left\{ \frac{1}{\sqrt{2}} \begin{pmatrix} 1 \\ -1 \end{pmatrix}, \frac{1}{\sqrt{5}} \begin{pmatrix} 2 \\ 1 \end{pmatrix} \right\}. \tag{3.64}$$

3.3.3.3 LEFT NULL SPACE $\mathcal{N}(\mathbf{A}^T) \subseteq \mathbb{R}^2$

This is the set of all 2-vectors mapped by $\mathbf{A}^T$ to $\mathbf{0} \in \mathbb{R}^3$,

$$\mathcal{N}(\mathbf{A}^T) = \left\{ \begin{pmatrix} \alpha \\ \beta \end{pmatrix} : \mathbf{A}^T \begin{pmatrix} \alpha \\ \beta \end{pmatrix} = \begin{pmatrix} 0 \\ 0 \\ 0 \end{pmatrix} \right\}. \tag{3.65}$$

Note that in this case $\mathcal{N}(\mathbf{A}^T)$ must be a zero-dimensional subspace of $\mathbb{R}^2$, the origin, because $\mathcal{R}(\mathbf{A})$ is two-dimensional subspace of $\mathbb{R}^2$, i.e., $\mathcal{R}(\mathbf{A}) = \mathbb{R}^2$. The salient equation for showing that this is so is

$$\mathbf{A}^T\mathbf{n}_l = \begin{pmatrix} 1 & -1 \\ 2 & 1 \\ 3 & 4 \end{pmatrix} \begin{pmatrix} \alpha \\ \beta \end{pmatrix} = \begin{pmatrix} 0 \\ 0 \\ 0 \end{pmatrix}. \tag{3.66}$$

We solve it using

$$\begin{aligned} \mathbf{E}_2\mathbf{E}_1\mathbf{A}^T &= \begin{pmatrix} 1 & 0 & 0 \\ 0 & 1 & 0 \\ 0 & -7 & 3 \end{pmatrix} \begin{pmatrix} 1 & 0 & 0 \\ -2 & 1 & 0 \\ -3 & 0 & 1 \end{pmatrix} \mathbf{A}^T \\ &= \begin{pmatrix} 1 & -1 \\ 0 & 3 \\ 0 & 0 \end{pmatrix} \begin{pmatrix} \alpha \\ \beta \end{pmatrix} = \begin{pmatrix} 0 \\ 0 \\ 0 \end{pmatrix}, \end{aligned} \tag{3.67}$$

which holds because on the right-hand side $\mathbf{E}_2\mathbf{E}_1\mathbf{0} = \mathbf{0}$. The second row (and first nontrivial one from the bottom up) indicates that $\beta = 0$. When substituted this into the top row, $\alpha = 0$ as well. Thus, the defining equation of $\mathcal{N}(\mathbf{A}^T)$ has only a trivial solution. That is, as predicted based on dimensionality, $\mathcal{N}(\mathbf{A}^T)$ is the zero-dimensional subspace of $\mathbb{R}^2$, $(0, 0)$.

3.3.3.4 ROW SPACE $\mathcal{R}(\mathbf{A}^T) \subseteq \mathbb{R}^3$

Because to get **U** from **A** we employed only elementary row operations on **A**'s rows, the row spaces of **U** and **A** are one and the same; $\mathcal{R}(\mathbf{A}^T)$ is invariant under row operations. Consequently,

$$\text{basis}\left[\mathcal{R}(\mathbf{A}^T)\right] = \left\{\frac{1}{\sqrt{14}}\begin{pmatrix}1\\2\\3\end{pmatrix}, \frac{1}{\sqrt{58}}\begin{pmatrix}0\\3\\7\end{pmatrix}\right\} \tag{3.68}$$

is a reasonable choice.

3.3.3.5 NULL SPACE $\mathcal{N}(\mathbf{A}) \subseteq \mathbb{R}^3$

This is the set of all 3-vectors mapped by **A** to $\mathbf{0} \in \mathbb{R}^2$,

$$\mathcal{N}(\mathbf{A}) = \left\{\begin{pmatrix}\alpha\\\beta\\\gamma\end{pmatrix} : \mathbf{A}\begin{pmatrix}\alpha\\\beta\\\gamma\end{pmatrix} = \begin{pmatrix}0\\0\end{pmatrix}\right\}. \tag{3.69}$$

To obtain this set, we solve

$$\mathbf{An} = \begin{pmatrix}1 & 2 & 3\\-1 & 1 & 4\end{pmatrix}\begin{pmatrix}\alpha\\\beta\\\gamma\end{pmatrix} = \begin{pmatrix}0\\0\end{pmatrix}, \tag{3.70}$$

reduced to

$$\mathbf{EAn} = \begin{pmatrix}1 & 0\\1 & 1\end{pmatrix}\begin{pmatrix}1 & 2 & 3\\-1 & 1 & 4\end{pmatrix}\begin{pmatrix}\alpha\\\beta\\\gamma\end{pmatrix} = \begin{pmatrix}0\\0\end{pmatrix} \tag{3.71}$$

(because $\mathbf{E0} = \mathbf{0}$), or

$$\begin{pmatrix}1 & 2 & 3\\0 & 3 & 7\end{pmatrix}\begin{pmatrix}\alpha\\\beta\\\gamma\end{pmatrix} = \begin{pmatrix}0\\0\end{pmatrix}. \tag{3.72}$$

The second equation states that $\gamma = -3\beta/7$, so the first becomes $\alpha = -2\beta - 3\gamma = -5\beta/7$, and

$$\mathbf{n} = -\frac{\beta}{7}\begin{pmatrix}5\\-7\\3\end{pmatrix}. \tag{3.73}$$

We verify this result by

$$\begin{pmatrix} 1 & 2 & 3 \\ 0 & 3 & 7 \end{pmatrix} \begin{pmatrix} 5 \\ -7 \\ 3 \end{pmatrix} = \begin{pmatrix} 0 \\ 0 \end{pmatrix}, \tag{3.74}$$

so

$$\text{basis}\left[\mathcal{N}(\boldsymbol{A})\right] = \frac{1}{\sqrt{83}} \begin{pmatrix} 5 \\ -7 \\ 3 \end{pmatrix}. \tag{3.75}$$

In summary:

$$\text{basis}\left[\mathcal{R}(\mathbf{A})\right] = \left\{ \frac{1}{\sqrt{2}} \begin{pmatrix} 1 \\ -1 \end{pmatrix}, \frac{1}{\sqrt{5}} \begin{pmatrix} 2 \\ 1 \end{pmatrix} \right\} \tag{3.76}$$

$$\text{basis}\left[\mathcal{N}(\mathbf{A}^T)\right] = \begin{pmatrix} 0 \\ 0 \end{pmatrix} \tag{3.77}$$

$$\text{basis}\left[\mathcal{R}(\mathbf{A}^T)\right] = \left\{ \frac{1}{\sqrt{14}} \begin{pmatrix} 1 \\ 2 \\ 3 \end{pmatrix}, \frac{1}{\sqrt{58}} \begin{pmatrix} 0 \\ 3 \\ 7 \end{pmatrix} \right\} \tag{3.78}$$

$$\text{basis}\left[\mathcal{N}(\mathbf{A})\right] = \left\{ \frac{1}{\sqrt{83}} \begin{pmatrix} 5 \\ -7 \\ 3 \end{pmatrix} \right\} \tag{3.79}$$

3.3.3.6 TEST THAT $\mathcal{R}(\mathbf{A})$ SPANS A'S COLUMNS

It is obvious that this basis spans **A**'s two left-most columns, but does it also span the third? First, it must, because **A**'s columns are from $\mathbb{R}^2$, and it only takes two linearly independent $\mathbb{R}^2$ vectors to span $\mathbb{R}^2$. Second, we know from the fact that $q = 2$ and not 1 that **A**'s two left-most columns are mutually linearly independent. If that is not enough, we can be more blindly formal, seeking $(\, \alpha \; \beta \,)^T$ satisfying

$$\begin{pmatrix} 1 & 2 \\ -1 & 1 \end{pmatrix} \begin{pmatrix} \alpha \\ \beta \end{pmatrix} = \begin{pmatrix} 3 \\ 4 \end{pmatrix}. \tag{3.80}$$

Premultiplying both sides by **E** gives

$$\begin{pmatrix} 1 & 0 \\ 1 & 1 \end{pmatrix} \begin{pmatrix} 1 & 2 \\ -1 & 1 \end{pmatrix} \begin{pmatrix} \alpha \\ \beta \end{pmatrix} = \begin{pmatrix} 1 & 0 \\ 1 & 1 \end{pmatrix} \begin{pmatrix} 3 \\ 4 \end{pmatrix}, \tag{3.81}$$

$$\begin{pmatrix} 1 & 2 \\ 0 & 3 \end{pmatrix} \begin{pmatrix} \alpha \\ \beta \end{pmatrix} = \begin{pmatrix} 3 \\ 7 \end{pmatrix}. \tag{3.82}$$

The second equation states that $\beta = \frac{7}{3}$, and the first that $\alpha = 3 - 2\beta = 3 - \frac{14}{3} = -\frac{5}{3}$. Thus,

$$\begin{pmatrix} \alpha \\ \beta \end{pmatrix} = \frac{1}{3}\begin{pmatrix} -5 \\ 7 \end{pmatrix}, \tag{3.83}$$

$$\frac{1}{3}\begin{pmatrix} 1 & 2 \\ -1 & 1 \end{pmatrix}\begin{pmatrix} -5 \\ 7 \end{pmatrix} = \frac{1}{3}\begin{pmatrix} 9 \\ 12 \end{pmatrix} = \begin{pmatrix} 3 \\ 4 \end{pmatrix}, \tag{3.84}$$

substantiating our earlier intuition.

3.3.3.7 TEST THAT $\mathcal{R}(\mathbf{A}) \perp \mathcal{N}(\mathbf{A}^T)$

Because $\mathcal{N}(\mathbf{A}^T)$ is spanned by $(\ 0\ 0\ )^T$, this test is also trivially passed; any 2-vector is orthogonal to $(\ 0\ 0\ )^T$ (because their dot product vanishes). Geometrically, this is clear: the dot product of **a** and **b** answers the question, What is the length of the projection of **a** on **b** (or of **b** on **a**)? Being that $(\ 0\ 0\ )^T$ is zero dimensional (a single point), no 2-vector can possibly have a nonzero projection on it.

3.3.3.8 TEST THAT $\mathcal{R}(\mathbf{A}) + \mathcal{N}(\mathbf{A}^T) = \mathbb{R}^2$

Because $\mathcal{N}(\mathbf{A}^T) = (\ 0\ 0\ )^T$, this test is really asking whether $\mathcal{R}(\mathbf{A}) = \mathbb{R}^2$. If our chosen basis$[\mathcal{R}(\mathbf{A})]$ spans $\mathbb{R}^2$, we should be able to solve

$$\begin{pmatrix} 1 & 2 \\ -1 & 1 \end{pmatrix}\begin{pmatrix} \alpha \\ \beta \end{pmatrix} = \begin{pmatrix} a \\ b \end{pmatrix} \tag{3.85}$$

with no restrictions on a and b (i.e., with no requisite relationship between a and b). With the following (leftmost) **E**,

$$\begin{pmatrix} 1 & 0 \\ 1 & 1 \end{pmatrix}\begin{pmatrix} 1 & 2 \\ -1 & 1 \end{pmatrix}\begin{pmatrix} \alpha \\ \beta \end{pmatrix} = \begin{pmatrix} 1 & 0 \\ 1 & 1 \end{pmatrix}\begin{pmatrix} a \\ b \end{pmatrix} \tag{3.86}$$

or

$$\begin{pmatrix} 1 & 2 \\ 0 & 3 \end{pmatrix}\begin{pmatrix} \alpha \\ \beta \end{pmatrix} = \begin{pmatrix} a \\ a+b \end{pmatrix}, \tag{3.87}$$

so $\beta = (a+b)/3$ and $\alpha = a - 2\beta = (a-2b)/3$. That is,

$$\frac{1}{3}\begin{pmatrix} 1 & 2 \\ -1 & 1 \end{pmatrix}\begin{pmatrix} a-2b \\ a+b \end{pmatrix} = \frac{1}{3}\begin{pmatrix} 3a \\ 3b \end{pmatrix} = \begin{pmatrix} a \\ b \end{pmatrix} \tag{3.88}$$

for any a and b, with no restrictions of the general form $a = a(b)$, as required. Thus, in this case $\mathcal{R}(\mathbf{A}) = \mathbb{R}^2$.

3.3.3.9 TEST THAT $\mathcal{R}(\mathbf{A}^T)$ SPANS $\mathbf{A}$'S ROWS

That **A**'s top row can be expressed in terms of our chosen basis$[\mathcal{R}(\mathbf{A}^T)]$ is obvious, as the latter contains the former. What about **A**'s second row? If basis$[\mathcal{R}(\mathbf{A}^T)]$ is to span **A**'s second row, we must be able to obtain (α, β) satisfying

$$\begin{pmatrix} 1 & 0 \\ 2 & 3 \\ 3 & 7 \end{pmatrix} \begin{pmatrix} \alpha \\ \beta \end{pmatrix} = \begin{pmatrix} -1 \\ 1 \\ 4 \end{pmatrix}. \tag{3.89}$$

The top equation requires that $\alpha = -1$. Substituted into the second equation, this yields $\beta = 1$. Since with these values the third row becomes the obviously true $-3 + 7 = 4$, the (α, β) we found indeed solve the problem. This shows that our chosen basis for $\mathcal{R}(\mathbf{A}^T)$ does span $\mathbf{A}$'s rows.

3.3.3.10 TEST THAT $\mathcal{R}(\mathbf{A}^T) \perp \mathcal{N}(\mathbf{A})$

Because

$$\begin{pmatrix} 1 & 2 & 3 \end{pmatrix} \begin{pmatrix} 5 \\ -7 \\ 3 \end{pmatrix} = 5 - 14 + 9 = 0 \tag{3.90}$$

and

$$\begin{pmatrix} 0 & 3 & 7 \end{pmatrix} \begin{pmatrix} 5 \\ -7 \\ 3 \end{pmatrix} = 0 - 21 + 21 = 0, \tag{3.91}$$

($\mathcal{R}(\mathbf{A}^T)$'s two spanning vectors are individually orthogonal to $\mathcal{N}(\mathbf{A})$'s single dimension), $\mathcal{R}(\mathbf{A}^T) \perp \mathcal{N}(\mathbf{A})$.

3.3.3.11 TEST THAT $\mathcal{R}(\mathbf{A}^T) + \mathcal{N}(\mathbf{A}) = \mathbb{R}^3$

If this holds, a set combining our basis[$\mathcal{R}(\mathbf{A}^T)$] and basis[$\mathcal{N}(\mathbf{A})$] should span $\mathbb{R}^3$. If it does, we should be able to solve

$$\begin{pmatrix} 1 & 0 & 5 \\ 2 & 3 & -7 \\ 3 & 7 & 3 \end{pmatrix} \begin{pmatrix} \alpha \\ \beta \\ \gamma \end{pmatrix} = \begin{pmatrix} a \\ b \\ c \end{pmatrix} \tag{3.92}$$

with no restrictions on (specified relationships between) a, b and c. Using

$$\mathbf{E}_1 = \begin{pmatrix} 1 & 0 & 0 \\ -2 & 1 & 0 \\ -3 & 0 & 1 \end{pmatrix} \quad \text{and} \quad \mathbf{E}_2 = \begin{pmatrix} 1 & 0 & 0 \\ 0 & 1 & 0 \\ 0 & -7 & 3 \end{pmatrix}, \tag{3.93}$$

our problem reduces to

$$\mathbf{E}_2\mathbf{E}_1 \begin{pmatrix} 1 & 0 & 5 \\ 2 & 3 & -7 \\ 3 & 7 & 3 \end{pmatrix} \begin{pmatrix} \alpha \\ \beta \\ \gamma \end{pmatrix} = \mathbf{E}_2\mathbf{E}_1 \begin{pmatrix} a \\ b \\ c \end{pmatrix} \tag{3.94}$$

or

$$\begin{pmatrix} 1 & 0 & 5 \\ 0 & 3 & -17 \\ 0 & 0 & 83 \end{pmatrix} \begin{pmatrix} \alpha \\ \beta \\ \gamma \end{pmatrix} = \begin{pmatrix} a \\ b - 2a \\ 5a - 7b + 3c \end{pmatrix}. \tag{3.95}$$

The bottom row indicates that

$$\gamma = \frac{5a - 7b + 3c}{83} = \frac{15a - 21b + 9c}{249} \tag{3.96}$$

and the second that

$$\begin{aligned} \beta &= \frac{b - 2a + 17\gamma}{3} = \frac{b - 2a}{3} + \frac{17}{3}\left(\frac{5a - 7b + 3c}{83}\right) \\ &= \frac{83b - 166a}{249} + \frac{85a - 119b + 51c}{249} \\ &= \frac{-81a - 36b + 51c}{249}. \end{aligned} \tag{3.97}$$

The top row then reads

$$\begin{aligned} \alpha = a - 5\gamma &= \frac{249a}{249} - 5\frac{15a - 21b + 9c}{249} \\ &= \frac{249a}{249} + \frac{-75a + 105b - 45c}{249} \\ &= \frac{174a + 105b - 45c}{249}. \end{aligned} \tag{3.98}$$

We check this result by ascertaining that

$$\frac{1}{249}\begin{pmatrix} 1 & 0 & 5 \\ 2 & 3 & -7 \\ 3 & 7 & 3 \end{pmatrix}\begin{pmatrix} 174a + 105b - 45c \\ -81a - 36b + 51c \\ 15a - 21b + 9c \end{pmatrix} = \begin{pmatrix} a \\ b \\ c \end{pmatrix} \tag{3.99}$$

holds. Since it does (carry this out!), $\mathcal{R}(\mathbf{A}^T) + \mathcal{N}(\mathbf{A}) = \mathbb{R}^3$.

3.3.4 Application to Systems of Linear Equations

Two of the fundamental vector spaces associated with a matrix—$\mathcal{R}(\mathbf{A})$ and $\mathcal{N}(\mathbf{A})$—are extremely pertinent in the context of a set of general (in general, not homogeneous) coupled linear algebraic equations, where $\mathbf{Ax} = \mathbf{b}$ ($\mathbf{A} \in \mathbb{R}^{M \times N}$) is a system of M equations in N unknowns. Two key questions the fundamental spaces will help us gain insights into are (1) whether there are enough equations (constraints) to fully determine $\mathbf{x}$, and, if not, what portion of it is unconstrained; and (2) whether the system and the rhs vector are consistent. The first question is answered by $\mathcal{N}(\mathbf{A})$, and the second by $\mathcal{R}(\mathbf{A})$.

Let's reconsider a familiar $\mathbf{A}$, but introduce a nonzero right-hand side;

$$\begin{pmatrix} 1 & 1 & 4 \\ 1 & 2 & 3 \\ 1 & 3 & 2 \\ 1 & 4 & 1 \end{pmatrix}\begin{pmatrix} \alpha \\ \beta \\ \gamma \end{pmatrix} = \begin{pmatrix} 8 \\ 7 \\ 6 \\ 5 \end{pmatrix} \equiv \mathbf{b}. \tag{3.100}$$

We start be reducing **A** to **U**, using the previously obtained elementary matrices,

$$\mathbf{E}_2\mathbf{E}_1\mathbf{A} = \begin{pmatrix} 1 & 0 & 0 & 0 \\ 0 & 1 & 0 & 0 \\ 0 & -2 & 1 & 0 \\ 0 & -3 & 0 & 1 \end{pmatrix}\begin{pmatrix} 1 & 0 & 0 & 0 \\ -1 & 1 & 0 & 0 \\ -1 & 0 & 1 & 0 \\ -1 & 0 & 0 & 1 \end{pmatrix}\mathbf{A} \tag{3.101}$$

$$= \begin{pmatrix} 1 & 0 & 0 & 0 \\ -1 & 1 & 0 & 0 \\ 1 & -2 & 1 & 0 \\ 2 & -3 & 0 & 1 \end{pmatrix}\mathbf{A} = \begin{pmatrix} \boxed{1} & 1 & 4 \\ 0 & \boxed{1} & -1 \\ 0 & 0 & 0 \\ 0 & 0 & 0 \end{pmatrix} \equiv \mathbf{U}.$$

With this **U**, we can next ask: Does this system have a solution? I.e., is

$$\begin{pmatrix} 8 \\ 7 \\ 6 \\ 5 \end{pmatrix} \in \mathcal{R}(\mathbf{A})? \tag{3.102}$$

We need basis[$\mathcal{R}(\mathbf{A})$]. Because this **A**'s corresponding **U** has two nontrivial pivots in columns 1 and 2, basis[$\mathcal{R}(\mathbf{A})$] comprises $\mathbf{a}_1$ and $\mathbf{a}_2$;

$$\mathcal{R}(\mathbf{A}) = \text{span}\left[\left\{\begin{pmatrix} 1 \\ 1 \\ 1 \\ 1 \end{pmatrix}, \begin{pmatrix} 1 \\ 2 \\ 3 \\ 4 \end{pmatrix}\right\}\right]. \tag{3.103}$$

We thus seek (α, β) satisfying

$$\begin{pmatrix} 1 & 1 \\ 1 & 2 \\ 1 & 3 \\ 1 & 4 \end{pmatrix}\begin{pmatrix} \alpha \\ \beta \end{pmatrix} = \begin{pmatrix} 8 \\ 7 \\ 6 \\ 5 \end{pmatrix}. \tag{3.104}$$

For this system to have a solution, the last two elements of $\mathbf{E}_2\mathbf{E}_1\mathbf{b}$ must vanish. They do (as we will see shortly), but let's instead employ a quick and dirty approach here. The first row states that $\beta = 8 - \alpha$. The second row reads $\alpha + 2\beta = \alpha + 2(8 - \alpha) = 16 - \alpha = 7$ or $\alpha = 9$, which means $\beta = -1$. The remaining rows conform, and

$$\begin{pmatrix} 1 & 1 \\ 1 & 2 \\ 1 & 3 \\ 1 & 4 \end{pmatrix}\begin{pmatrix} 9 \\ -1 \end{pmatrix} = \begin{pmatrix} 8 \\ 7 \\ 6 \\ 5 \end{pmatrix}, \tag{3.105}$$

which means that this **b** is a linear combination of **A**'s columns; $\mathbf{b} \in \mathcal{R}(\mathbf{A})$. This is by no means general: **b** need not be from $\mathcal{R}(\mathbf{A})$. But in this case we deliberately chose a **b** from **A**'s column space, so this system has a solution.

To continue with the solution procedure and preserve the conditions of the equations, we must operate on the rhs vector in a consistent manner:

$$\mathbf{E}_2\mathbf{E}_1\mathbf{b} = \begin{pmatrix} 1 & 0 & 0 & 0 \\ -1 & 1 & 0 & 0 \\ 1 & -2 & 1 & 0 \\ 2 & -3 & 0 & 1 \end{pmatrix} \begin{pmatrix} 8 \\ 7 \\ 6 \\ 5 \end{pmatrix} = \begin{pmatrix} 8 \\ -1 \\ 0 \\ 0 \end{pmatrix} \equiv \mathbf{d}. \tag{3.106}$$

Thus, we have transformed the original $\mathbf{Ax} = \mathbf{b}$ to an equivalent, but readily solvable, problem, $\mathbf{Ux} = \mathbf{d}$, by premultiplying both sides by $\mathbf{E}_2\mathbf{E}_1$,

$$\mathbf{E}_2\mathbf{E}_1\mathbf{Ax} = \mathbf{Ux} = \begin{pmatrix} 1 & 1 & 4 \\ 0 & 1 & -1 \\ 0 & 0 & 0 \\ 0 & 0 & 0 \end{pmatrix} \begin{pmatrix} \alpha \\ \beta \\ \gamma \end{pmatrix} = \mathbf{E}_2\mathbf{E}_1\mathbf{b} = \begin{pmatrix} 8 \\ -1 \\ 0 \\ 0 \end{pmatrix} = \mathbf{d}. \tag{3.107}$$

Let's take a brief detour now and find the $3 - q$ vectors $\{\boldsymbol{n}_i\}_{i=1}^{3-q}$ that span $\mathcal{N}(\mathbf{A})$ by considering an alternative, homogeneous, rhs. Noting that since $3 - q = 1$ there is only one such spanning vector and the index is unnecessary (we seek only one vector, call it $\mathbf{n}$), and denoting $\mathbf{n} = (\, n_1 \; n_2 \; n_3 \,)^T$, row 2 states that $n_2 = n_3$. Row 1 then states that $n_1 = -5n_2$, so $\mathbf{n} = \xi(-5 \; 1 \; 1\,)^T$, where ξ is any scalar.

Returning next to the nonhomogeneous case, the bottom two equations tell us nothing, because they are satisfied identically for any $\mathbf{x}$ (which, in turn, holds because $\mathbf{b} \in \mathcal{R}(\mathbf{A})$). The second equation says that $\beta - \gamma = -1$, or $\gamma = \beta + 1$. The first equation, $\alpha = 8 - \beta - 4\gamma$, can be combined with $\gamma = \beta + 1$ to yield $\alpha = 4 - 5\beta$. That is,

$$\mathbf{x} = \begin{pmatrix} 4 - 5\beta \\ \beta \\ \beta + 1 \end{pmatrix} = \beta \begin{pmatrix} -5 \\ 1 \\ 1 \end{pmatrix} + \begin{pmatrix} 4 \\ 0 \\ 1 \end{pmatrix} = \mathbf{x}_h + \mathbf{x}_p, \tag{3.108}$$

with any β, satisfies $\mathbf{Ax} = \mathbf{b}$ perfectly (and equally) well, where $\mathbf{x}_h$ and $\mathbf{x}_p$ are the solution's homogeneous and particular parts. The particular solution, $\mathbf{x}_p$, solves the equations, conforming to the particular rhs we were given,

$$\begin{pmatrix} 1 & 1 & 4 \\ 1 & 2 & 3 \\ 1 & 3 & 2 \\ 1 & 4 & 1 \end{pmatrix} \begin{pmatrix} 4 \\ 0 \\ 1 \end{pmatrix} = \begin{pmatrix} 4+4 \\ 4+3 \\ 4+2 \\ 4+1 \end{pmatrix} = \begin{pmatrix} 8 \\ 7 \\ 6 \\ 5 \end{pmatrix}, \tag{3.109}$$

as required. The solution's homogeneous part $\mathbf{x}_h$ gives a direction in $\mathbb{R}^3$ along which any vector can be added to $\mathbf{x}$ without affecting the solution's ability to reproduce the rhs, because

$$\begin{pmatrix} 1 & 1 & 4 \\ 1 & 2 & 3 \\ 1 & 3 & 2 \\ 1 & 4 & 1 \end{pmatrix} \begin{pmatrix} -5 \\ 1 \\ 1 \end{pmatrix} = \begin{pmatrix} -5+1+4 \\ -5+2+3 \\ -5+3+2 \\ -5+4+1 \end{pmatrix} = \begin{pmatrix} 0 \\ 0 \\ 0 \\ 0 \end{pmatrix}. \tag{3.110}$$

That is, vectors parallel to $\mathbf{x}_h$ are mapped by $\mathbf{A}$ to $\mathbf{0} \in \mathbb{R}^4$, so we can add any of them to the solution without violating the constraints. This can be true only if $\mathbf{x}_h \in \mathcal{N}(\mathbf{A})$, which can be readily verified by comparing $\mathbf{x}_h$ to our previously obtained basis$[\mathcal{N}(\mathbf{A})]$. That the solution admits a nontrivial, one-dimensional, homogeneous part is yet another way of saying that $\mathbf{A}$'s rank is 2, not the full possible 3.

Let's contrast the above situation with

$$\mathbf{Ax} = \begin{pmatrix} 1 & 1 & 4 \\ 0 & 2 & 3 \\ -1 & 3 & 2 \\ 0 & 4 & 1 \end{pmatrix} \begin{pmatrix} \alpha \\ \beta \\ \gamma \end{pmatrix} = \begin{pmatrix} 7 \\ 5 \\ 3 \\ 5 \end{pmatrix} = \mathbf{b}. \tag{3.111}$$

With

$$\mathbf{E}_3\mathbf{E}_2\mathbf{E}_1 = \begin{pmatrix} 1 & 0 & 0 & 0 \\ 0 & 1 & 0 & 0 \\ 0 & 0 & 0 & 1 \\ 0 & 0 & 1 & 0 \end{pmatrix} \begin{pmatrix} 1 & 0 & 0 & 0 \\ 0 & 1 & 0 & 0 \\ 0 & -2 & 1 & 0 \\ 0 & -2 & 0 & 1 \end{pmatrix} \begin{pmatrix} 1 & 0 & 0 & 0 \\ 0 & 1 & 0 & 0 \\ 1 & 0 & 1 & 0 \\ 0 & 0 & 0 & 1 \end{pmatrix}, \tag{3.112}$$

$$\mathbf{E}_3\mathbf{E}_2\mathbf{E}_1\mathbf{A} = \begin{pmatrix} \boxed{1} & 1 & 4 \\ 0 & \boxed{2} & 3 \\ 0 & 0 & \boxed{-5} \\ 0 & 0 & 0 \end{pmatrix} \equiv \mathbf{U} \tag{3.113}$$

and

$$\mathbf{E}_3\mathbf{E}_2\mathbf{E}_1\mathbf{b} = \begin{pmatrix} 7 \\ 5 \\ -5 \\ 0 \end{pmatrix} \equiv \mathbf{d}. \tag{3.114}$$

We solve $\mathbf{Ux} = \mathbf{d}$ for $\mathbf{x} = (\,\alpha\,\beta\,\gamma\,)^T$ using back-substitution. The third equation indicates that $\gamma = 1$, with which the second, $2\beta + 3\gamma = 5$, becomes $\beta = 1$. Finally, the first equation yields $\alpha = 2$, and

$$\mathbf{x} = \begin{pmatrix} 2 \\ 1 \\ 1 \end{pmatrix} = \mathbf{x}_p, \tag{3.115}$$

with $\mathbf{x}_h = (\,0\;0\;0\,)^T$; $\mathbf{x}$ is completely determined, with no lingering indeterminacy. This is because this $\mathbf{A}$ is full rank, $q = 3 = \min(4, 3)$, which is true, in turn, because its columns are linearly independent, so no nontrivial combination of the columns vanishes.

Let's discuss one last example,

$$\mathbf{Ax} = \begin{pmatrix} 1 & 1 & 0 & 2 \\ 1 & 2 & -1 & 3 \\ 1 & 3 & -2 & 4 \\ 1 & 4 & -3 & 5 \\ 2 & 3 & -1 & 5 \end{pmatrix} \begin{pmatrix} \alpha \\ \beta \\ \gamma \\ \delta \end{pmatrix} = \begin{pmatrix} 1 \\ 4 \\ 7 \\ 10 \\ 5 \end{pmatrix} = \mathbf{b}. \tag{3.116}$$

With

$$\mathbf{E}_2\mathbf{E}_1 = \begin{pmatrix} 1 & 0 & 0 & 0 & 0 \\ 0 & 1 & 0 & 0 & 0 \\ 0 & -2 & 1 & 0 & 0 \\ 0 & -3 & 0 & 1 & 0 \\ 0 & -1 & 0 & 0 & 1 \end{pmatrix} \begin{pmatrix} 1 & 0 & 0 & 0 & 0 \\ -1 & 1 & 0 & 0 & 0 \\ -1 & 0 & 1 & 0 & 0 \\ -1 & 0 & 0 & 1 & 0 \\ -2 & 0 & 0 & 0 & 1 \end{pmatrix}, \tag{3.117}$$

$$\mathbf{E}_2\mathbf{E}_1\mathbf{A}\mathbf{x} = \begin{pmatrix} \boxed{1} & 1 & 0 & 2 \\ 0 & \boxed{1} & -1 & 1 \\ 0 & 0 & 0 & 0 \\ 0 & 0 & 0 & 0 \\ 0 & 0 & 0 & 0 \end{pmatrix} \mathbf{x} \equiv \mathbf{U}\mathbf{x} = \mathbf{E}_2\mathbf{E}_1\mathbf{b} = \begin{pmatrix} 1 \\ 3 \\ 0 \\ 0 \\ 0 \end{pmatrix} \equiv \mathbf{d}. \tag{3.118}$$

As previously emphasized, that $\mathbf{d}$'s bottom 3 rows vanish is an indication that $\mathbf{b} \in \mathcal{R}(\mathbf{A})$. But the structure of the solution is more interesting. Row 2 yields $\beta = 3 + \gamma - \delta$ and row 1 $\alpha = -2 - \gamma - \delta$. Thus,

$$\begin{aligned} \mathbf{x} = \begin{pmatrix} -2 - \gamma - \delta \\ 3 + \gamma - \delta \\ \gamma \\ \delta \end{pmatrix} &= \begin{pmatrix} -2 \\ 3 \\ 0 \\ 0 \end{pmatrix} + \gamma \begin{pmatrix} -1 \\ 1 \\ 1 \\ 0 \end{pmatrix} + \delta \begin{pmatrix} -1 \\ -1 \\ 0 \\ 1 \end{pmatrix} \\ &= \mathbf{x}_p + \mathbf{x}_h^1 + \mathbf{x}_h^2, \end{aligned} \tag{3.119}$$

where $\mathbf{x}_h^1$ and $\mathbf{x}_h^2$ are homogeneous solution vectors jointly spanning the plane in $\mathbb{R}^4$ that is $\mathbf{A}$'s null space. That $\dim[\mathcal{N}(\mathbf{A})] = 2$ is, of course, expected, given that $\mathbf{x} \in \mathbb{R}^4$ and $q = 2$, so $\mathbf{A}$'s nullity (see section 3.3.1) must be $4 - 2 = 2$.

Our introduction to the fundamental vector spaces associated with $\mathbf{A} \in \mathbb{R}^{M \times N}$ involved in $\mathbf{A}\mathbf{x} = \mathbf{b}$ can be summarized as follows:

1. **A**'s *column space* $\mathcal{R}(\mathbf{A})$ comprises all $\mathbb{R}^M$ vectors that are linear combinations of **A**'s columns. Its dimension is **A**'s rank q. If $\mathbf{b} \in \mathcal{R}(\mathbf{A})$, the system can have a solution; if $\mathbf{b} \notin \mathcal{R}(\mathbf{A})$, the system has no solution (it is inconsistent).
2. **A**'s *left null space* $\mathcal{N}(\mathbf{A}^T)$ comprises all $\mathbb{R}^M$ vectors that $\mathbf{A}^T$ maps onto $\mathbf{0} \in \mathbb{R}^N$. Its dimension is $M - q$. Vectors from this space will contribute nothing to the inverse problem of obtaining **x** from observed **b**.
3. **A**'s *row space* $\mathcal{R}(\mathbf{A}^T)$ comprises all $\mathbb{R}^N$ vectors that are linear combinations of **A**'s rows. Its dimension is q.
4. **A**'s *null space* $\mathcal{N}(\mathbf{A})$ comprises all $\mathbb{R}^N$ vectors that **A** maps onto $\mathbf{0} \in \mathbb{R}^M$. Its dimension is $N - q$, and it corresponds to the homogeneous solution.

3.3.5 Numerical Considerations in Determining q

The number q of independent columns or rows of a matrix is a very fundamental question, one that can be answered in a number of ways (all of which, of course, will lead to the same conclusion!). It is also a question that is somewhat ambiguous in practice. To see why, recall that linear dependence has to do with parallel vectors. In simple examples, vectors are either mutually parallel or not; there is no in between. Noisy data matrices, on the other hand, can have columns that are almost parallel, but not quite. Then, the question is a more subtle one. The condition number of a matrix (the ratio of the largest and smallest singular values, see chapter 5) is meant to answer the question, but in a somewhat subjective way. We need to know more about matrices to fully appreciate this dilemma.

For now, we conclude this brief discussion by noting that when $\mathbf{A}$ is full rank, $q = \min(M, N)$. Otherwise $[q < \min(M, N)]$, $\mathbf{A}$ is rank deficient; there exists a nontrivial null space. If $q = M = N$ and ($\mathbf{A}$ is both square and full rank), $\mathbf{A}$ has an inverse, denoted by $\mathbf{A}^{-1}$, satisfying

$$\mathbf{A}^{-1}\mathbf{A} = \mathbf{A}\,\mathbf{A}^{-1} = \mathbf{I}_N,$$

where $\mathbf{I}_N$, the $N \times N$ identity matrix, is the N-dimensional generalization of 1:

$$\mathbf{I}_{ij} = \begin{cases} 1 & i = j \\ 0 & i \neq j, \end{cases}$$

leaving intact $\mathbb{R}^N$ vectors it operates on, changing neither their direction nor their magnitude.

3.4 Gram-Schmidt Orthogonalization

In various situations (notably when forming the bases for the fundamental spaces of an analyzed matrix), we have a set of linearly independent but nonorthogonal vectors $\{\mathbf{x}_1, \mathbf{x}_2, \ldots, \mathbf{x}_N\}$ we wish to turn into an alternative set of vectors, $\{\hat{\mathbf{q}}_1, \hat{\mathbf{q}}_2, \ldots, \hat{\mathbf{q}}_N\}$, that are mutually orthonormal,

$$\hat{\mathbf{q}}_i^T \hat{\mathbf{q}}_j = \begin{cases} 1, & i = j \\ 0, & i \neq j \end{cases} \tag{3.120}$$

Achieving this orthonormalization is the purpose of the Gram-Schmidt procedure.

Let's start with the simple example:

$$\mathbf{x}_1 = \begin{pmatrix} 1 \\ 1 \\ 1 \end{pmatrix}, \quad \mathbf{x}_2 = \begin{pmatrix} 2 \\ 1 \\ 1 \end{pmatrix}, \quad \mathbf{x}_3 = \begin{pmatrix} 1 \\ 2 \\ 1 \end{pmatrix}. \tag{3.121}$$

In general, it would make sense to ensure before launching the orthogonalization procedure of $\mathbf{x}_i \in \mathbb{R}^M$, $i = [1, N]$ that the original vectors are linearly independent, by showing that

$$\begin{pmatrix} \mathbf{x}_1 & \mathbf{x}_2 & \cdots & \mathbf{x}_N \end{pmatrix} \begin{pmatrix} \alpha_1 \\ \alpha_2 \\ \vdots \\ \alpha_N \end{pmatrix} = \mathbf{0} \in \mathbb{R}^M \tag{3.122}$$

admits only the trivial solution, $\alpha_i = 0$ for $1 \le i \le N$. Let's skip this step in this, and subsequent, made up, clean examples.

Obtaining the first element of $\{\hat{\mathbf{q}}_i\}$ is simple, involving nothing more than a simple normalization,

$$\hat{\mathbf{q}}_1 = \frac{1}{\sqrt{\mathbf{x}_1^T \mathbf{x}_1}} \mathbf{x}_1 = \frac{1}{\sqrt{1+1+1}} \begin{pmatrix} 1 \\ 1 \\ 1 \end{pmatrix} = \frac{1}{\sqrt{3}} \begin{pmatrix} 1 \\ 1 \\ 1 \end{pmatrix}. \tag{3.123}$$

The second vector, $\mathbf{q}_2$, is the original second vector, $\mathbf{x}_2$, minus its projection on the first orthonormal basis vector, $\hat{\mathbf{q}}_1$,

$$\mathbf{q}_2 = \mathbf{x}_2 - (\mathbf{x}_2^T \hat{\mathbf{q}}_1) \hat{\mathbf{q}}_1 \tag{3.124}$$

$$= \begin{pmatrix} 2 \\ 1 \\ 1 \end{pmatrix} - \left[\begin{pmatrix} 2 & 1 & 1 \end{pmatrix} \frac{1}{\sqrt{3}} \begin{pmatrix} 1 \\ 1 \\ 1 \end{pmatrix} \right] \frac{1}{\sqrt{3}} \begin{pmatrix} 1 \\ 1 \\ 1 \end{pmatrix} \tag{3.125}$$

$$= \begin{pmatrix} 2 \\ 1 \\ 1 \end{pmatrix} - \frac{4}{3} \begin{pmatrix} 1 \\ 1 \\ 1 \end{pmatrix} = \frac{1}{3} \begin{pmatrix} 2 \\ -1 \\ -1 \end{pmatrix}. \tag{3.126}$$

Because $\|\mathbf{q}_2\| = (\sqrt{4+1+1})/3 = \sqrt{6}/3$, this $\mathbf{q}_2$ readily yields

$$\hat{\mathbf{q}}_2 = \frac{\mathbf{q}_2}{\|\mathbf{q}_2\|} = \frac{1}{3} \begin{pmatrix} 2 \\ -1 \\ -1 \end{pmatrix} \frac{3}{\sqrt{6}} = \frac{1}{\sqrt{6}} \begin{pmatrix} 2 \\ -1 \\ -1 \end{pmatrix}. \tag{3.127}$$

Finally, $\mathbf{q}_3$ is the original $\mathbf{x}_3$ minus its projections on the first and second orthonormal basis vectors $\hat{\mathbf{q}}_1$ and $\hat{\mathbf{q}}_2$,

$$\mathbf{q}_3 = \mathbf{x}_3 - (\mathbf{x}_3^T \hat{\mathbf{q}}_1) \hat{\mathbf{q}}_1 - (\mathbf{x}_3^T \hat{\mathbf{q}}_2) \hat{\mathbf{q}}_2 \tag{3.128}$$

$$= \begin{pmatrix} 1 \\ 2 \\ 1 \end{pmatrix} - \frac{\begin{pmatrix} 1 & 2 & 1 \end{pmatrix} \begin{pmatrix} 1 \\ 1 \\ 1 \end{pmatrix}}{3} \begin{pmatrix} 1 \\ 1 \\ 1 \end{pmatrix} - \frac{\begin{pmatrix} 1 & 2 & 1 \end{pmatrix} \begin{pmatrix} 2 \\ -1 \\ -1 \end{pmatrix}}{6} \begin{pmatrix} 2 \\ -1 \\ -1 \end{pmatrix}$$

$$= \begin{pmatrix} 1 \\ 2 \\ 1 \end{pmatrix} - \frac{4}{3} \begin{pmatrix} 1 \\ 1 \\ 1 \end{pmatrix} + \frac{1}{6} \begin{pmatrix} 2 \\ -1 \\ -1 \end{pmatrix} = \frac{1}{6} \begin{pmatrix} 6 \\ 12 \\ 6 \end{pmatrix} - \frac{8}{6} \begin{pmatrix} 1 \\ 1 \\ 1 \end{pmatrix} + \frac{1}{6} \begin{pmatrix} 2 \\ -1 \\ -1 \end{pmatrix}$$

$$= \frac{1}{2} \begin{pmatrix} 0 \\ 1 \\ -1 \end{pmatrix}$$

from which

$$\hat{\mathbf{q}}_3 = \frac{1}{\sqrt{2}} \begin{pmatrix} 0 \\ 1 \\ -1 \end{pmatrix} \tag{3.129}$$

follows.

It is, in general, prudent to test the results, by making sure that

$$\hat{\mathbf{q}}_i^T \hat{\mathbf{q}}_j = \begin{cases} 1 & i = j \quad \text{(unit norm)} \\ 0 & i \neq j \quad \text{(orthogonality)} \end{cases} \tag{3.130}$$

as required. This criterion is met by the above set $\{\hat{\mathbf{q}}_i\}$.

Let's address one other particular example, from $\mathbb{R}^4$,

$$\mathbf{x}_1 = \begin{pmatrix} 1 \\ 2 \\ 3 \\ 4 \end{pmatrix}, \quad \mathbf{x}_2 = \begin{pmatrix} 1 \\ 6 \\ 12 \\ 4 \end{pmatrix}, \quad \mathbf{x}_3 = \begin{pmatrix} 3 \\ 7 \\ 11 \\ 0 \end{pmatrix}. \tag{3.131}$$

The first of the set $\{\hat{\mathbf{q}}_i\}$ is again the simple renormalization

$$\hat{\mathbf{q}}_1 = \frac{1}{\sqrt{\mathbf{x}_1^T \mathbf{x}_1}} \mathbf{x}_1 = \frac{1}{\sqrt{1+4+9+16}} \begin{pmatrix} 1 \\ 2 \\ 3 \\ 4 \end{pmatrix}. \tag{3.132}$$

The second vector is

$$\begin{aligned} \mathbf{q}_2 &= \mathbf{x}_2 - \left(\mathbf{x}_2^T \hat{\mathbf{q}}_1\right) \hat{\mathbf{q}}_1 \\ &= \begin{pmatrix} 1 \\ 6 \\ 12 \\ 4 \end{pmatrix} - \left[\begin{pmatrix} 1 & 6 & 12 & 4 \end{pmatrix} \frac{1}{\sqrt{30}} \begin{pmatrix} 1 \\ 2 \\ 3 \\ 4 \end{pmatrix} \right] \frac{1}{\sqrt{30}} \begin{pmatrix} 1 \\ 2 \\ 3 \\ 4 \end{pmatrix} \\ &= \begin{pmatrix} 1 \\ 6 \\ 12 \\ 4 \end{pmatrix} - \frac{65}{30} \begin{pmatrix} 1 \\ 2 \\ 3 \\ 4 \end{pmatrix} = \frac{1}{6} \begin{pmatrix} 6 \\ 36 \\ 72 \\ 24 \end{pmatrix} - \frac{1}{6} \begin{pmatrix} 13 \\ 26 \\ 39 \\ 52 \end{pmatrix} = \frac{1}{6} \begin{pmatrix} -7 \\ 10 \\ 33 \\ -28 \end{pmatrix}, \end{aligned} \tag{3.133}$$

from which we derive

$$\hat{\mathbf{q}}_2 = \frac{\mathbf{q}}{\|\mathbf{q}_2\|} \approx \begin{pmatrix} -0.16 \\ 0.22 \\ 0.73 \\ -0.62 \end{pmatrix}. \tag{3.134}$$

The last basis vector is

$$\mathbf{q}_3 = \mathbf{x}_3 - \left(\mathbf{x}_3^T \hat{\mathbf{q}}_1\right) \hat{\mathbf{q}}_1 - \left(\mathbf{x}_3^T \hat{\mathbf{q}}_2\right) \hat{\mathbf{q}}_2. \tag{3.135}$$

Sparing you the straightforward yet laborious arithmetic,

$$\mathbf{q}_3 \approx \begin{pmatrix} 2.76 \\ 1.63 \\ -0.72 \\ -0.96 \end{pmatrix} \implies \hat{\mathbf{q}}_3 = \frac{\mathbf{q}_3}{\|\mathbf{q}_3\|} \approx \begin{pmatrix} 0.81 \\ 0.48 \\ -0.21 \\ -0.28 \end{pmatrix}. \tag{3.136}$$

In conclusion, the Gram-Schmidt procedure is

```
start with linearly independent {x_i}_{i=1}^N
for each i from 1 to N
```

$$\mathbf{q}_i = \mathbf{x}_i - \sum_{j=1}^{i-1} (\mathbf{x}_i^T \hat{\mathbf{q}}_j)\hat{\mathbf{q}}_j$$

$$\hat{\mathbf{q}}_i = \mathbf{q}_i / \|\mathbf{q}_i\|$$

```
end
```

The procedure transforms an original, linearly independent but not orthonormal set $\{\mathbf{x}_i\}$ to an equivalent set $\{\mathbf{q}_i\}$ whose vectors are all mutually orthogonal and have unit norms. Now, when we express an arbitrary vector from the space spanned by these alternative sets (in the above two cases $\mathbb{R}^3$ and $\mathbb{R}^4$, respectively) as a linear combination of the orthonormal vectors $\hat{\mathbf{q}}_i$, the projections (the weights) are independent of one another, and their squared elements' sum is exactly the squared norm of the represented vector, not more. This is the reason orthonormalization of the basis vectors is very useful.

Let's demonstrate this point by representing $\mathbf{y} = (\ 1\ -1\ 1\ )^T$ in various bases. First, in the Cartesian basis $(\{\hat{\mathbf{i}}, \hat{\mathbf{j}}, \hat{\mathbf{k}}\})$, the norm of the coefficient vector $\mathbf{c}_c$ (with which $(\ \hat{\mathbf{i}}\ \hat{\mathbf{j}}\ \hat{\mathbf{k}}\ )\mathbf{c}_c = \mathbf{y}$) is the same as $\mathbf{y}$'s norm, $\|\mathbf{c}_c\| = \|\mathbf{y}\| = \sqrt{3} \approx 1.73$ (because $\mathbf{I}^{-1} = \mathbf{I}$). The equality of represented vector norm to that of the representation coefficient vector holds in representing any vector in terms of any orthonormal basis. For example, it holds for representing $\mathbf{y}$ in terms of the orthonormal set obtained earlier in this section from $\mathbf{x}_{1,2,3}$, $\mathbf{Q} = (\ \hat{\mathbf{q}}_1\ \hat{\mathbf{q}}_2\ \hat{\mathbf{q}}_3\ )$ (eqs. 3.132, 3.134, and 3.136), where $\mathbf{c}_q \approx (\ 0.58\ 0.82\ -1.41\ )^T$ but $\|\mathbf{c}_q\| = \|\mathbf{y}\| = \sqrt{3}$ still holds. By contrast, in terms of $\mathbf{Q}$'s nonorthonormal forebearer $\mathbf{X} = (\ \hat{\mathbf{x}}_1\ \hat{\mathbf{x}}_2\ \hat{\mathbf{x}}_3\ )$ obtained by normalizing $\mathbf{x}_{1,2,3}$ (eq. 3.121), $\|\mathbf{c}_x\| \approx 7.14$, more than 4 times larger than $\|\mathbf{c}_q\|$ or $\|\mathbf{c}_c\|$.

To gain further insights into the representation norm question, let's consider representing $\mathbf{y}$ in terms of

$$\mathbf{Z} = (\ \hat{\mathbf{z}}_1\ \hat{\mathbf{z}}_2\ \hat{\mathbf{z}}_3\ ) \tag{3.137}$$

derived from

$$\mathbf{z}_1 = \begin{pmatrix} 1 \\ 1 \\ 1 \end{pmatrix}, \mathbf{z}_2 = \begin{pmatrix} 1 \\ 1 \\ 1-\varepsilon \end{pmatrix}, \mathbf{z}_3 = \begin{pmatrix} 1+\varepsilon \\ 1 \\ 1 \end{pmatrix}, \tag{3.138}$$

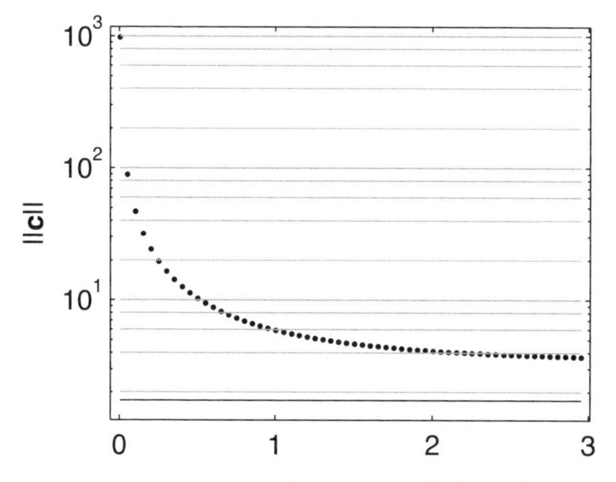

Figure 3.4. Representation coefficient vector, $\|\mathbf{c}\|$, as a function of ε for the representation problem discussed in section 3.4 and defined in eqs. 3.137–3.138. Because of its wide numerical range (~3 orders of magnitude), the vertical axis is represented in $\log_{10}$. The norm of the represented vector $\mathbf{y}$, also the minimal possible coefficient vector norm, is given by the thick horizontal black (also lowermost) line at $\min_{\varepsilon}(\|\mathbf{c}\|) \equiv \|\mathbf{y}\| \approx 10^{0.24}$. Thin gray horizontal grid lines are at $10^{2/10}$ increments.

where ε is a parameter we vary over $[5 \cdot 10^{-3}, 3]$ in increments of $5 \cdot 10^{-2}$ (for a total of 60 values). Formally, this set is a basis for $\mathbb{R}^3$. As ε gets smaller, however, the angles between the three vectors become very small as the vectors become nearly mutually parallel (e.g., for $\varepsilon = 5 \cdot 10^{-3}$, the angles between the vectors are all near 0.13°). Figure 3.4 displays the norm of the representation coefficient vector, $\|\mathbf{c}\|$, as a function of ε. Clearly, as $\varepsilon \to 0$ (as we move from right to left in fig. 3.4), the representation coefficient vector norm increases dramatically, demonstrating the increasing suboptimality of the respective basis and thus the importance of the Gram-Schmidt orthonormalization process.

3.5 Summary

An $M \times N$ matrix is associated with four fundamental spaces. The *column space* is the set of all M-vectors that are linear combinations of the columns. If the matrix has M independent columns (its rank is M, which requires that $N \geq M$), then the column space is $\mathbb{R}^M$; otherwise (if the rank is less than M), the column space is a subspace of $\mathbb{R}^M$. Also in $\mathbb{R}^M$ is the *left null space*, the set of all M-vectors that the matrix's transpose maps to the zero N-vector. The column and left null spaces are mutually orthogonal (any vector from one is orthogonal to any vector from the other) and combine (direct sum) to $\mathbb{R}^M$.

The *row space* is the set of all N-vectors that are linear combinations of the rows. If the matrix has N independent rows (its rank is N, which requires that $M \geq N$), then the row space is $\mathbb{R}^N$; otherwise (if the rank is less than N), the row space is a subspace of $\mathbb{R}^N$. Also in $\mathbb{R}^N$ is the *null space*, the set of all N-vectors that the matrix maps to the zero M-vector. The row and null spaces are also mutually orthogonal and combine (direct sum) to $\mathbb{R}^N$.

The linear system $\mathbf{Ax} = \mathbf{b}$ has an exact solution if and only if $\mathbf{b}$ is from $\mathbf{A}$'s column space. Otherwise, the system has no exact solution. (But, as we see in chapter 9, it can be transformed into the alternative $\mathbf{Ax} = \mathbf{d}$ system, where $\mathbf{d}$ is "as close to $\mathbf{b}$ as possible" in the *least-squares* sense.)

If $\mathbf{b}$ is from $\mathbf{A}$'s column space, and the system admits an exact solution, if $\mathbf{A}$ has less than N independent columns (which can arise for both $M > N$ and $N > M$), the exact solution is not unique because there exists a nonempty null space from which any vector can be added to the solution.

FOUR

Introduction to Eigenanalysis

4.1 Preface

EIGENANALYSIS AND ITS NUMEROUS offsprings form the suite of algebraic operations most important and relevant to data analysis, as well as to dynamical systems, modeling, numerical analysis, and related key branches of applied mathematics. This chapter introduces, and places in a broader context, the algebraic operation of eigen-decomposition.

To have eigen-decomposition, a matrix must be square. Yet data matrices are very rarely square. The direct relevance of eigen-decomposition to data analysis is therefore limited. Indirectly, however, generalized eigenanalysis is enormously important to studying data matrices, as we will see later. Because of the centrality of generalized eigenanalysis to data matrices, and because those generalizations (notably the singular value decomposition, chapter 5), build, algebraically and logically, on eigenanalysis itself, it makes sense to discuss eigenanalysis at some length.

4.1.1 Background

Matrices are higher-dimensional generalization of real numbers, in general altering both direction and magnitude of vectors on which they operate. To be sure, the real numbers have direction too; it's just always the same, forward or backward along the real line. The thick gray horizontal line in fig. 4.1 represents the number 3. Next, we take 3×2, shown by the slightly thinner black horizontal line. The magnitude changed (from 3 to 6), but the direction was conserved; both are along the real line. Thus, any real number can be represented as $\alpha\hat{\mathbf{r}}$, where α is the length, while $\hat{\mathbf{r}} = (1)$ can be thought of as the real numbers' only spanning 1-vector (where only one is needed because real numbers all share the same direction).

In higher dimensions ($\mathbb{R}^{\geq 2}$), both magnitude and direction can change, as fig. 4.1 shows. The other two lines of fig. 4.1 are the $\mathbb{R}^2$ vectors

$$\begin{pmatrix} 1 \\ 2 \end{pmatrix} \quad \text{and} \quad \begin{pmatrix} 1 & 2 \\ 3 & 0 \end{pmatrix}\begin{pmatrix} 1 \\ 2 \end{pmatrix} = \begin{pmatrix} 5 \\ 3 \end{pmatrix} \tag{4.1}$$

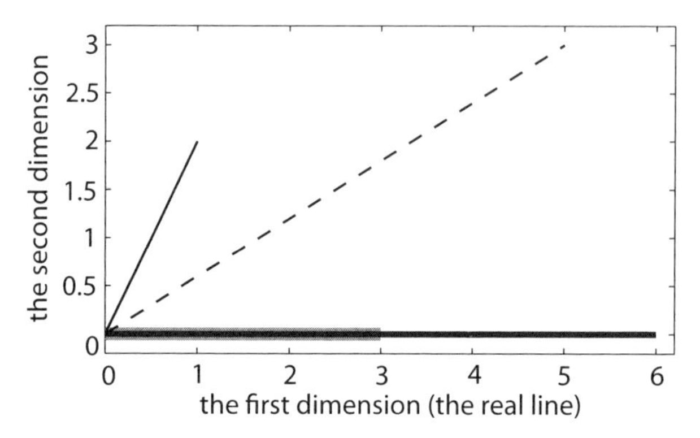

Figure 4.1. Demonstration of direction and magnitude of vectors and matrices.

(in thin solid and dashed lines, respectively). Clearly, the matrix premultiplication altered both magnitude and direction of $(1\ 2)^T$. (But this doesn't have to be so; just recall the identity matrix!)

The change in direction is easy to understand—it stems from the multiplication rules and depends on the elements of the matrices involved. But what about the magnitude? This is determined by one of the most fundamental properties of square matrices, their eigenvalues. (The direction, too, can be cast in these terms, being governed by the eigenvectors, which we will get to a bit later.)

Because eigenanalysis (the identification of eigenvalues and eigenvectors) applies only to square matrices, in this chapter all matrices are square. If you are not persuaded by this chapter's preface, and still wonder why we want to bother with square matrices when data matrices are almost always rectangular, there exists a more specific and straightforward answer: because the principal tool of multidimensional data analysis is the (square symmetric) covariance matrix. But we are getting a bit ahead of ourselves.

4.2 Eigenanalysis Introduced

To get a handle on eigenvalues, it's best to start with an example. Consider the temporal evolution (system state through time t) of an ecosystem comprising two species, $x(t)$ and $y(t)$. Let's further envision that, left alone, each species multiplies at a given rate, and that the species do not interact,

$$\frac{dx(t)}{dt} = ax(t) \tag{4.2}$$

$$\frac{dy(t)}{dt} = by(t) \tag{4.3}$$

or, in vector form (suppressing explicit t dependence for neatness),

$$\frac{d}{dt}\begin{pmatrix} x \\ y \end{pmatrix} = \begin{pmatrix} a & 0 \\ 0 & b \end{pmatrix}\begin{pmatrix} x \\ y \end{pmatrix}. \tag{4.4}$$

However, the species do interact; let's assume they compete over the same food source. Then, the less of x in the system, the happier y gets, and vice versa. The above system of linear ODEs (ordinary differential equations) must therefore be modified to take note of the competition,

$$\frac{dx(t)}{dt} = ax(t) - cy(t) \tag{4.5}$$

$$\frac{dy(t)}{dt} = by(t) - dx(t) \tag{4.6}$$

(You surely realize that this is a grotesque oversimplification of competition, I just need a system of linear ODEs . . .) Now the system's state, $\mathbf{x} = (\, x\, y \,)^T$, evolves according to

$$\frac{d}{dt}\begin{pmatrix} x \\ y \end{pmatrix} \equiv \frac{d\mathbf{x}}{dt} = \begin{pmatrix} a & -c \\ -d & b \end{pmatrix}\begin{pmatrix} x \\ y \end{pmatrix} \equiv A\mathbf{x}, \tag{4.7}$$

where the state's first element x is easily distinguished from the full state $\mathbf{x}$.

Just like in the scalar case, we can always try a solution and check whether it satisfies the equation. Let's choose, then

$$x(t) = \alpha e^{\lambda t} \qquad \text{and} \qquad y(t) = \beta e^{\lambda t}, \tag{4.8}$$

with amplitudes α and β and a timescale λ describing the temporal evolution of both species. Substituting the solutions into the 2 scalar equations (eqs. 4.5 and 4.6), we get

$$\alpha\lambda e^{\lambda t} = a\alpha e^{\lambda t} - c\beta e^{\lambda t} \tag{4.9}$$

$$\beta\lambda e^{\lambda t} = -d\alpha e^{\lambda t} + b\beta e^{\lambda t}. \tag{4.10}$$

The exponential is common to all terms, and is nonzero. Hence, the equations can be divided by it, yielding

$$\alpha\lambda = a\alpha - c\beta \tag{4.11}$$

$$\beta\lambda = -d\alpha + b\beta \tag{4.12}$$

or, in vector form,

$$\lambda\mathbf{e} \equiv \lambda\begin{pmatrix} \alpha \\ \beta \end{pmatrix} = \begin{pmatrix} a & -c \\ -d & b \end{pmatrix}\begin{pmatrix} \alpha \\ \beta \end{pmatrix} = \mathbf{Ae}. \tag{4.13}$$

Let's examine this equation. We are looking for scalar–vector pair $(\lambda, \mathbf{e})$ (or $(\lambda_i, \mathbf{e}_i)$ pairs) satisfying $\mathbf{Ae}_i = \lambda_i \mathbf{e}_i$. That is, we seek vectors whose direction is invariant under premultiplication by $\mathbf{A}$ and whose magnitude thus changes by a factor λ_i, so that $\|\mathbf{A}\mathbf{e}_i\|/\|\mathbf{e}_i\| = \lambda_i$.

Since both sides of $\lambda\mathbf{e} = \mathbf{A}\mathbf{e}$ premultiply $\mathbf{e}$, we next strive to combine them. However, recall that we cannot simply add the scalar λ to the matrix $\mathbf{A}$, as this is dimensionally impossible. What we can do, however, is

$$\left[\begin{pmatrix} a & -c \\ -d & b \end{pmatrix} - \lambda \begin{pmatrix} 1 & 0 \\ 0 & 1 \end{pmatrix}\right]\begin{pmatrix} \alpha \\ \beta \end{pmatrix} = \begin{pmatrix} 0 \\ 0 \end{pmatrix} \tag{4.14}$$

or

$$(\mathbf{A} - \lambda\mathbf{I})\mathbf{e} = \mathbf{0}. \tag{4.15}$$

This is the central equation of this discussion.

Let's write the equation explicitly,

$$\begin{pmatrix} a - \lambda & -c \\ -d & b - \lambda \end{pmatrix}\begin{pmatrix} \alpha \\ \beta \end{pmatrix} = \begin{pmatrix} 0 \\ 0 \end{pmatrix}, \tag{4.16}$$

i.e., the vector we are looking for is from the null space of $\mathbf{B} \equiv (\mathbf{A} - \lambda\mathbf{I})$. For $\mathbf{B}$ to have a nontrivial null space, it must be singular. One way to check whether a matrix is singular or not is to evaluate its determinant; if $\det(\mathbf{A} - \lambda\mathbf{I}) = 0$, $\mathbf{B}$ has a nontrivial null space, as required. The determinant of a 2×2

$$\mathbf{D} = \begin{pmatrix} a & b \\ c & d \end{pmatrix} \quad \text{is} \quad \det(\mathbf{D}) = \begin{vmatrix} a & b \\ c & d \end{vmatrix} = ad - bc, \tag{4.17}$$

while the determinant of

$$\mathbf{D} = \begin{pmatrix} a & b & c \\ d & e & f \\ g & h & i \end{pmatrix} \in \mathbb{R}^{3\times 3} \tag{4.18}$$

is

$$\det(\mathbf{D}) = \begin{vmatrix} a & b & c \\ d & e & f \\ g & h & i \end{vmatrix} = a(ei - fh) - b(di - fg) + c(dh - eg). \tag{4.19}$$

Formulae for higher-dimensional determinants can be readily found in linear algebra textbooks, but you can also figure out (with some mental gymnastics) the rule from the above.

The determinant of $\mathbf{A} - \lambda\mathbf{I}$ yields $\mathbf{A}$'s characteristic polynomial. For an $\mathbf{A} \in \mathbb{R}^{N\times N}$ matrix, the characteristic polynomial has N roots, $\mathbf{A}$'s eigenvalues. For the above 2-species competition scenario,

$$\det(\mathbf{A} - \lambda\mathbf{I}) = \lambda^2 - (a+b)\lambda + ab - cd, \tag{4.20}$$

with roots

$$\lambda_{1,2} = \frac{(a+b) \pm \sqrt{(a+b)^2 - 4(ab-cd)}}{2}, \tag{4.21}$$

the eigenvalues.

If you are a bit mystified by the determinant, notice that the characteristic polynomial can be derived directly from the requirement for a nontrivial null space for $\mathbf{A} - \lambda\mathbf{I}$ (eq. 4.16). Carrying out the left-hand-side product in eq. 4.16, we get

$$\alpha(a-\lambda) - \beta c = 0 \quad \text{and} \quad -\alpha d + \beta(b-\lambda) = 0$$

or

$$\alpha = \frac{\beta c}{a-\lambda} \quad \text{and} \quad \alpha = \frac{\beta(b-\lambda)}{d}. \tag{4.22}$$

Equating the two expressions for α and dividing by β, we get

$$\frac{c}{a-\lambda} = \frac{b-\lambda}{d} \implies (a-\lambda)(b-\lambda) - cd = 0, \tag{4.23}$$

which is the characteristic polynomial.

Either way, for each (eigenvalue, eigenvector) pair, we solve the equation

$$(\mathbf{A} - \lambda_i\mathbf{I})\mathbf{e}_i = \mathbf{0}, \qquad i = 1,2,\ldots,N. \tag{4.24}$$

Since we have already established that $(\mathbf{A} - \lambda_i\mathbf{I})$ has a nontrivial null space (λ_i were chosen to ensure that), nontrivial $\mathbf{e}_i$ must exist.

Let's consider a numerical example for the 2-species system

$$\mathbf{A} = \begin{pmatrix} 1 & -\frac{3}{2} \\ -\frac{1}{2} & 2 \end{pmatrix}. \tag{4.25}$$

The 1 in position (1, 1) means that species x, when unmolested by y, grows exponentially with an e-folding timescale of 1 in whatever time units we employ (the e-folding timescale is the time it takes an exponentially growing/decaying entity x to grow/decay by a factor of e, so for an exponentially growing x, an e-folding of 1 means that $x(t+1)/x(t) = \mathrm{e} \approx 2.72$). For species y the corresponding number is 2 (given by element (2, 2)). That is, y's biology enables it to exploit its available resources for expansion at twice x's rate. The off-diagonal elements mean that the species are affected by the competition differently; species x is rather sensitive to the fierce competition species y puts up (the $-\frac{3}{2}$ in position (1, 2)), while species y is less easily perturbed by the presence of species x. So much for population dynamics 101.

Let's eigenanalyze $\mathbf{A}$, starting with

$$\det\left(\mathbf{A} - \lambda\mathbf{I}\right) = \begin{vmatrix} 1-\lambda & -\frac{3}{2} \\ -\frac{1}{2} & 2-\lambda \end{vmatrix} = \left(1-\lambda\right)\left(2-\lambda\right) - \frac{3}{4}, \tag{4.26}$$

which yields the characteristic equation $\lambda^2 - 3\lambda + \frac{5}{4} = 0$, with roots (1, 5)/2. Solving $\mathbf{A} - \lambda_1\mathbf{I} = \mathbf{0}$ with the first root ($\lambda_1 = \frac{1}{2}$) yields

$$\frac{1}{2}\begin{pmatrix} 1 & -3 \\ -1 & 3 \end{pmatrix}\begin{pmatrix} \alpha \\ \beta \end{pmatrix} = \begin{pmatrix} 0 \\ 0 \end{pmatrix} \implies \begin{pmatrix} \alpha \\ \beta \end{pmatrix} = \beta\begin{pmatrix} 3 \\ 1 \end{pmatrix} \tag{4.27}$$

with unconstrained β, so $\mathbf{e}_1 = (\ 3\ 1\ )^T$. To ascertain that $\mathbf{Ae}_1 = \lambda_1\mathbf{e}_1$ is indeed satisfied when $\lambda_1 = \frac{1}{2}$ and $\mathbf{e}_1 = (\ 3\ 1\ )^T$, we evaluate the left-hand side

$$\underbrace{\begin{pmatrix} 1 & -\frac{3}{2} \\ -\frac{1}{2} & 2 \end{pmatrix}}_{\mathbf{A}}\underbrace{\begin{pmatrix} 3 \\ 1 \end{pmatrix}}_{\mathbf{e}_1} = \begin{pmatrix} 3 - \frac{3}{2} \\ -\frac{3}{2} + 2 \end{pmatrix} = \begin{pmatrix} \frac{3}{2} \\ \frac{1}{2} \end{pmatrix}, \tag{4.28}$$

which is indeed $\lambda_1\mathbf{e}_1$, as required.

Solving $\mathbf{A} - \lambda_2\mathbf{I} = \mathbf{0}$ with $\lambda_2 = \frac{5}{2}$ yields

$$\frac{1}{2}\begin{pmatrix} -3 & -3 \\ -1 & -1 \end{pmatrix}\begin{pmatrix} \alpha \\ \beta \end{pmatrix} = \begin{pmatrix} 0 \\ 0 \end{pmatrix} \implies \begin{pmatrix} \alpha \\ \beta \end{pmatrix} = \beta\begin{pmatrix} -1 \\ 1 \end{pmatrix} \tag{4.29}$$

with unconstrained β, so $\mathbf{e}_2 = (\ -1\ 1\ )^T$. Since the test

$$\underbrace{\begin{pmatrix} 1 & -\frac{3}{2} \\ -\frac{1}{2} & 2 \end{pmatrix}}_{\mathbf{A}}\underbrace{\begin{pmatrix} -1 \\ 1 \end{pmatrix}}_{\mathbf{e}_2} = \begin{pmatrix} -1 - \frac{3}{2} \\ \frac{1}{2} + 2 \end{pmatrix} = \frac{5}{2}\begin{pmatrix} -1 \\ 1 \end{pmatrix} = \lambda_2\mathbf{e}_2 \tag{4.30}$$

is also satisfied, our problem is solved. We can write down victoriously the complete solution to

$$\frac{d}{dt}\begin{pmatrix} x \\ y \end{pmatrix} = \frac{1}{2}\begin{pmatrix} 2 & -3 \\ -1 & 4 \end{pmatrix}\begin{pmatrix} x \\ y \end{pmatrix} \tag{4.31}$$

as the superposition of the two pure exponentials governing the evolution of the two modes (the two eigenvalue/eigenvector pairs)

$$\mathbf{x}(t) = a_1 \exp\left(\frac{1}{2}t\right)\mathbf{e}_1 + a_2 \exp\left(\frac{5}{2}t\right)\mathbf{e}_2, \tag{4.32}$$

with amplitudes a_1 and a_2 determined from the initial conditions. Let's pick $x(0) = y(0) = 1000$, in which case

$$\begin{pmatrix} 1000 \\ 1000 \end{pmatrix} = a_1\begin{pmatrix} 3 \\ 1 \end{pmatrix} + a_2\begin{pmatrix} -1 \\ 1 \end{pmatrix} \implies a_1 = a_2 = 500, \tag{4.33}$$

yielding finally

$$\mathbf{x}(t) = 500\left[\exp\left(\frac{1}{2}t\right)\begin{pmatrix} 3 \\ 1 \end{pmatrix} + \exp\left(\frac{5}{2}t\right)\begin{pmatrix} -1 \\ 1 \end{pmatrix}\right]. \tag{4.34}$$

The left panels of fig. 4.2 show the results of the system over one-half time unit.

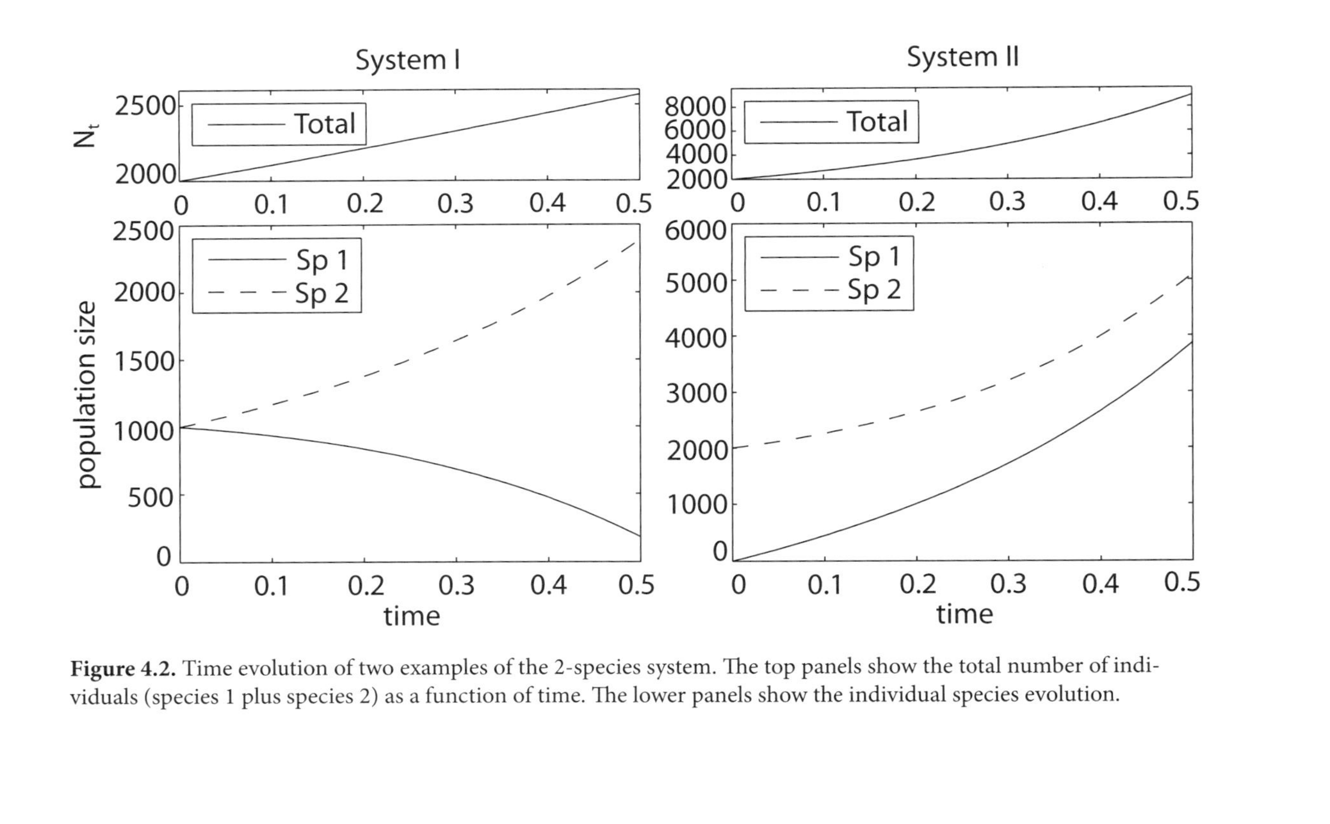

Figure 4.2. Time evolution of two examples of the 2-species system. The top panels show the total number of individuals (species 1 plus species 2) as a function of time. The lower panels show the individual species evolution.

The right panels correspond to the different system whose governing matrix is

$$\mathbf{A} = \begin{pmatrix} 1 & -2 \\ -2 & 1 \end{pmatrix}, \tag{4.35}$$

i.e., a system in which both species exhibit the same inherent growth rate (the diagonal elements) and equal susceptibility to competition with the other species (the off-diagonal elements). The characteristic polynomial is

$$(1 - \lambda)^2 - 4 = 0, \tag{4.36}$$

with roots $\lambda_1 = 3$ and $\lambda_1 = -1$. The corresponding eigenvectors are

$$\lambda_1 = 3, \mathbf{e}_1 = \begin{pmatrix} 1 \\ -1 \end{pmatrix} \quad \text{and} \quad \lambda_2 = -1, \mathbf{e}_2 = \begin{pmatrix} 1 \\ 1 \end{pmatrix}. \tag{4.37}$$

The usual tests yield

$$\mathbf{Ae}_1 = \begin{pmatrix} 1 & -2 \\ -2 & 1 \end{pmatrix}\begin{pmatrix} 1 \\ -1 \end{pmatrix} = 3\begin{pmatrix} 1 \\ -1 \end{pmatrix} = \lambda_1\mathbf{e}_1 \tag{4.38}$$

$$\mathbf{Ae}_2 = \begin{pmatrix} 1 & -2 \\ -2 & 1 \end{pmatrix}\begin{pmatrix} 1 \\ 1 \end{pmatrix} = -1\begin{pmatrix} 1 \\ 1 \end{pmatrix} = \lambda_2\mathbf{e}_2, \tag{4.39}$$

as required. Let's choose the same initial total number of individuals, 2000, which dictates the complete solution

$$\mathbf{x}(t) = 1000\left[\mathbf{e}^{3t}\begin{pmatrix} 1 \\ -1 \end{pmatrix} + e^{-t}\begin{pmatrix} 1 \\ 1 \end{pmatrix}\right]. \tag{4.40}$$

As the right panels of fig. 4.2 show, the small difference in the growth rate of the fastest growing mode (the largest eigenvalue and its corresponding eigenvector) is sufficient to give very different time behavior from that of the system in the previous example. Note that it is often useful (and customary) to normalize the eigenvectors (which I did not do above).

We can normalize next the eigenvectors to unit norm and form an eigenvector matrix $\mathbf{E}$ (not to be confused with elementary row operation matrices of Gaussian elimination) whose columns are the normalized eigenvectors

$$\mathbf{E} = \frac{1}{\sqrt{2}}\begin{pmatrix} 1 & 1 \\ -1 & 1 \end{pmatrix}. \tag{4.41}$$

Next, we obtain $\mathbf{E}$'s inverse

$$\mathbf{E}^{-1} = \frac{1}{\sqrt{2}}\begin{pmatrix} 1 & -1 \\ 1 & 1 \end{pmatrix}. \tag{4.42}$$

With these matrices,

$$\mathbf{E}^{-1}\mathbf{AE} = \frac{1}{\sqrt{2}}\begin{pmatrix} 1 & -1 \\ 1 & 1 \end{pmatrix}\begin{pmatrix} 1 & -2 \\ -2 & 1 \end{pmatrix}\frac{1}{\sqrt{2}}\begin{pmatrix} 1 & 1 \\ -1 & 1 \end{pmatrix} \tag{4.43}$$

$$= \frac{1}{2}\begin{pmatrix} 1 & -1 \\ 1 & 1 \end{pmatrix}\begin{pmatrix} 1 & -2 \\ -2 & 1 \end{pmatrix}\begin{pmatrix} 1 & 1 \\ -1 & 1 \end{pmatrix} = \begin{pmatrix} 3 & 0 \\ 0 & -1 \end{pmatrix} = \mathbf{\Lambda}, \tag{4.44}$$

where $\mathbf{\Lambda}$ is the eigenvalue matrix, with the eigenvalues along the diagonal. For matrices with a full set of (N) eigenvectors (some matrices, discussed below, have fewer eigenvectors than N, and are thus nondiagonalizable), this is one of the many important aspects of eigenanalysis:

$$\mathbf{E}^{-1}\mathbf{A}\mathbf{E} = \mathbf{\Lambda} \quad \text{or} \quad \mathbf{A} = \mathbf{E}\mathbf{\Lambda}\mathbf{E}^{-1}. \tag{4.45}$$

If the eigenvectors are orthonormal, this can be simplified even further to

$$\mathbf{E}^{T}\mathbf{A}\mathbf{E} = \mathbf{\Lambda} \quad \text{or} \quad \mathbf{A} = \mathbf{E}\mathbf{\Lambda}\mathbf{E}^{T} \tag{4.46}$$

because for any orthonormal

$$\mathbf{D} = (\mathbf{d}_1 \quad \mathbf{d}_2 \quad \cdots \quad \mathbf{d}_N), \qquad \mathbf{d}_i^T\mathbf{d}_j = \begin{cases} 1, & 1 = j \\ 0, & 1 \neq j' \end{cases} \tag{4.47}$$

$\mathbf{D}^T\mathbf{D} = \mathbf{D}\mathbf{D}^T = \mathbf{I}$.

Let's look at some interesting and revealing examples.

- Failure to diagonalize: With

  $$\mathbf{A} = \begin{pmatrix} 0 & 3 \\ 0 & 0 \end{pmatrix},$$

  the characteristic polynomial is $(-\lambda)^2 = 0$, with $\lambda_{1,2} = 0$. The corresponding eigenvectors satisfy

  $$\begin{pmatrix} 0 & 3 \\ 0 & 0 \end{pmatrix}\begin{pmatrix} \alpha \\ \beta \end{pmatrix} = \begin{pmatrix} 0 \\ 0 \end{pmatrix},$$

  which yields $\mathbf{e}_1 = \mathbf{e}_2 = (\ 1\ 0\ )^T$. In this case there are not enough eigenvectors to form $\mathbf{E}$ of the necessary dimension (2×2), and diagonalization fails. Note that this is not because of $\lambda_{1,2} = 0$, or even because $\lambda_1 = \lambda_2$; it is the twice-repeated eigenvalue whose algebraic multiplicity (the number of times it is repeated) is 2, but whose geometric multiplicity (the dimension of the subspace spanned by the corresponding eigenvectors) fails to achieve the required 2 ($\mathbf{A} - \lambda\mathbf{I}$ has only a one-dimensional null space). When the number of linearly independent eigenvectors is smaller than N, we can still form $\mathbf{E}$, but it will be rank deficient and thus not invertible. If $\mathbf{E}$ is singular, we cannot proceed with the representation $\mathbf{A} = \mathbf{E}\mathbf{\Lambda}\mathbf{E}^{-1}$, so $\mathbf{A}$ is not diagonalizable.

  To further clarify failure to diagonalize, let's briefly consider

  $$\mathbf{A} = \begin{pmatrix} 2 & 0 \\ 2 & 2 \end{pmatrix}, \tag{4.48}$$

  which gives rise to

$$|\mathbf{A} - \lambda \mathbf{I}| = (2-\lambda)^2 = 0, \qquad \Longrightarrow \qquad \lambda_1 = \lambda_2 = 2. \tag{4.49}$$

Solving next $(\mathbf{A} - 2\mathbf{I})\mathbf{e}_i = \mathbf{0}$,

$$\begin{pmatrix} 0 & 0 \\ 2 & 0 \end{pmatrix} \mathbf{e}_i = \begin{pmatrix} 0 \\ 0 \end{pmatrix} \quad \Longrightarrow \quad \mathbf{e}_{1,2} = \begin{pmatrix} 0 \\ \alpha \end{pmatrix} \tag{4.50}$$

for any α. Since $\dim[\mathcal{N}(\mathbf{A} - \lambda \mathbf{I})] = 1$, $\mathbf{E}$ is singular, and thus this $\mathbf{A}$ is not diagonalizable.

- Things are great when $\mathbf{A}$ yields an orthonormal $\mathbf{E}$, as with, e.g.,

$$\mathbf{A} = \begin{pmatrix} 1 & 1 \\ 1 & 1 \end{pmatrix}. \tag{4.51}$$

The characteristic polynomial is $(1-\lambda)^2 - 1 = 0$, with $\lambda_1 = 0$ and $\lambda_2 = 2$. The eigenvector equations give

$$\begin{pmatrix} 1 & 1 \\ 1 & 1 \end{pmatrix}\begin{pmatrix} \alpha \\ \beta \end{pmatrix} = \begin{pmatrix} 0 \\ 0 \end{pmatrix} \quad \Longrightarrow \quad \mathbf{e}_1 = \begin{pmatrix} 1 \\ -1 \end{pmatrix} \tag{4.52}$$

$$\begin{pmatrix} 1 & 1 \\ 1 & 1 \end{pmatrix}\begin{pmatrix} \alpha \\ \beta \end{pmatrix} = 2\begin{pmatrix} \alpha \\ \beta \end{pmatrix} \quad \Longrightarrow \quad \mathbf{e}_2 = \begin{pmatrix} 1 \\ 1 \end{pmatrix}, \tag{4.53}$$

which we normalize and use to construct

$$\mathbf{E} = \frac{1}{\sqrt{2}}\begin{pmatrix} 1 & -1 \\ 1 & 1 \end{pmatrix}. \tag{4.54}$$

Now recall that, because this $\mathbf{E}$ is orthonormal, $\mathbf{E}^T\mathbf{E} = \mathbf{E}\mathbf{E}^T = \mathbf{I}$, $\mathbf{E}$'s transpose is also its inverse. Very convenient. It's worth noting that for the eigenvectors of $\mathbf{A}$ to for an orthonormal set, $\mathbf{A}\mathbf{A}^T = \mathbf{A}^T\mathbf{A}$ must hold.[1]

- Powers of a matrix: Consider the square of the above

$$\mathbf{A} = \begin{pmatrix} 1 & 1 \\ 1 & 1 \end{pmatrix}, \tag{4.55}$$

$$\mathbf{A}^2 = \mathbf{A}\mathbf{A} = \begin{pmatrix} 1 & 1 \\ 1 & 1 \end{pmatrix}\begin{pmatrix} 1 & 1 \\ 1 & 1 \end{pmatrix} = \begin{pmatrix} 2 & 2 \\ 2 & 2 \end{pmatrix}. \tag{4.56}$$

The characteristic equation $(2-\lambda)^2 - 4 = 0$ yields $\lambda_1 = 0$ and $\lambda_2 = 4$, the square of $\mathbf{A}$'s eigenvalues. The eigenvectors are

$$\begin{pmatrix} 2 & 2 \\ 2 & 2 \end{pmatrix}\begin{pmatrix} \alpha \\ \beta \end{pmatrix} = \begin{pmatrix} 0 \\ 0 \end{pmatrix} \quad \Longrightarrow \quad \mathbf{e}_1 = \begin{pmatrix} 1 \\ -1 \end{pmatrix} \tag{4.57}$$

[1] For a discussion of the profoundly important effects of dynamical matrices failing to meet this criterion, see, e.g., Farrell, B. F. (1982) The initial growth of disturbances in a baroclinic flow. *J. Atmos. Sci.* **39**, 1663–1686; Farrell, B. F. and P. J. Ioannou (1996) Generalized stability theory, part I: autonomous operators. *J. Atmos. Sci.* **53**, 2025–2040; or Trefethen, L. N. (2005) *Spectra and Pseudospectra: The Behavior of Nonnormal Matrices and Operators*, Princeton University Press, Princeton, NJ, 624 pp., ISBN-13: 978-0691119465.

$$\begin{pmatrix} 2 & 2 \\ 2 & 2 \end{pmatrix}\begin{pmatrix} \alpha \\ \beta \end{pmatrix} = 4\begin{pmatrix} \alpha \\ \beta \end{pmatrix} \quad \Longrightarrow \quad \mathbf{e}_2 = \begin{pmatrix} 1 \\ 1 \end{pmatrix}, \tag{4.58}$$

just like **A**'s. This too is a general property of diagonalizable matrices; their eigenspace is invariant under raising to an arbitrary power, while the eigenvalues are raised to the required power.

4.3 Eigenanalysis as Spectral Representation

One of the most important aspects of eigen-decomposition of data matrices is that it affords a spectral representation of the data. Consequently, it is appropriate to discuss briefly eigenanalysis as spectral representation. In anticipation of later sections, and, in particular, ones addressing the singular value decomposition (SVD, chapter 5), let's now get to know one of this play's key characters, the spectrum.

4.3.1 The Spectrum: First Encounter

There are various definitions of the spectrum. For real matrices, the *spectral theorem* states that if $\mathbf{S} \in \mathbb{R}^{N\times N}$ is symmetric ($\mathbf{S} = \mathbf{S}^T$), there exists an orthonormal matrix $\mathbf{Q} \in \mathbb{R}^{N\times N}$ and a diagonal matrix $\mathbf{D} \in \mathbb{R}^{N\times N}$ such that $\mathbf{S} = \mathbf{QDQ}^T$ or, equivalently, $\mathbf{D} = \mathbf{Q}^T\mathbf{SQ}$. Then, **S**'s *spectrum* is the set $\{d_i\}_{i=1}^N$, **D**'s N diagonal elements. In **S**'s spectral representation, the full information in **S** is equivalently held by the N columns of **Q**, $\{\hat{\mathbf{q}}_i\}_{i=1}^N$, and the N numbers $\{d_i\}_{i=1}^N$,

$$\mathbf{S} = d_1\hat{\mathbf{q}}_1\hat{\mathbf{q}}_1^T + d_2\hat{\mathbf{q}}_2\hat{\mathbf{q}}_2^T + \cdots + d_{N-1}\hat{\mathbf{q}}_{N-1}\hat{\mathbf{q}}_{N-1}^T + d_N\hat{\mathbf{q}}_N\hat{\mathbf{q}}_N^T. \tag{4.59}$$

Beyond the above narrow definition, and inclusive of it, the key importance of spectral decomposition and spectral representation is the split of the full information being spectrally represented into distinct orthonormal patterns ($\{\hat{\mathbf{q}}_i\}_{i=1}^N$ above), which in some cases are chosen by the analyst and are thus considered "known," and the spectrum, which reflects the relative importance of each of the patterns in making up the full information. Even in cases when the patterns are also determined by the analyzed data (as in the above **S** determined $\{\hat{\mathbf{q}}_i\}_{i=1}^N$), the split provides many advantages, as discussed in the remainder of this section.

In the context of real scalar data, the spectrum is a set of N numbers $\{\phi_i\}_{i=1}^N$ that jointly amount to an alternative—and entirely equivalent—way of representing an original set of N numbers, $\{v_i\}_{i=1}^N$. Importantly, the transformation is reversible and symmetric, $\{\phi_i\}_{i=1}^N \Longleftrightarrow \{v_i\}_{i=1}^N$. Let's look at some examples that give this definition specificity most suitable for this book's focus and that motivate the need for such alternative representations.

We start with a simple $\mathbb{R}^3$ example. Suppose we took three measurements, $v_1 = 1$, $v_2 = -1$, and $v_3 = 2$, which we hold in the data vector

$$\mathbf{v} = \begin{pmatrix} 1 \\ -1 \\ 2 \end{pmatrix}. \tag{4.60}$$

Since $\mathbf{v}$ is a 3-vector, it represents three degrees of freedom, three independent choices to make while constructing $\mathbf{v}$: the values of v_1, v_2, and v_3. One way to construct this data vector is therefore to simply put the three measurements in the vector's three slots, 1 in v_1, -1 in v_2, and 2 in v_3. But there are alternative ways, and while perhaps initially less straightforward, they may be more desirable under some circumstances. For example, we can represent $\mathbf{v}$ in the Cartesian $\mathbb{R}^3$ basis $(\hat{\mathbf{i}}, \hat{\mathbf{j}}, \hat{\mathbf{k}})$,

$$\mathbf{v} = \phi_1\hat{\mathbf{i}} + \phi_2\hat{\mathbf{j}} + \phi_3\hat{\mathbf{k}}. \tag{4.61}$$

In this case the values are the same ($\phi_i = v_i$, $i = 1, 2, 3$) while their meaning—ϕ_i is the weight you need to give the ith basis vector in order to reproduce $\mathbf{v}$—is slightly different. This divergence of interpretations is even clearer when we choose a different basis to span $\mathbb{R}^3$. For example, with

$$\mathbf{s}_1 = \begin{pmatrix} 1 \\ 0 \\ 1 \end{pmatrix}, \quad \mathbf{s}_2 = \begin{pmatrix} 1 \\ 0 \\ -1 \end{pmatrix} \quad \text{and} \quad \mathbf{s}_3 = \begin{pmatrix} 0 \\ 1 \\ 1 \end{pmatrix}, \tag{4.62}$$

$$\text{span}(\{\mathbf{s}_1, \mathbf{s}_2, \mathbf{s}_3\}) = \mathbb{R}^3 \tag{4.63}$$

still holds, but now $\phi_1 = 2$, $\phi_2 = -1$ and $\phi_3 = -1$, because these are the loadings the above $\mathbf{s}_i$ basis vectors require for satisfying

$$\phi_1\mathbf{s}_1 + \phi_2\mathbf{s}_2 + \phi_3\mathbf{s}_3 = \mathbf{v}. \tag{4.64}$$

In this example, therefore, $\{\phi_1 = 2, \phi_2 = -1, \phi_3 = -1\} \Longleftrightarrow \{v_1 = 1, v_2 = -1, v_3 = 2\}$, so that $\phi_i = v_i$—unique to the Cartesian basis and most emphatically not general—no longer holds. Notice the reversibility: given the basis $\{\mathbf{s}_i\}_{i=1}^3$, constructing $\mathbf{v}$ means choosing $\{\phi_i\}_{i=1}^3$, and from $\{\phi_i\}_{i=1}^3$, $\{v_i\}_{i=1}^3$ can be readily retrieved: $\{\phi_i\}_{i=1}^3 \Longleftrightarrow \{v_i\}_{i=1}^3$.

The completeness of the spanning set is absolutely essential for the reversibility; if $\{\mathbf{s}_i\}$ is incomplete, the $\{v_i\} \Longleftrightarrow \{\phi_i\} \Longleftrightarrow \{v_i\}$ transformation entails loss of information. In the above case, with $\mathbf{s}_i \in \mathbb{R}^3$, if $\text{span}\{\mathbf{s}_1, \mathbf{s}_2, \mathbf{s}_3\} = \mathcal{S} \subset \mathbb{R}^3$, $\{v_i\} \Longrightarrow \{\phi_i\} \Longrightarrow \{v_i\}$ preserves the part of $\mathbf{v}$ that projects on the $\mathbb{R}^3$ subspace spanned by $\{\mathbf{s}_i\}$ (i.e., the part of $\mathbf{v}$ from $\mathcal{S}$) while annihilating $\mathbf{v}$'s part from the $\mathbb{R}^3$ subspace orthogonal to $\mathcal{S}$. This can be demonstrated by modifying slightly the above example to

$$\begin{aligned} \mathcal{S} &= \text{span}\left(\left\{\begin{pmatrix} 1 \\ 0 \\ 1 \end{pmatrix}, \begin{pmatrix} 1 \\ 0 \\ -1 \end{pmatrix}, \begin{pmatrix} 0 \\ 0 \\ 2 \end{pmatrix}\right\}\right) \\ &= \text{span}\left(\{\mathbf{s}_1, \mathbf{s}_2, \mathbf{s}_3\}\right), \end{aligned} \tag{4.65}$$

in which $\mathbf{s}_3 = \mathbf{s}_1 - \mathbf{s}_2$, so that $\mathcal{S} \subset \mathbb{R}^3$ is a plane in $\mathbb{R}^3$. Now the equation for $\mathbf{\Phi} = (\phi_1, \phi_2, \phi_3)$ is

$$\mathbf{S\Phi} = \begin{pmatrix} 1 & 1 & 0 \\ 0 & 0 & 0 \\ 1 & -1 & 2 \end{pmatrix} \begin{pmatrix} \phi_1 \\ \phi_2 \\ \phi_3 \end{pmatrix} = \begin{pmatrix} 1 \\ -1 \\ 2 \end{pmatrix} = \mathbf{v}, \tag{4.66}$$

which is obviously an impossibility because of row 2. In a feat of trickery that will become somewhat clearer below and entirely clear after we introduce both regression and singular value decomposition, let's do the best we can. That is, instead of accepting full defeat, let's solve the closest problem we can, by brushing aside this impossibility and solving instead

$$\begin{pmatrix} 1 & 1 & 0 \\ 1 & -1 & 2 \end{pmatrix} \begin{pmatrix} \phi_1 \\ \phi_2 \\ \phi_3 \end{pmatrix} = \begin{pmatrix} 1 \\ 2 \end{pmatrix}. \tag{4.67}$$

This yields

$$\mathbf{\Phi} = \frac{1}{2} \begin{pmatrix} 3 \\ -1 \\ 0 \end{pmatrix} + \xi \begin{pmatrix} -1 \\ 1 \\ 1 \end{pmatrix} = \mathbf{\Phi}_p + \mathbf{\Phi}_h, \tag{4.68}$$

where $\mathbf{\Phi}_p$ and $\mathbf{\Phi}_h$ are the solution's particular and homogeneous parts, and ξ is unconstrained (arbitrary). Because $\mathbf{S\Phi}_h$ vanishes and can thus have no relevance to any right-hand side, we need only examine

$$\mathbf{S\Phi}_p = \begin{pmatrix} 1 \\ 0 \\ 2 \end{pmatrix}. \tag{4.69}$$

The vector $\mathbf{S\Phi}_p$ is the projection of $\mathbf{v}$ on $\mathcal{S} = \operatorname{span}(\mathbf{s}_1, \mathbf{s}_2)$:

$$\frac{\mathbf{v}^T\mathbf{s}_1}{\mathbf{s}_1^T\mathbf{s}_1}\mathbf{s}_1 + \frac{\mathbf{v}^T\mathbf{s}_2}{\mathbf{s}_2^T\mathbf{s}_2}\mathbf{s}_2 = \frac{3}{2}\begin{pmatrix} 1 \\ 0 \\ 1 \end{pmatrix} - \frac{1}{2}\begin{pmatrix} 1 \\ 0 \\ -1 \end{pmatrix} = \begin{pmatrix} 1 \\ 0 \\ 2 \end{pmatrix}, \tag{4.70}$$

which is also the reverse transformation, $\{\phi_i\} \Longrightarrow \{v_i\}$. In turn, the missing part, $\mathbf{v} - (1\ 0\ 2)^T = (0\ -1\ 0)^T$, is the projection of $\mathbf{v}$ on $\mathcal{S}_\perp$, the $\mathbb{R}^3$ subspace orthogonal to to $\mathcal{S}$, i.e., orthogonal to both $\mathbf{s}_1$ and $\mathbf{s}_2$. Since $\dim(\mathcal{S}_\perp) = 1$ ($\mathcal{S}_\perp$ is one dimensional), it is spanned by a single vector, $\mathbf{n} = (n_1\ n_2\ n_3)^T$, which must satisfy $\mathbf{n}^T\mathbf{s}_1 = \mathbf{n}^T\mathbf{s}_2 = 0$. This yields $\hat{\mathbf{n}} = (\ 0\ 1\ 0\ )^T$, and

$$\left(\mathbf{v}^T\hat{\mathbf{n}}\right)\hat{\mathbf{n}} = -1\begin{pmatrix} 0 \\ 1 \\ 0 \end{pmatrix} = \begin{pmatrix} 0 \\ -1 \\ 0 \end{pmatrix} = \mathbf{v} - \begin{pmatrix} 1 \\ 0 \\ 2 \end{pmatrix}. \tag{4.71}$$

Thus, our spectral representation

$$\left\{v_1 = 1, v_2 = -1, v_3 = 3,\right\} \Longrightarrow$$

$$\{\phi_1 = \tfrac{3}{2} - \xi, \phi_2 = -\tfrac{1}{2} + \xi, \phi_3 = \xi\}$$

is only imperfectly reversible, yielding the altered reconstructed $\mathbf{v}\{v_1^a = 1, v_2^a = 0, v_3^a = 2\}$, decidedly not $\mathbf{v}$. This failure is entirely attributable to the fact that $\mathcal{S} \subset \mathbb{R}^3$ instead of the requisite $\mathcal{S} = \mathbb{R}^3$.

In summary, as long as $\{\mathbf{s}_i\}$ is complete (as long as it is a basis for $\mathbb{R}^N$), we can take any original set of N scalars $\{v_i\}_{i=1}^N$ and recast them as an equivalent but different set of N numbers, $\{\phi_i\}_{i=1}^N$ and collectively the spectrum of $\{v_i\}_{i=1}^N$, the amplitudes of the N basis vectors that make up the data. Knowing the basis vectors (in the examples above $\{\mathbf{s}_i\}$) and the amplitudes $\{\phi_i\}_{i=1}^N$ (the spectrum) allows us to fully reconstruct the data. Eigenanalysis is a form of a spectrum.

4.3.2 Utility of Spectral Representations

At this point you probably understand, at least in principle, what the spectrum is. But why we need it, why would we want to represent $\{v_i\}_{i=1}^N$ as $\{\phi_i\}_{i=1}^N$, must be rather mysterious to some. The answer is actually simple: The spectral representation may be desirable because the basis functions can be chosen to have certain properties that will afford clean separation of various physical workings of the studied system that are otherwise hard to distinguish.

As an example, consider a synthetic signal $\mathbf{d} = (d_0\ d_2\ \cdots\ d_{100})^T$ where $d_i = d(x_i)$ and

$$d_i = 6\cos\left(\frac{6\pi x_i}{N}\right) + 5\cos\left(\frac{8\pi x_i}{N}\right) + 4\cos\left(\frac{10\pi x_i}{N}\right) + 2\cos\left(\frac{20\pi x_i}{N}\right) + n_i; \tag{4.72}$$

$N = 101$ is the signal length and n_i is the ith realization of low-amplitude random noise centered on zero. This signal is shown in fig. 4.3a. Let's span (slightly imperfectly) the relevant space, $\mathbb{R}^{101}$, with the following (nearly) complete set:

$$\hat{\mathbf{b}}_j = \sqrt{\frac{2}{101}}\cos\left(\frac{j2\pi x}{N}\right), \quad \hat{\mathbf{b}}_{j+50} = \sqrt{\frac{2}{101}}\sin\left(\frac{j2\pi x}{N}\right) \tag{4.73}$$

with $1 \le j \le 50$ and $\mathbf{x} = (x_0\ x_1\ \cdots\ x_{100})^T \in \mathbb{R}^{101}$, which we place in $\mathbf{B} \in \mathbb{R}^{101\times 100}$. This basis—a close sibling of the celebrated real Fourier basis, with the appealing

$$\hat{\mathbf{b}}_i^T\hat{\mathbf{b}}_j = \begin{cases} 1 & \text{for} \quad i = j \\ 0 & \text{for} \quad i \neq j \end{cases} \tag{4.74}$$

quality—does not quite fully span $\mathbb{R}^{101}$ (it comprises only 100 vectors). Nevertheless, $\mathbf{d}$ is constructed in such a way $[(\mathbf{d} - \mathbf{n}) \in \mathcal{R}(\mathbf{B})]$ that only the noise ($\mathbf{n}$, with individual elements n_i), but not the structured signal $\mathbf{d} - \mathbf{n}$, has a nonzero projection on $\mathbf{B}$'s left null space, so the structured part of the signal should be fully reconstructable. Next we derive the coefficients of $\mathbf{d}$'s spectral representation in this basis,

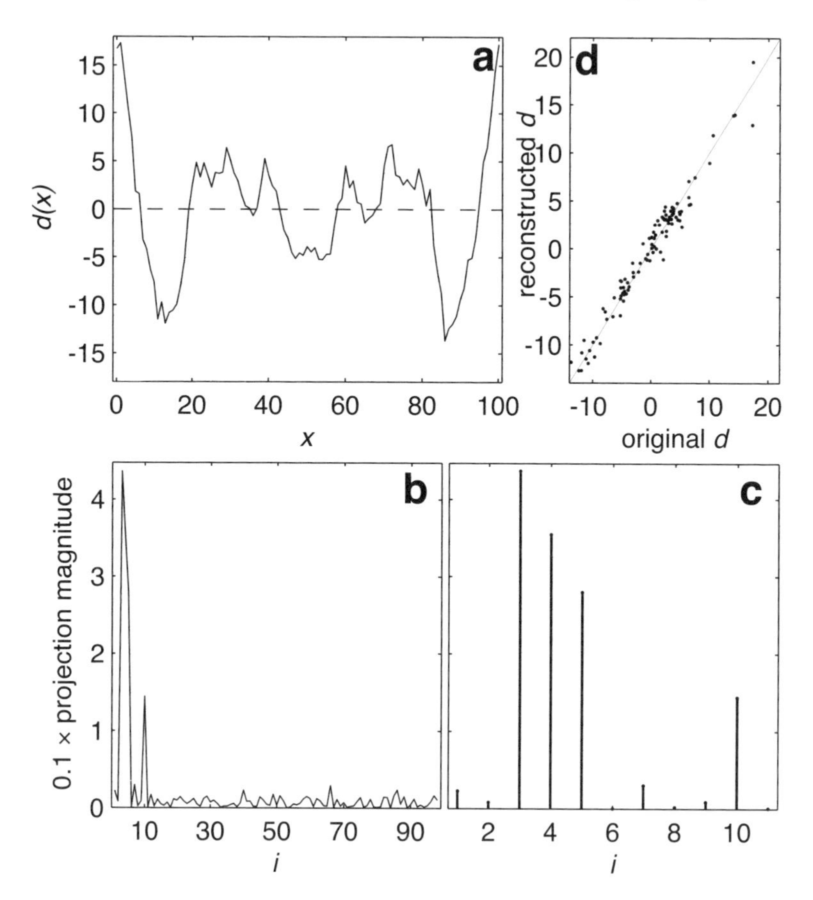

Figure 4.3. Example demonstrating proprieties of spectral representation. Panel a shows the noise-contaminated synthetic signal $\mathbf{d} = d(x)$ discussed in the text. Panel b shows the projection magnitude of $\mathbf{d}$ on the $\mathbb{R}^{101}$ spanning set described in the text. Panel c is a blowup of panel b close to the origin. Panel d addresses the quality and imperfection of reconstructing $\mathbf{d}$ from its spectral representation.

$$\mathbf{\Phi} = |\mathbf{B}^T\mathbf{d}|, \tag{4.75}$$

where the absolute value applies element-wise, and here $\{d_i\}$ plays the role of $\{v_i\}$. This is the step yielding $\{\phi_i\}$, what we previously denoted generally as $\{v_i\} \Longrightarrow \{\phi_i\}$. The $\{\phi_i\}$ coefficients are shown in fig. 4.3b, and their meaningful subset is emphasized in panel c. From fig. 4.3b it is clear that all the sine terms ($51 \leq i \leq 100$) span nothing but noise, as expected from $\mathbf{d}$'s structure (eq. 4.72). It is also clear, from fig. 4.3c, that the only significantly nonzero projections occur for $i = 3, 4, 5, 10$. This confirms the adequacy of $\mathbf{d}$'s spectral representation, because, e.g.,

$$\hat{\mathbf{b}}_3 \propto \cos\left(\frac{6\pi\mathbf{x}}{N}\right), \tag{4.76}$$

$\mathbf{d}$'s leading term (eq. 4.72). The decreasing amplitudes of subsequent terms in eq. 4.72 is similarly captured by the relative magnitudes of ϕ_3, ϕ_4, ϕ_5, and ϕ_{10}, as required.

With $\{\phi_i\}$ thus obtained,

$$\mathbf{d} = \mathbf{B}\boldsymbol{\Phi} + \mathbf{n}_r = \sum_{i=1}^{100} \phi_i \mathbf{b}_i + \mathbf{n}_r = \hat{\mathbf{d}} + \mathbf{n}_r, \tag{4.77}$$

where $\hat{\mathbf{d}}$ is the reconstructed signal, shown in fig. 4.3d, and $\mathbf{n}_r \in \mathcal{N}(\mathbf{B}^T)$ is not quite $\mathbf{n}$ but rather its residual, $\mathbf{n}$'s part orthogonal to $\mathcal{R}(\mathbf{B})$ ($\mathbf{n}$'s other part being the collective contribution of the small, irregular bumps in fig. 4.3b,c).

This example highlights the utility of spectral representation. Visual examination of the noise-contaminated signal $\mathbf{d}$ (fig. 4.3a) offers little insight into the dominant signals that collectively make up $\mathbf{d}$; while $\mathbf{d}$ surely goes up and down and is even vaguely symmetrical about its midpoint ($i \approx 50$), one would be hard pressed to suggest dominance of particular frequencies. Upon spectral decomposition (fig. 4.3b,c), however, the dominant frequencies are readily visible. If those frequencies also have a simple association (e.g., if their reciprocals are ~365 days or ~24 hours), the analyst can better appreciate $\mathbf{d}$'s physical origins. In such cases, which are often realized in the analysis of actual data, the algebraic machinery facilitates mechanistic understanding, science's principal objective.

4.3.3 Eigen-decomposition as Spectral Representation

Accepting significant loss of generality for clarity of presentation, let's consider a very restrictive $\mathbf{A}$, symmetric and full rank, which assures a full set of orthogonal eigenvectors. Assuming the eigenvectors have all been normalized and placed in $\mathbf{E} = (\hat{\mathbf{e}}_1 \; \hat{\mathbf{e}}_2 \cdots \hat{\mathbf{e}}_N) \in \mathbb{R}^{N\times N}$, $\mathbf{A}$'s $\mathbf{E}^T = \mathbf{E}^{-1}$. With these stipulations in place,

$$\mathbf{A} = \mathbf{E}\boldsymbol{\Lambda}\mathbf{E}^T \tag{4.78}$$

$$= \begin{pmatrix} \vdots & \vdots & & \vdots \\ \hat{\mathbf{e}}_1 & \hat{\mathbf{e}}_2 & \cdots & \hat{\mathbf{e}}_N \\ \vdots & \vdots & & \vdots \end{pmatrix} \begin{pmatrix} \lambda_1 & & & \\ & \lambda_2 & & \\ & & \ddots & \\ & & & \lambda_N \end{pmatrix} \begin{pmatrix} \cdots & \hat{\mathbf{e}}_1^T & \cdots \\ \cdots & \hat{\mathbf{e}}_2^T & \cdots \\ & \vdots & \\ \cdots & \hat{\mathbf{e}}_N^T & \cdots \end{pmatrix}$$

$$= \begin{pmatrix} \vdots & \vdots & & \vdots \\ \lambda_1\hat{\mathbf{e}}_1 & \lambda_2\hat{\mathbf{e}}_2 & \cdots & \lambda_N\hat{\mathbf{e}}_N \\ \vdots & \vdots & & \vdots \end{pmatrix} \begin{pmatrix} \cdots & \hat{\mathbf{e}}_1^T & \cdots \\ \cdots & \hat{\mathbf{e}}_2^T & \cdots \\ & \vdots & \\ \cdots & \hat{\mathbf{e}}_N^T & \cdots \end{pmatrix}$$

$$= \lambda_1(\hat{\mathbf{e}}_1\hat{\mathbf{e}}_1^T) + \lambda_2(\hat{\mathbf{e}}_2\hat{\mathbf{e}}_2^T) + \cdots + \lambda_N(\hat{\mathbf{e}}_N\hat{\mathbf{e}}_N^T), \tag{4.79}$$

where the ith term is the product of the ith eigenvalue λ_i and the rank 1 $N \times N$ matrix $\mathbf{E}_i = \hat{\mathbf{e}}_i\hat{\mathbf{e}}_i^T$. This is a spectral representation of $\mathbf{A}$. The action of $\mathbf{A}$ on any $\mathbf{x} \in \mathbb{R}^N$ vector it premultiplies is broken down into N distinct actions corresponding to the N elements in the sum (4.79). For example, in the earlier population dynamics-motivated example (eq. 4.35) of

$$\mathbf{A} = \begin{pmatrix} 1 & -2 \\ -2 & 1 \end{pmatrix}, \quad \mathbf{E} = \frac{1}{\sqrt{2}}\begin{pmatrix} 1 & 1 \\ -1 & 1 \end{pmatrix} \quad \text{and} \quad \begin{matrix} \lambda_1 & = & 3 \\ \lambda_2 & = & -1 \end{matrix}, \tag{4.80}$$

$$\mathbf{Ax} = \frac{3}{2}\left[\begin{pmatrix} 1 \\ -1 \end{pmatrix}\begin{pmatrix} 1 & -1 \end{pmatrix}\right]\mathbf{x} - \frac{1}{2}\left[\begin{pmatrix} 1 \\ 1 \end{pmatrix}\begin{pmatrix} 1 & 1 \end{pmatrix}\right]\mathbf{x} \tag{4.81}$$

$$= \underbrace{\frac{3}{2}\begin{pmatrix} 1 & -1 \\ -1 & 1 \end{pmatrix}}_{\lambda_1\mathbf{E}_1}\mathbf{x} - \underbrace{\frac{1}{2}\begin{pmatrix} 1 & 1 \\ 1 & 1 \end{pmatrix}}_{\lambda_2\mathbf{E}_2}\mathbf{x}. \tag{4.82}$$

Mode 1 (the $(\lambda_1, \hat{\mathbf{e}}_1)$ pair) contributes to $\mathbf{Ax}$'s total action $3(x_1 - x_2\ x_2 - x_1)^T/2$ (where $\mathbf{x} = (x_1\ x_2)^T$), while mode 2's contribution is $-(x_1 + x_2\ x_1 + x_2)^T/2$. The sum of the individual mode contributions is the full action of $\mathbf{A}$, in this case

$$\frac{3}{2}\begin{pmatrix} x_1 - x_2 \\ x_2 - x_1 \end{pmatrix} - \frac{1}{2}\begin{pmatrix} x_1 + x_2 \\ x_1 + x_2 \end{pmatrix} = \begin{pmatrix} x_1 - 2x_2 \\ x_2 - 2x_1 \end{pmatrix} = \mathbf{Ax}, \tag{4.83}$$

as required. What is interesting and important is the way the individual modal contributions sum to $\mathbf{A}$'s full action. Mode i's full contribution—$\lambda_i\mathbf{E}_i\mathbf{x}$—has two parts. The scalar amplification due to multiplication by λ_i is straightforward: all else being equal, the larger $|\lambda_i|$, the larger the modal contribution.

The second part, $\mathbf{E}_i\mathbf{x}$, is slightly trickier. First, since $\mathbf{E}_i = \hat{\mathbf{e}}_i\hat{\mathbf{e}}_i^T$ and $|\hat{\mathbf{e}}_i| = 1$ by construction, $\mathbf{E}_i$ is rank 1, as mentioned above, with a single unit eigenvalue and vanishing remaining $N - 1$ eigenvalues. More generally, $\mathbf{E}_i = \hat{\mathbf{e}}_i\hat{\mathbf{e}}_i^T$ are a special case of orthogonal projection matrices, with $\mathbf{E}_i$ projecting vectors it premultiplies onto $\hat{\mathbf{e}}_i$. For example, the action of the 2×2 $\mathbf{E}_1$ in the population dynamics problem discussed above (based on eq. 4.35), on an arbitrary $\mathbf{x} = (x_1\ x_2)^T$ is

$$\mathbf{E}_1\mathbf{x} = \frac{1}{2}\begin{pmatrix} 1 & -1 \\ -1 & 1 \end{pmatrix}\begin{pmatrix} x_1 \\ x_2 \end{pmatrix} = \frac{1}{2}\begin{pmatrix} x_1 - x_2 \\ x_2 - x_1 \end{pmatrix}, \tag{4.84}$$

as shown above. By comparison, the direct projection of $\mathbf{x}$ on $\hat{\mathbf{e}}_1$ is

$$\mathbf{x}_{\hat{\mathbf{e}}_1} = (\hat{\mathbf{e}}_1^T\mathbf{x})\hat{\mathbf{e}}_1 = \left[\frac{1}{\sqrt{2}}\begin{pmatrix} 1 & -1 \end{pmatrix}\begin{pmatrix} x_1 \\ x_2 \end{pmatrix}\right]\frac{1}{\sqrt{2}}\begin{pmatrix} 1 \\ -1 \end{pmatrix}$$

$$= \frac{x_1 - x_2}{2}\begin{pmatrix} 1 \\ -1 \end{pmatrix} = \frac{1}{2}\begin{pmatrix} x_1 - x_2 \\ x_2 - x_1 \end{pmatrix}, \tag{4.85}$$

obviously the same.

Because, each $\mathbf{E}_i$'s rank is 1, premultiplying a vector by it can at most preserve the vector's magnitude, not increase it. But, clearly, if $\mathbf{e}_i$ and $\mathbf{x}$ are orthogonal, one's projection on the other vanishes. For an arbitrary $\mathbf{x}$, therefore, $0 \le \|\mathbf{E}_i\mathbf{x}\|/\|\mathbf{x}\| \le 1$, with orthogonal $\mathbf{x}$ and $\hat{\mathbf{e}}_i$ yielding $\|\mathbf{E}_i\mathbf{x}\|/\|\mathbf{x}\| = 0$, parallel $\mathbf{x}$ and $\hat{\mathbf{e}}_i$ yielding $\|\mathbf{E}_i\mathbf{x}\|/\|\mathbf{x}\| = 1$, and other vectors forming an angle $0 < \mathrm{mod}(\theta, \pi/2) < \pi/2$ with $\hat{\mathbf{e}}_i$ yielding intermediate values. That is, premultiplication by $\mathbf{E}_i$ reduces or preserves—but never amplifies—the norm of a vector. That latter role is reserved exclusively for λ_i.

And what about the suppression of $\|\mathbf{x}\|$ by $\mathbf{E}_i$? Can an $\mathbf{x}$ simply be out of luck because it is orthogonal to a particular $\hat{\mathbf{e}}_i$? No, not really. It can certainly be mapped by a particular mode—to which it is orthogonal—to zero; no doubt about that. But recall that an $\mathbf{A}$ that meets the stringent criteria specified at the beginning of this section has a complete set of orthogonal eigenvectors, and $\{\hat{\mathbf{e}}_i\}_{i=1}^N$ fully span $\mathbb{R}^N$. Consequently, if $\mathbf{x}$ is annihilated by mode i, it simply has to project favorably on at least one other mode (and if it is only one, then $\mathbf{x}$ must itself be an eigenvector).

In the more general case of asymmetrical $\mathbf{A}$ for which $\mathbf{E}^T \neq \mathbf{E}^{-1}$, the modal description of $\mathbf{A}$'s action on $\mathbf{x}$, $\mathbf{A}\mathbf{x} = \sum_{i=1}^N \lambda_i \hat{\mathbf{e}}_i \hat{\mathbf{e}}_i^T \mathbf{x}$, is lost. Yet, provided $\mathbf{E}$ is invertible (which, in general, need not be true, but is guaranteed here by the assumption of nontrivial eigenvalues), the modal nature of the action is still apparent, if not as beautifully: $\mathbf{A}\mathbf{x} = \sum_{i=1}^N a_i \lambda_i \hat{\mathbf{e}}_i$, where $\mathbf{a} = \mathbf{E}^{-1}\mathbf{x} = (\, a_1 \; a_2 \cdots a_N \,)^T$. If $\mathbf{A}$ has even a single repeated eigenvalue with algebraic multiplicity exceeding its geometric multiplicity, $\mathbf{E}$ is singular, and $\mathbf{A}\mathbf{x}$'s eigen-representation is lost. Note also that, even if $\mathbf{E}$ is invertible, for nonnormal $\mathbf{A}$s (ones failing to satisfy $\mathbf{A}\mathbf{A}^T = \mathbf{A}^T\mathbf{A}$), the eigenvectors are, in general, not mutually orthogonal, $\hat{\mathbf{e}}_i^T \hat{\mathbf{e}}_j \neq 0$ for at least some $(i, j \neq i)$ pairs, so most of the appeal of the modal representation is lost despite the existence of $\mathbf{E}^{-1}$.

4.3.4 Eigenfunctions and Eigenvectors, Spectra of Linear Operators

While data matrices are in general rectangular, there is an extremely important and broad class of square matrices: numerical operators representing linear operators in finite (vector, as opposed to function) spaces. Because of these matrices prominence, and most importantly because their analysis will help us gain further insights into eigenanalysis, let's examine some differential operators, starting with continuous ones and progressing to their finite (discrete) counterparts.

A linear operator $\mathcal{L}$ can be thought of as a functional, a function of a function, taking in an input function and carrying out some linear manipulation of the function; $\mathcal{L}[f(x)]$ is a linear operator operating on the function $f(x)$, which we assume here for simplicity is one dimensional, depending on x alone. For example, $\mathcal{L}$ can be ∂_x or ∂_{xx}, returning $\partial_x f(x)$ or $\partial_{xx} f(x)$, respectively. By analogy to matrix eigenvectors, continuous linear operators have eigenfunctions satisfying $\mathcal{L}[e(x)] = \lambda e(x)$. For the two $\mathcal{L}$ examples above, this means

$$\mathcal{L}[g(x)] = \partial_x g(x) = \lambda g(x) \Longrightarrow (\partial_x - \lambda)g(x) = 0 \tag{4.86}$$

and

$$\mathcal{L}[h(x)] = \partial_{xx} h(x) = \lambda h(x) \Longrightarrow (\partial_{xx} - \lambda)h(x) = 0. \tag{4.87}$$

In each of these cases, we are thus looking for a suite of functions that when x differentiated once or twice, as dictated by $\mathcal{L}$, will reproduce themselves times a scalar constant. Given differentiation properties of exponential and trigonometric functions, this means that $g(x) = ae^{\lambda x}$ and $h(x) = a\cos(\alpha x) + b\sin(\alpha x)$ with $\lambda = -\alpha^2$. Thus, the eigenfunctions of $\mathcal{L} = \partial_x$ and $\mathcal{L} = \partial_{xx}$ are exponentials and trigonometric waves.

Let's focus on $\mathcal{L} = \partial_{xx}$ and assume the studied physical problem includes vanishing boundary values $h(0) = h(L) = 0$. At $x = 0$, h's sine term vanishes, so $a = 0$. At $x = L$, $h(L) = b\sin(\alpha L) = 0$ or $\alpha = n\pi/L$, where $n = 0, 1, 2, \ldots$ is any of the natural numbers. Since we ran out of boundary conditions and b is still unconstrained, let's set $b = 1$, so

$$h_n(x) = \sin\left(\frac{n\pi x}{L}\right), \qquad \lambda_n = -\left(\frac{n\pi}{L}\right)^2, \tag{4.88}$$

with $h_0(x) = \lambda_0 = 0$, $\lambda_{n\geq 1}$ becoming more negative with increasing n, and h_n comprises one half wave more than h_{n-1} for all n. Figure 4.4 shows the three leading nontrivial eigenfunctions of $\mathcal{L} = \partial_{xx}$ with vanishing boundary conditions.

Now let's examine the discrete case, in which the continuous $x = [0, L]$ (over a physical domain extending between $x = 0$ on the left and $x = L$ on the right) is replaced by the finite set $\{x_i\}_{i=1}^N$ (where for simplicity we assume $\delta x \equiv x_{i+1} - x_i$ is uniform and i independent, so that $x_i = i\delta x$ for $1 \leq i \leq N$),

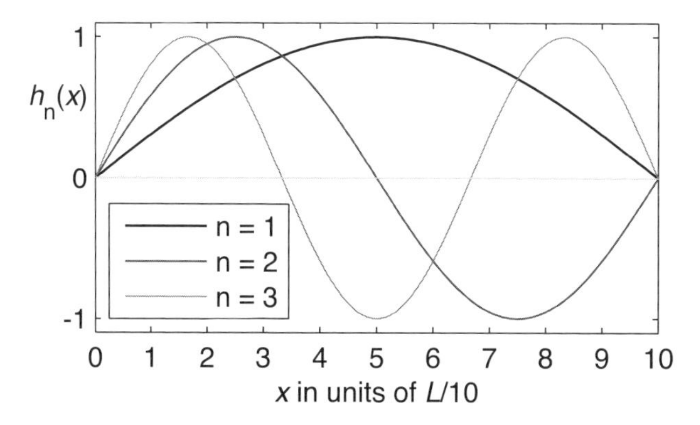

Figure 4.4. The three leading nontrivial eigenfunctions of $\mathcal{L} = \partial_{xx}$ with vanishing boundary conditions. The lines get thinner and lighter as n increases.

and the continuous $h(x)$ is replaced by the finite set $\{h(x_1), h(x_2), \ldots, h(x_N)\} \equiv \{h^{(1)}, h^{(2)}, \ldots, h^{(N)}\}$, with—to conform to the vanishing boundary values we chose in the continuous case—the known boundary values $h^{(0)} = h^{(L)} = 0$.

Choosing $N = 200$ and applying twice the differential approximation $\partial_x h|^{(i+1/2)} \approx (h^{(i+1)} - h^{(i)})/\delta x$ yields

$$\begin{aligned} \left.\frac{\partial^2 h}{\partial x^2}\right|_i &\approx \frac{1}{\delta x}\left(\left.\frac{\partial h}{\partial x}\right|^{(i+\frac{1}{2})} - \left.\frac{\partial h}{\partial x}\right|^{(i-\frac{1}{2})}\right) \\ &\approx \frac{1}{\delta x}\left(\frac{h^{(i+1)} - h^{(i)}}{\delta x} - \frac{h^{(i)} - h^{(i-1)}}{\delta x}\right) \\ &= \frac{1}{\delta x^2}\left(h^{(i+1)} - 2h^{(i)} + h^{(i-1)}\right). \end{aligned} \tag{4.89}$$

Near the left and right boundaries, this becomes

$$\left.\frac{\partial^2 h}{\partial x^2}\right|^{(1)} \approx \frac{h^{(2)} - 2h^{(1)}}{\delta x^2} \quad \text{and} \quad \left.\frac{\partial^2 h}{\partial x^2}\right|^{(N)} \approx \frac{h^{(N-1)} - 2h^{(N)}}{\delta x^2}, \tag{4.90}$$

respectively. With these choices, the matrix operator that is the discrete analog of ∂_{xx} is therefore the tridiagonal

$$\mathbf{D}_{xx} = \frac{1}{\delta x^2}\begin{pmatrix} -2 & 1 & & & & & \\ 1 & -2 & 1 & & & & \\ 0 & 1 & -2 & 1 & & & \\ & & & \ddots & & & \\ & & & 1 & -2 & 1 & \\ & & & & 1 & -2 & 1 \\ & & & & & 1 & -2 \end{pmatrix} \in \mathbb{R}^{N\times N}. \tag{4.91}$$

Figure 4.5 shows numerically obtained eigenvectors and eigenvalues of $\mathbf{D}_{xx}$.

The analytic and numerical results (figs. 4.4 and 4.5) clearly compare favorably. The leading three eigenfunctions (h_{1-3}, fig. 4.4) are virtually identical, to within unimportant scaling, to $\mathbf{D}_{xx}$'s leading eigenvectors (fig. 4.5), as are the eigenvalues. This close correspondence deteriorates rapidly after $n \approx 50$, because at this high spatial frequency (large number of crests or troughs within the $[0, L]$ interval), each wave (eigenvector of the discrete operator) is resolved by too few grid points to be adequately represented.

This dual (analytic/numerical) example illuminates the spectral nature of eigen-decomposition in an unusually revealing light. Because $\mathbf{D}_{xx}$ is symmetrical, $\mathbf{E}^T\mathbf{E} = \mathbf{I}$, i.e., $\mathbf{D}_{xx}$'s eigenvectors form an orthonormal complete set, in the shown example a spanning set for $\mathbb{R}^{200}$. As a consequence, the fate of an initial structure $\mathbf{s}$ subject to repeated premultiplication by $\mathbf{D}_{xx}$ ($\mathbf{D}_{xx}\mathbf{s}$, $\mathbf{D}_{xx}\mathbf{D}_{xx}\mathbf{s} = \mathbf{D}_{xx}^2\mathbf{s}$, and so on) depends on the part of $\mathbf{D}_{xx}$'s eigenspace on which $\mathbf{s}$ projects. If it projects on

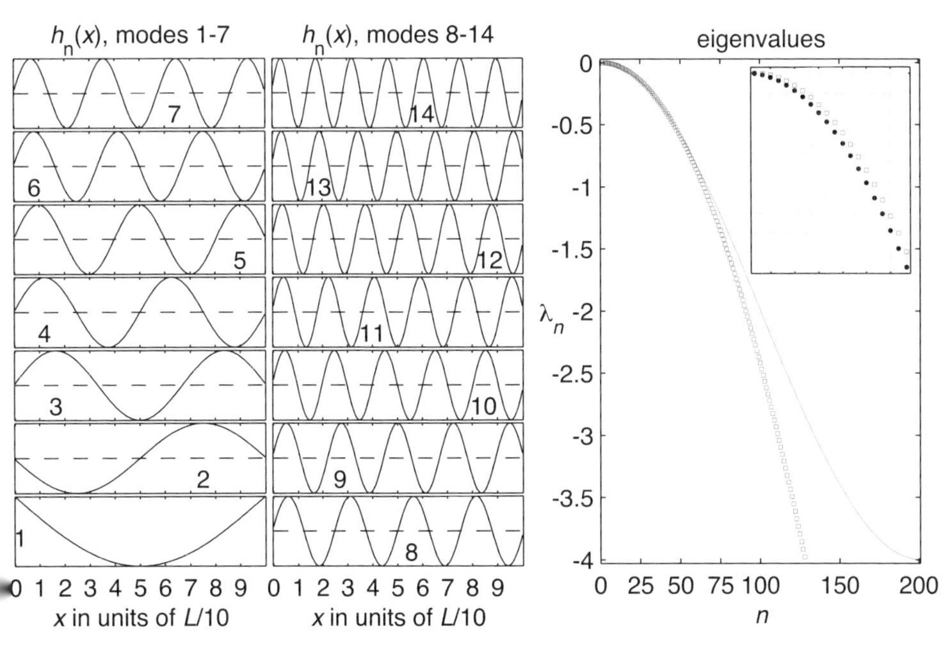

Figure 4.5. Leading eigenvectors, $h_n(x)$, $n = [1, 14]$ (left panels) and the full eigenspectrum (right panel; λ_n, with open gray squares showing the analytic eigenvalues, $-(n\pi/L)^2$) of the differential operator matrix $\mathbf{D}_{xx}$ with vanishing boundary conditions. The 20 leading numerical and analytic eigenvalues are magnified in the upper right inset (where gray lines replace tick marks at $n = 2, 4, \dots, 20$ and $\lambda_n = 0, -0.01, \dots, -0.1$).

leading $\mathbf{e}_i$s (where what "leading" means depends on the rate of eigenspectrum falloff with i), $\|\mathbf{D}_{xx}\mathbf{s}\| / \|\mathbf{s}\|$ decreases much more slowly than when $\mathbf{s}$ projects on trailing ($i \to N$) $\mathbf{e}_i$s. Let's take the extreme contrast cases of $\mathbf{s}_1 = \hat{\mathbf{e}}_1$ and $\mathbf{s}_2 = \hat{\mathbf{e}}_N$, still assuming $\mathbf{e}_i^T\mathbf{e}_j = 0$ for $i \neq j$ and 1 for $i = j$ (as satisfied by $\mathbf{D}_{xx}$'s eigenvectors).

In the $\mathbf{s}_1 = \hat{\mathbf{e}}_1$ case,

$$\mathbf{E}^T\mathbf{s}_1 = \begin{pmatrix} 1 \\ 0 \\ \vdots \\ 0 \end{pmatrix} \quad \text{so that} \quad \mathbf{D}_{xx}\mathbf{s}_1 = \lambda_1\hat{\mathbf{e}}_1 \tag{4.92}$$

by appeal to eq. 4.79. Conversely, in the $\mathbf{s}_2 = \hat{\mathbf{e}}_N$ case,

$$\mathbf{E}^T\mathbf{s}_2 = \begin{pmatrix} 0 \\ \vdots \\ 0 \\ 1 \end{pmatrix} \quad \text{so that} \quad \mathbf{D}_{xx}\mathbf{s}_2 = \lambda_N\hat{\mathbf{e}}_N. \tag{4.93}$$

Therefore, the magnitude of the initial signal after n repeated applications of the operator satisfies, using $\mathbf{s}_1$ as an example,

$$\|\mathbf{D}_{xx}^n \mathbf{s}_1\| = \|\lambda_1^n \hat{\mathbf{e}}_1\| = |\lambda_1|^n \|\hat{\mathbf{e}}_1\| = |\lambda_1|^n, \tag{4.94}$$

and therefore

$$\frac{\|\mathbf{D}_{xx}^n \mathbf{s}_1\|}{\|\mathbf{D}_{xx}^n \mathbf{s}_2\|} = \left(\frac{|\lambda_1|}{|\lambda_N|}\right)^n, \tag{4.95}$$

because $\|\hat{\mathbf{e}}_i\| = 1$ for any i; the magnitude ratio of the two vectors subject to repeated application of the operator is the ratio of the corresponding eigenvalue magnitudes to the nth power.

In less extreme cases, and in most realistic situations, the initial vectors excite more than one mode (project on more than one $\hat{\mathbf{e}}_i$), and the action of premultiplication is a modal sum. If, e.g., $\mathbf{s}$ projects on modes 1–3, in the first application,

$$\mathbf{D}_{xx}\mathbf{s} = \mathbf{E}\Lambda\mathbf{E}^T\mathbf{s} = \begin{pmatrix} \vdots & & \vdots \\ \lambda_1 \hat{\mathbf{e}}_1 & \cdots & \lambda_N \hat{\mathbf{e}}_N \\ \vdots & & \vdots \end{pmatrix} \begin{pmatrix} \mathbf{s}^T\hat{\mathbf{e}}_1 \\ \mathbf{s}^T\hat{\mathbf{e}}_2 \\ \mathbf{s}^T\hat{\mathbf{e}}_3 \\ 0 \\ \vdots \\ 0 \end{pmatrix}$$

$$= \sum_{i=1}^{3} \lambda_i (\mathbf{s}^T\hat{\mathbf{e}}_i)\hat{\mathbf{e}}_i, \tag{4.96}$$

again emphasizing the modal, spectral, nature of premultiplying a vector by a matrix when the matrix is expressed in terms of it eigenstructure. Note that the chosen projection on successive modes (1–3) is nonessential and unimportant and is simply used here for succinctness of the presentation permitting a continuous rightmost sum in eq. 4.96, which would otherwise need to be an explicit sum of individual nonsuccessive terms. If $\mathbf{E}$'s columns form an orthonormal complete set, as we have assumed throughout this discussion, then in subsequent applications of the matrix each mode evolves entirely independently of all others, growing or decaying ($|\lambda_i| > 1$ and $|\lambda_i| < 1$, respectively) as $|\lambda_i|^n$, i.e., exclusively based on its eigenvalue. It should be emphasized that for nonnormal matrices, whose $\mathbf{E}$s do not satisfy $\mathbf{E}^T\mathbf{E} = \mathbf{I}$, this property is emphatically not true.

This discussion is distilled into a handful of simple examples in fig. 4.6. All four signals $\mathbf{s}_1$–$\mathbf{s}_4$ are of unit norm (i.e., are $\hat{\mathbf{s}}_1 - \hat{\mathbf{s}}_4$), derived from

$$\mathbf{s}_1 = \sin\left(\frac{2\pi x}{N}\right) \tag{4.97}$$

$$\mathbf{s}_2 = \sin\left(\frac{40\pi x}{N}\right) \tag{4.98}$$

$$\mathbf{s}_3 = \sin\left(\frac{2\pi x}{N}\right) + \sin\left(\frac{3\pi x}{N}\right) \tag{4.99}$$

$$\mathbf{s}_4 = \sin\left(\frac{30\pi x}{N}\right) + \sin\left(\frac{38\pi x}{N}\right), \tag{4.100}$$

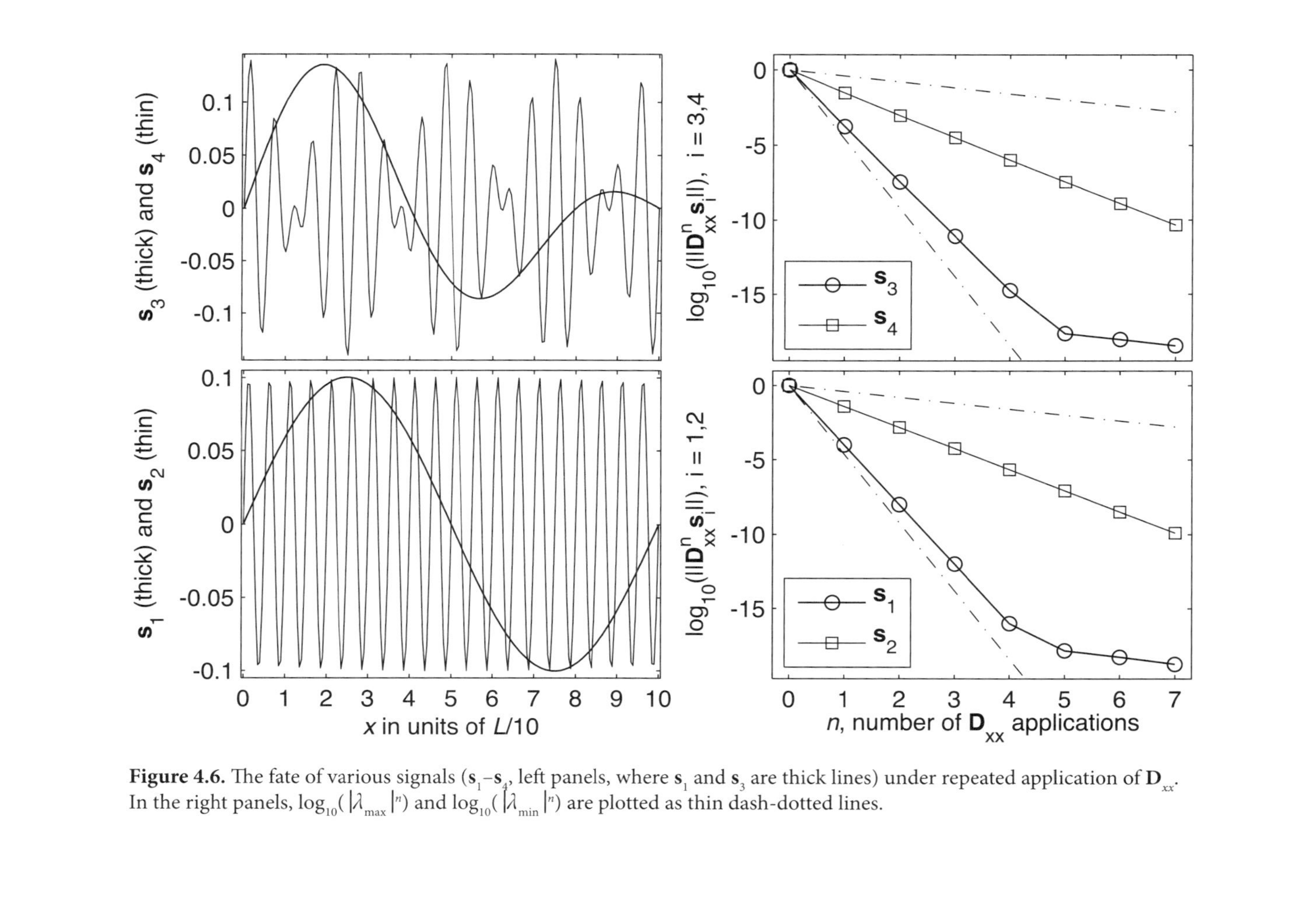

Figure 4.6. The fate of various signals ($\mathbf{s}_1$–$\mathbf{s}_4$, left panels, where $\mathbf{s}_1$ and $\mathbf{s}_3$ are thick lines) under repeated application of $\mathbf{D}_{xx}$. In the right panels, $\log_{10}(|\lambda_{\max}|^n)$ and $\log_{10}(|\lambda_{\min}|^n)$ are plotted as thin dash-dotted lines.

shown in fig. 4.6's left panels. All are eigenfunctions of ∂_{xx}, but not necessarily exactly eigenvectors of $\mathbf{D}_{xx}$ (fig. 4.5). The signals show the various discussed cases; $\mathbf{s}_1$ and $\mathbf{s}_2$ contain a single x frequency each, while $\mathbf{s}_3$ and $\mathbf{s}_4$ blend two frequencies each, two long waves in $\mathbf{s}_3$ and two short ones in $\mathbf{s}_4$. Figure 4.6's right panels show the dramatically higher effect of $\mathbf{D}_{xx}$ on the long waves, pure (bottom) or compound (top). After 7 $\mathbf{D}_{xx}$ applications, the remaining short wave magnitudes are 8–9 orders of magnitude higher than the long waves' (while the magnitude of the difference depends on the chosen δx, the relative differences do not). Note the flattening of the decay rate, which occurs as the norm approaches machine accuracy, in the shown case $\mathcal{O}(10^{-16})$.

The significance and spectral nature of this result is best appreciated in the context of a physical problem in which $\mathbf{D}_{xx}$ features prominently. Let's therefore consider a toy problem of a massive, rapid deluge on the ocean, occurring in two distinct spatial patterns. In the first, rainfall is centered in the middle of the domain, with symmetrically decreasing intensity away from the domain center (a crude pedagogically motivated simulation of a particularly stubborn warm front). While "piling up" of fresh water over the ocean surface may violate some readers' intuitive expectation, it is, in fact, what actually happens in the real ocean when the rain is rapid and widespread enough. The initial ($t = 0$) upward ocean surface displacement associated with this rainfall pattern is modeled as

$$\eta_{1,0}(x) \equiv \eta_1(x, t = 0) = \exp\left[-\frac{(2x - L)^2}{8L}\right], \tag{4.101}$$

with $L = 200$.

In the second scenario, the rain is more "patchy," forming the initial surface displacement

$$\tilde{\eta}_2(x, t = 0) = \sum_{i=1}^{7}\left[\frac{8 - |2i - 8|}{10}\right]\exp\left[-8\frac{(x + 20 - 30i)^2}{L}\right], \tag{4.102}$$

which we normalize to have the same x mean as $\eta_{1,0}$, $\eta_{2,0}(x) \equiv \eta_2(x, t = 0) = \tilde{\eta}_2(x, 0)\bar{\eta}_{1,0}/\bar{\tilde{\eta}}_{2,0}$ (where the overbar denotes x averaging, and the normalization simulates the same domain total rainfall, so the difference is strictly due to the different spatial patterns). Note that η_1 and η_2 are deliberately chosen to not be eigenvectors of ∂_{xx}, so as to excite (project on) multiple modes, as fig. 4.7 shows. The two initial surface displacements $\eta_1(x, 0)$ and $\eta_2(x, 0)$ are shown in black ($t = 0$) in fig. 4.8's lower panels.

Of course, as time goes on, the initial surface displacements erode under dissipation. A crude way to model this process is

$$\frac{\partial \eta}{\partial t} = k\frac{\partial^2 \eta}{\partial x^2}, \tag{4.103}$$

where k is the diffusion coefficient with units of m^2 s^{-1}. Assuming both surface displacement patterns evolve according to eq. 4.103, and discretizing η in space (x) again but not time, this becomes

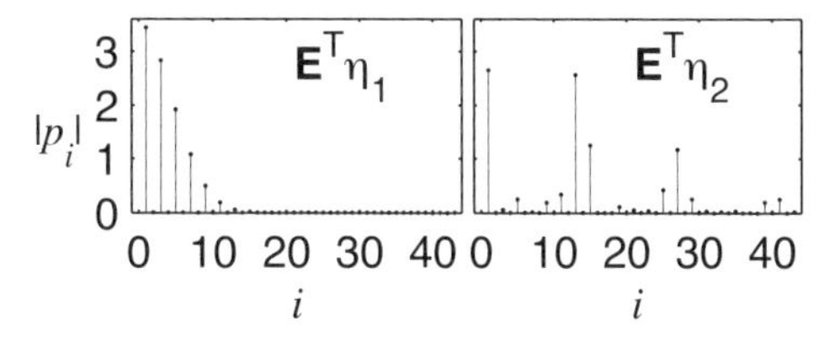

Figure 4.7. Projection magnitudes on **P**'s eigenvectors of $\vec{\eta}_1$ and $\vec{\eta}_2$ discussed in the text.

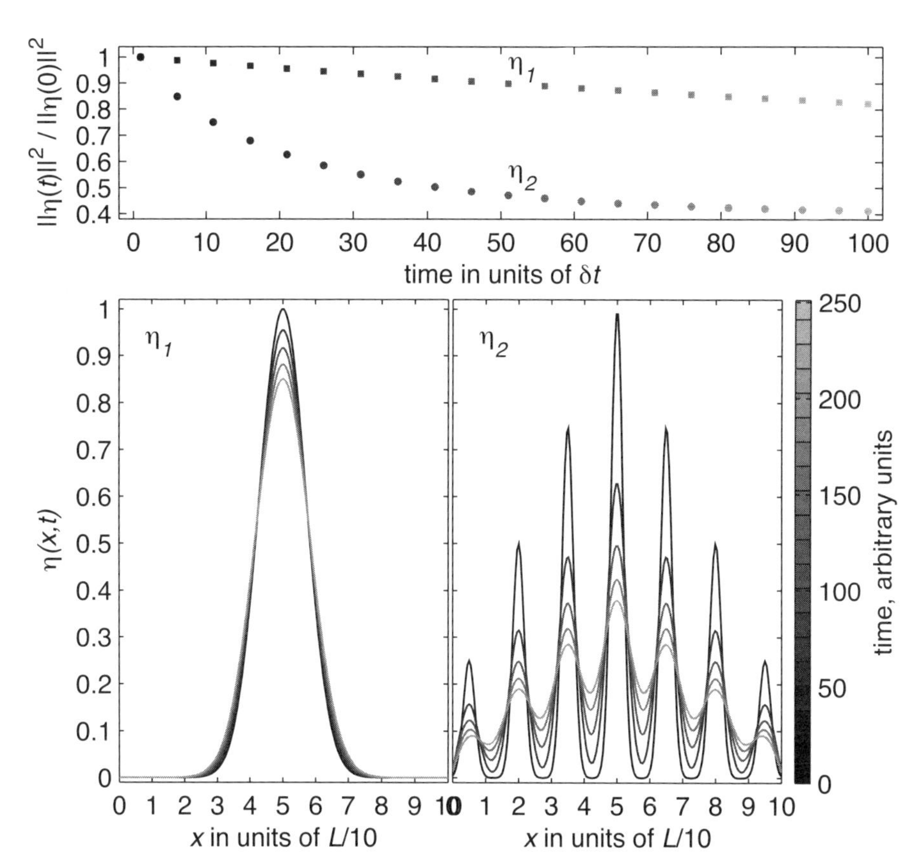

Figure 4.8. Time evolution (indicated by gray shades) of the two initial ocean surface elevation displacement patterns discussed in the text, from initial states shown in the lower panels in black. Subsequent cross sections of the displacement patterns are shown in progressively lighter gray shades. The evolution of the squared norms of the two signals under dissipation is shown in the upper panel as a fraction of the respective initial value, with solid curves showing the actual numerical results (using fourth-order Runge-Kutta scheme) and colored squares showing approximations based on eq. 4.106, i.e., on assuming $\vec{\eta}_\tau = (\mathbf{I} + \tilde{\mathbf{D}}_{xx})^{N_t}\vec{\eta}_0$.

$$\frac{\partial \vec{\eta}}{\partial t} = k\mathbf{D}_{xx}\vec{\eta}. \tag{4.104}$$

Devising a suitable approximation for the tendency (left-hand derivative) term is often much harder than the space derivative approximation, requiring more elaborate discretization schemes (e.g., in fig. 4.8, I use an explicit fourth-order Runge-Kutta method[2]). For presentation purposes only, let's pretend that

$$\frac{\vec{\eta}_{t+\delta t} - \vec{\eta}_t}{\delta t} \approx k\mathbf{D}_{xx}\vec{\eta}_t \qquad \text{or} \qquad \vec{\eta}_{t+\delta t} - \vec{\eta}_t \approx \tilde{\mathbf{D}}_{xx}\vec{\eta}_t, \tag{4.105}$$

with a finite time step δt and with $\tilde{\mathbf{D}}_{xx} \equiv k\delta t\mathbf{D}_{xx}$, is a good enough approximation of eq. 4.103. While numerically shaky in practice, this form is a reasonable way to represent the evolution symbolically, and it makes clear that the larger $\tilde{\mathbf{D}}_{xx}\vec{\eta}$, the more $\vec{\eta}$ will change in time. Propagating an initial signal $\vec{\eta}_0$ by $\tau = N_t\delta t$ time units and rearranging, this becomes

$$\vec{\eta}_\tau \approx \vec{\eta} + \tilde{\mathbf{D}}_{xx}\vec{\eta}_0 = \left(\mathbf{I} + \tilde{\mathbf{D}}_{xx}\right)^{N_t}\vec{\eta}_0 = \mathbf{P}^{N_t}\vec{\eta}_0, \tag{4.106}$$

where $\mathbf{P} = \mathbf{I} + \tilde{\mathbf{D}}_{xx}$ is the propagator. Referring back to fig. 4.6's right panels, it is easy to see that long waves—which project most strongly on the less damped subspace of $\tilde{\mathbf{D}}_{xx}$'s eigenspace (the subspace spanned by $\tilde{\mathbf{D}}_{xx}$'s leading eigenvectors)—evolve slowly (their characteristic $\tilde{\mathbf{D}}_{xx}\vec{\eta}$ are small), while short waves—which project mostly on the subspace of $\tilde{\mathbf{D}}_{xx}$'s eigenspace spanned by trailing eigenvectors with strongly negative eigenvalues, and whose $\tilde{\mathbf{D}}_{xx}\vec{\eta}$ are thus large—rapidly dissipate, as fig. 4.8 shows.

A reasonable way to characterize initial signals in terms of the degree to which they are altered by the propagator is to consider the squared magnitude after N_t propagator applications,

$$\frac{\vec{\eta}_\tau^T\vec{\eta}_\tau}{\vec{\eta}_0^T\vec{\eta}_0} = \frac{\left[\mathbf{P}^{N_t}\vec{\eta}_0\right]^T\mathbf{P}^{N_t}\vec{\eta}_0}{\vec{\eta}_0^T\vec{\eta}_0} = \frac{\vec{\eta}_0^T\mathbf{P}^{N_t T}\mathbf{P}^{N_t}\vec{\eta}_0}{\vec{\eta}_0^T\vec{\eta}_0}. \tag{4.107}$$

Introducing $\mathbf{P} = \mathbf{E}\Lambda\mathbf{E}^T$, the symmetric propagator's eigen-representation, $\mathbf{P}^{N_t}\vec{\eta}_0 = \left(\mathbf{E}\Lambda\mathbf{E}^T\right)^{N_t}\vec{\eta}_0 = \mathbf{E}\Lambda^{N_t}\mathbf{E}^T\vec{\eta}_0$ (because $\mathbf{E}^T\mathbf{E} = \mathbf{I}$), so that

$$\frac{\vec{\eta}_\tau^T\vec{\eta}_\tau}{\vec{\eta}_0^T\vec{\eta}_0} = \frac{\vec{\eta}_0^T\mathbf{E}\Lambda^{N_t}\mathbf{E}^T\mathbf{E}\Lambda^{N_t}\mathbf{E}^T\vec{\eta}_0}{\vec{\eta}_0^T\vec{\eta}_0} = \frac{\vec{\eta}_0^T\mathbf{E}\Lambda^{2N_t}\mathbf{E}^T\vec{\eta}_0}{\vec{\eta}_0^T\vec{\eta}_0}, \tag{4.108}$$

essentially a spectrally weighted mean decay rate. To better understand the modally averaged decay rate interpretation of $\vec{\eta}_\tau^T\vec{\eta}_\tau/\left(\vec{\eta}_0^T\vec{\eta}_0\right)$, let

[2] See, e.g., section 16.1, pp. 704–708, in Press, W. H., S. A. Teukolsky, W. T. Vetterling and B. P. Flannery, 1992: *Numerical Recipes in Fortran: The Art of Scientific Computing*, 2nd edition, 963 pp.

$$\mathbf{p} := \mathbf{E}^T \vec{\eta}_0 = \begin{pmatrix} \mathbf{e}_1^T \vec{\eta}_0 \\ \mathbf{e}_2^T \vec{\eta}_0 \\ \vdots \\ \mathbf{e}_N^T \vec{\eta}_0 \end{pmatrix} \tag{4.109}$$

denote the set of $\vec{\eta}_0$'s projections on the N eigenvectors; note that $\vec{\eta}_0^T \mathbf{E} = \mathbf{p}^T$ and $\mathbf{p}^T \mathbf{p} = \vec{\eta}_0^T \vec{\eta}_0$ because $\mathbf{E}^T \mathbf{E} = \mathbf{I}$. Consequently,

$$\frac{\vec{\eta}_\tau^T \vec{\eta}_\tau}{\vec{\eta}_0^T \vec{\eta}_0} = \frac{\vec{\eta}_0^T \mathbf{E} \boldsymbol{\Lambda}^{2N_t} \mathbf{E}^T \vec{\eta}_0}{\vec{\eta}_0^T \vec{\eta}_0} = \frac{\mathbf{p}^T \boldsymbol{\Lambda}^{2N_t} \mathbf{p}}{\mathbf{p}^T \mathbf{p}}$$

$$= \frac{\sum_{i=1}^N p_i^2 \lambda_i^{2N_t}}{\sum_{j=1}^N p_j^2} = \left(\frac{1}{\sum_{j=1}^N p_j^2} \right) \sum_{i=1}^N p_i^2 \lambda_i^{2N_t}, \tag{4.110}$$

i.e., the weighted mean of the eigenvalues to the $2N_t$th power, with the ith weight being $p_i^2 / \sum_{j=1}^N p_j^2 = p_i^2 / \|\vec{\eta}_0\|^2$.

The evolution of $\vec{\eta}_1$ and $\vec{\eta}_2$ under a particular $\tilde{\mathbf{D}}_{xx}$ (i.e., with specific k, δx, and δt choices) is shown in fig. 4.8. The above predictions, which are based on spectral decomposition of the initial signals, are clearly borne out quantitatively. Over the shown span, during which $\mathbf{P}$ was applied 100 times, the short wave signal lost very nearly twice as much of its initial squared norm as its long wave counterpart. Referring back to fig. 4.7, this disparity is now readily understood in terms of the distinct subspaces of the propagator's eigenspace on which each of the $\vec{\eta}$s project.

4.4 Summary

The eigenvalues λ_i and eigenvectors $\mathbf{e}_i$ of a square $(N \times N)$ $\mathbf{A}$ satisfy the key equation of this chapter,

$$\mathbf{A}\mathbf{e}_i = \lambda_i \mathbf{e}_i.$$

The eigenvectors corresponding to nonzero eigenvalues describe the directions in $\mathbb{R}^N$ along which premultiplication by $\mathbf{A}$ changes at most the magnitude, but not the direction, of the premultiplied vector. If $\lambda_i \neq 1$ for some i, that magnitude change is λ_i.

Not all matrices have a full set of distinct eigenvectors. If they do, It is customary and useful to place the eigenvectors in a single matrix,

$$\mathbf{E} = \begin{pmatrix} \mathbf{e}_1 & \cdots & \mathbf{e}_N \end{pmatrix},$$

and the eigenvalues in a diagonal matrix,

$$\boldsymbol{\Lambda} = \begin{pmatrix} \lambda_1 & & \\ & \ddots & \\ & & \lambda_N \end{pmatrix},$$

with which

$$\mathbf{A} = \mathbf{E}\boldsymbol{\Lambda}\mathbf{E}^{-1}.$$

If $\mathbf{A}$ is normal (or self-adjoint, satisfying $\mathbf{A}\mathbf{A}^T = \mathbf{A}^T\mathbf{A}$), this simplified further, to

$$\mathbf{A} = \mathbf{E}\boldsymbol{\Lambda}\mathbf{E}^T,$$

because for a normal $\mathbf{A}$, $\mathbf{E}\mathbf{E}^T = \mathbf{E}^T\mathbf{E} = \mathbf{I}$, so $\mathbf{E}^{-1} = \mathbf{E}^T$. If $\mathbf{A}$ is viewed as an operator, the set $\{\lambda_i\}_{i=1}^N$ is collectively known as the operator's spectrum.

FIVE

The Algebraic Operation of SVD

IN THE PRECEDING CHAPTER we discussed the eigenvalue/eigenvector diagonalization of a matrix. Perhaps the biggest problem for this to be very useful in data analysis is the restriction to square matrices. We have already emphasized time and again that data matrices, unlike dynamical operators, are rarely square. The algebraic operation of the singular value decomposition, SVD, is the answer. Note the distinction between the data analysis method widely known as SVD and the actual algebraic machinery. The former uses the latter, but isn't the latter! In this chapter, I describe the method, postponing the discussion of the analytic tool to the book's second part.

5.1 SVD Introduced

Any real $\mathbf{A} \in \mathbb{R}^{M \times N}$ can be represented as the product of three very special matrices:

$$\mathbf{A} = \mathbf{U}\mathbf{\Sigma}\mathbf{V}^T, \tag{5.1}$$

where $\mathbf{U} \in \mathbb{R}^{M \times M}$, $\mathbf{\Sigma} \in \mathbb{R}^{M \times N}$, and $\mathbf{V} \in \mathbb{R}^{N \times N}$, with $\{\hat{\mathbf{u}}_i\}_{i=1}^{M}$ and $\{\hat{\mathbf{v}}_i\}_{i=1}^{N}$ (**U**'s and **V**'s columns) being $\mathbf{A}\mathbf{A}^T$'s and $\mathbf{A}^T\mathbf{A}$'s eigenvectors, respectively. Because $\mathbf{A}\mathbf{A}^T$ and $\mathbf{A}^T\mathbf{A}$ are both square and symmetric quadratics, their eigenvalues are real and nonnegative, and their eigenvectors form complete orthonormal spanning sets for $\mathbb{R}^M$ and $\mathbb{R}^N$, respectively.

To show why the SVD representation exists, let's start by noting that, by virtue of being $\mathbf{A}^T\mathbf{A}$'s and $\mathbf{A}\mathbf{A}^T$'s eigenvectors, $\{\hat{\mathbf{v}}_i\}$ and $\{\hat{\mathbf{u}}_i\}$ satisfy

$$\mathbf{A}^T\mathbf{A}\hat{\mathbf{v}}_i = \lambda_{vi}\hat{\mathbf{v}}_i, \qquad i = [1, N], \tag{5.2}$$

$$\mathbf{A}\mathbf{A}^T\hat{\mathbf{u}}_i = \lambda_{ui}\hat{\mathbf{u}}_i, \qquad i = [1, M] \tag{5.3}$$

mode-wise. Allowing for rank deficiency, $q \leq \min(M, N)$, and premultiplying (5.2) by **A** gives

$$\mathbf{A}\mathbf{A}^T(\mathbf{A}\hat{\mathbf{v}}_i) = \lambda_{vi}(\mathbf{A}\hat{\mathbf{v}}_i), \qquad i = [1, q], \tag{5.4}$$

which shows that $\hat{\mathbf{u}}_i = \mathbf{A}\hat{\mathbf{v}}_i / \|\mathbf{A}\hat{\mathbf{v}}_i\|$ are unit norm eigenvectors of $\mathbf{A}\mathbf{A}^T$, with the corresponding eigenvalues λ_{vi}. That is, $\lambda_{vi} = \lambda_{ui} = \lambda_i$, the two sets of nontrivial eigenvalues are one and the same (which makes sense, as $\mathbf{A}^T\mathbf{A}$ and $\mathbf{A}\mathbf{A}^T$ contain essentially the same information). To get the scaling factor $\|\mathbf{A}\hat{\mathbf{v}}_i\|$ for $\hat{\mathbf{u}}$, we premultiply (5.2) by $\hat{\mathbf{v}}^T$,

$$\hat{\mathbf{v}}_i^T \mathbf{A}^T \mathbf{A} \hat{\mathbf{v}}_i \equiv \|\mathbf{A}\hat{\mathbf{v}}_i\|^2 = \lambda_i \hat{\mathbf{v}}_i^T \hat{\mathbf{v}}_i = \lambda_i, \qquad i = [1, N], \tag{5.5}$$

so

$$\|\mathbf{A}\hat{\mathbf{v}}_i\| = \sqrt{\lambda_i}, \qquad i = [1, N], \tag{5.6}$$

$$\hat{\mathbf{u}}_i = \frac{\mathbf{A}\hat{\mathbf{v}}_i}{\|\mathbf{A}\hat{\mathbf{v}}_i\|} = \frac{\mathbf{A}\hat{\mathbf{v}}_i}{\sqrt{\lambda_i}}, \qquad i = [1, q], \tag{5.7}$$

and eq. 5.4 reads

$$\mathbf{A}\mathbf{A}^T \hat{\mathbf{u}}_i = \lambda_i \frac{\mathbf{A}\hat{\mathbf{v}}_i}{\sqrt{\lambda_i}} = \sqrt{\lambda_i}\,\mathbf{A}\hat{\mathbf{v}}_i, \qquad i = [1, q]. \tag{5.8}$$

Despite the limited i range above, the symmetry of $\mathbf{A}^T\mathbf{A}$ and $\mathbf{A}\mathbf{A}^T$ guarantees a full set of eigenvectors. By employing Gram-Schmidt orthonormalization we can therefore extend $\{\hat{\mathbf{u}}_i\}$ and $\{\hat{\mathbf{v}}_i\}$ from q to M and N vectors, respectively, thus fully spanning $\mathbb{R}^M$ and $\mathbb{R}^N$, and construct $\mathbf{U} \in \mathbb{R}^{M \times M}$ and $\mathbf{V} \in \mathbb{R}^{N \times N}$ satisfying

$$\mathbf{U}\mathbf{U}^T = \mathbf{U}^T\mathbf{U} = \mathbf{I}_M \qquad \text{and} \qquad \mathbf{V}\mathbf{V}^T = \mathbf{V}^T\mathbf{V} = \mathbf{I}_N, \tag{5.9}$$

where $\mathbf{I}_M$ and $\mathbf{I}_N$ are the $M \times M$ and $N \times N$ identity matrices, irrespective of q. Let us also arrange, with no loss of generality, the singular values, $\sigma_i \equiv \sqrt{\lambda_i}$, $I = [1, q]$, the square root of the nonzero eigenvalues that $\mathbf{A}^T\mathbf{A}$ and $\mathbf{A}\mathbf{A}^T$ share, along the diagonal of $\boldsymbol{\Sigma} \in \mathbb{R}^{M \times N}$, and reshuffle the columns of $\mathbf{U}$ and $\mathbf{V}$ accordingly. With these extensions, eq. 5.8 becomes

$$\mathbf{A}\mathbf{A}^T\mathbf{U} = \mathbf{A}\mathbf{V}\boldsymbol{\Sigma}^T \qquad \Longrightarrow \qquad \mathbf{A}\mathbf{A}^T = \mathbf{A}\mathbf{V}\boldsymbol{\Sigma}^T\mathbf{U}^T \tag{5.10}$$

or

$$\mathbf{A}\left(\mathbf{A}^T - \mathbf{V}\boldsymbol{\Sigma}^T\mathbf{U}^T\right) = 0 \in \mathbb{R}^{M \times M}, \tag{5.11}$$

which, in general, holds only if

$$\mathbf{A}^T = \mathbf{V}\boldsymbol{\Sigma}^T\mathbf{U}^T. \tag{5.12}$$

Taking the transpose of both sides, we finally get the SVD representation,

$$\mathbf{A} = \mathbf{U}\boldsymbol{\Sigma}\mathbf{V}^T. \tag{5.13}$$

The matrix $\boldsymbol{\Sigma}$ and the relationship of the SVD representation to $\mathbf{A}$'s fundamental spaces require further discussion. Recall that $\boldsymbol{\Sigma}$ is diagonal but rectangular, with $\mathbf{A}$'s q nonzero singular values, $\sigma_i = \sqrt{\lambda_i}$, arranged in descending numerical order along its diagonal. It can be viewed as comprising three matrices, depending on M, N, and q, as fig. 5.1 shows. While both $\mathbf{N}$ and $\mathbf{N}_l$ comprise only zero elements,

$$\mathbf{D} = \begin{pmatrix} \sigma_1 & & & \\ & \sigma_2 & & \\ & & \ddots & \\ & & & \sigma_q \end{pmatrix} \in \mathbb{R}^{q \times q} \tag{5.14}$$

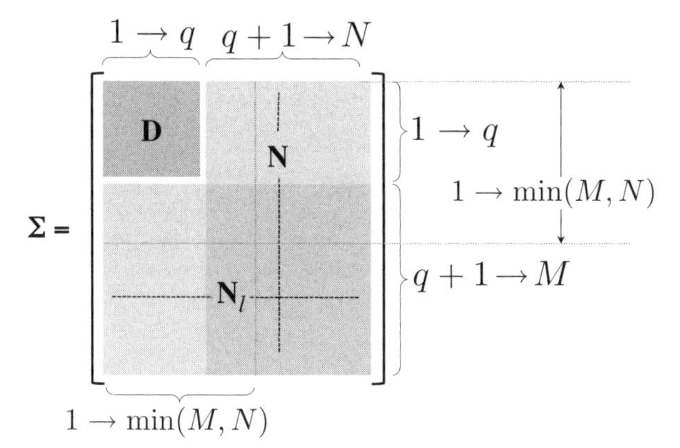

Figure 5.1. The structure of $\boldsymbol{\Sigma}$ in the SVD representation of a rank q $\mathbf{A}\in\mathbb{R}^{M\times N}$. The submatrices are $\mathbf{D}\in\mathbb{R}^{q\times q}$, corresponding to $\mathcal{R}(\mathbf{A})$ and $\mathcal{R}(\mathbf{A}^T)$; $\mathbf{N}_l\in\mathbb{R}^{(M-q)\times N}$—which exists only if $q<M$—corresponding to $\mathcal{N}(\mathbf{A}^T)$; and $\mathbf{N}\in\mathbb{R}^{M\times(N-q)}$—which exists only if $q<N$—corresponding to $\mathcal{N}(\mathbf{A})$.

is the diagonal matrix of nonzero singular values (with zero elements suppressed above), corresponding to $\mathcal{R}(\mathbf{A})$ and $\mathcal{R}(\mathbf{A}^T)$. Its maximum size is $\min(M, N)\times\min(M, N)$, which is realized only if $\mathbf{A}$ is full rank. In that case, $\mathbf{D}$ occupies $\boldsymbol{\Sigma}$'s full left end (if $M<N$) or top (if $M>N$).

Since $\mathbf{A}$'s rank is $q\leq\min(M, N)$, the remainder (which for a full rank $\mathbf{A}$ exists in either $\mathbf{U}$ (if $M>N=q$) or $\mathbf{V}$ (if $N>M=q$)) must represent a null space. Assuming $\mathbf{A}$ is full rank, when $M>N$, $\mathbf{U}$ is larger than $\mathbf{V}$ and $\mathbf{D}$'s N diagonal elements are nonzero, but $\{\mathbf{u}_i\}_{i=N+1}^{M}$ span a nonempty left null space comprising all $\mathbb{R}^M$ vectors orthogonal to $\mathbf{A}$'s columns. (Because such an $\mathbf{A}$ has only $N<M$ independent columns, there is no reason to even hope its columns fully span $\mathbb{R}^M$!) If, on the other hand (still assuming full rank), $M<N$ so $\mathbf{V}$ is larger than $\mathbf{U}$, all $\mathbf{D}$'s M diagonal elements are nonzero, but $\{\mathbf{v}_i\}_{i=M+1}^{N}$ span a nonempty null space comprising all $\mathbb{R}^N$ vectors orthogonal to $\mathbf{A}$'s rows. (Because now $\mathbf{A}$ has $q=M<N$ independent columns, its columns fully span $\mathbb{R}^M$, but only a subspace of $\mathbb{R}^N$.) It is because of these considerations that the limits in eqs. 5.4, 5.7, and 5.8 are $[1, q]$ despite the fact that there are more columns than this limit in either $\mathbf{U}$, $\mathbf{V}$ or both (for the cases $M>N=q$, $N>M=q$, and $q<(M, N)$, respectively).

Figure 5.2 represents schematically the SVD of three $\mathbf{A}\in\mathbb{R}^{M\times N}$ examples satisfying (top) $q=N<M$, (middle) $q<N<M$, and (bottom) $q<M<N$. When $q=N<M$ (top), $\mathbf{u}_{1\to N}$ and $\mathbf{u}_{N+1\to M}$ span $\mathbf{A}$'s column and left null spaces in $\mathbb{R}^M$, $\mathbf{v}_{1\to N}$ span $\mathbf{A}$'s row space in $\mathbb{R}^N$, and this $\mathbf{A}$ has only a trivial null space in $\mathbb{R}^N$ (i.e., $\mathcal{R}(\mathbf{A}^T)=\mathbb{R}^N$). For $q<N<M$ (middle) and $q<M<N$ (bottom), $\mathbf{u}_{1\to q}$ and

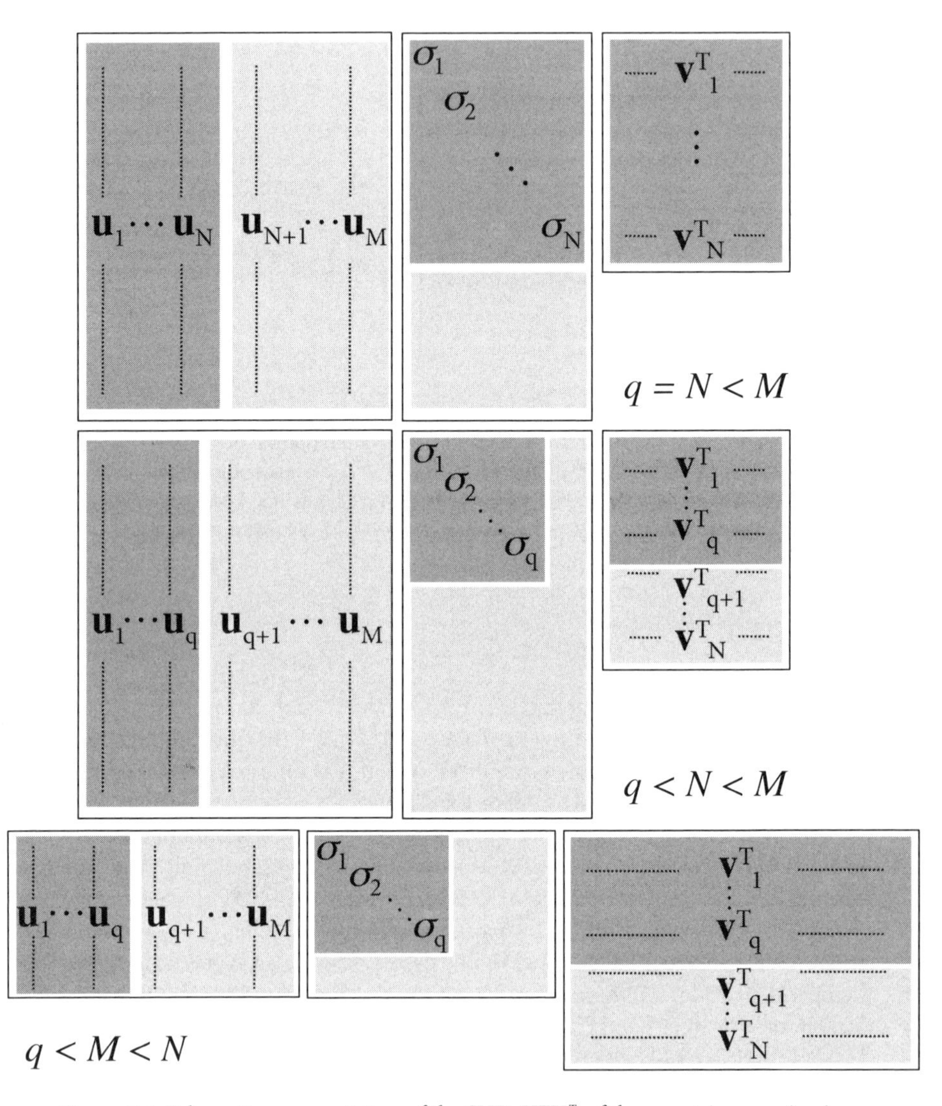

Figure 5.2. Schematic representation of the SVD, $\mathbf{U\Sigma V}^T$, of three matrices varying in the relationships between q, M, and N. In all cases $\mathbf{u}_i \in \mathbb{R}^M$, $\mathbf{U} \in \mathbb{R}^{M \times M}$, $\mathbf{v}_i \in \mathbb{R}^N$, $\mathbf{V} \in \mathbb{R}^{N \times N}$, and $\mathbf{\Sigma} \in \mathbb{R}^{M \times N}$. As indicated, the schematics represent matrices satisfying $q = N < M$, $q < N < M$, and $q < M < N$. Darker and lighter shades of gray indicate column and left null spaces in $\mathbf{U}$ and row and null spaces in $\mathbf{V}$, respectively. The same gray shading convention also holds for $\mathbf{\Sigma}$, where darker gray corresponds to both $\mathcal{R}(\mathbf{A})$ and $\mathcal{R}(\mathbf{A}^T)$.

$\mathbf{u}_{q+1\to M}$ span $\mathbf{A}$'s column and left null spaces in $\mathbb{R}^M$, and $\mathbf{v}_{1\to q}$ and $\mathbf{v}_{q+1\to N}$ span $\mathbf{A}$'s row and null spaces in $\mathbb{R}^N$.

Note the important expansion that follows from $\mathbf{A} = \mathbf{U\Sigma V}^T$:

$$\mathbf{A} = \sum_{i=1}^{q} \sigma_i \mathbf{u}_i \mathbf{v}_i^T = \sum_{i=1}^{q} \sigma_i \begin{pmatrix} \vdots \\ \mathbf{u}_i \\ \vdots \end{pmatrix} \begin{pmatrix} \cdots & \mathbf{v}_i^T & \cdots \end{pmatrix} \tag{5.15}$$
$$= \mathbf{M}_1 + \mathbf{M}_2 + \cdots + \mathbf{M}_q$$

where the unit rank $\mathbf{M}_i \in \mathbb{R}^{M\times N}$ are $\mathbf{A}$'s modes. This spectral representation of $\mathbf{A}$ as a series of modes is among SVD's most important aspects, as discussed later.

Consistent with our earlier derivation of the SVD, from $\mathbf{A} = \mathbf{U\Sigma V}^T$ and the orthonormality of $\mathbf{U}$ and $\mathbf{V}$ it follows that

$$\mathbf{AA}^T = \mathbf{U\Sigma V}^T\mathbf{V\Sigma}^T\mathbf{U}^T = \mathbf{U\Sigma}_M^2\mathbf{U}^T, \tag{5.16}$$

$$\mathbf{A}^T\mathbf{A} = \mathbf{V\Sigma}^T\mathbf{U}^T\mathbf{U\Sigma V}^T = \mathbf{V\Sigma}_N^2\mathbf{V}^T, \tag{5.17}$$

with diagonal $\mathbf{\Sigma}_M^2 \in \mathbb{R}^{M\times M}$ and $\mathbf{\Sigma}_N^2 \in \mathbb{R}^{N\times N}$ whose $[1, q]$ diagonal elements are the squared singular values (i.e., the eigenvalues of both $\mathbf{AA}^T$ and $\mathbf{A}^T\mathbf{A}$).

The orthonormality of $\mathbf{U}$ and $\mathbf{V}$ also leads straightforwardly to biorthogonality relations between $\mathbf{U}$ and $\mathbf{V}$,

$$\mathbf{A} = \mathbf{U\Sigma V}^T \;\Rightarrow\; \mathbf{AV} = \mathbf{U\Sigma V}^T\mathbf{V} = \mathbf{U\Sigma}, \tag{5.18}$$

$$\mathbf{U}^T\mathbf{A} = \mathbf{U}^T\mathbf{U\Sigma V}^T = \mathbf{\Sigma V}^T \;\Rightarrow\; \mathbf{A}^T\mathbf{U} = \mathbf{V\Sigma}^T, \tag{5.19}$$

which we have encountered and used in slightly different guises in the derivation of the SVD. Let's examine more closely the two relations. The first one is

$$\mathbf{AV} = \mathbf{A}\begin{pmatrix} \vdots & & \vdots & \vdots & & \vdots \\ \mathbf{v}_1 & \cdots & \mathbf{v}_q & \mathbf{v}_{q+1} & \cdots & \mathbf{v}_N \\ \vdots & & \vdots & \vdots & & \vdots \end{pmatrix} = \mathbf{U\Sigma}$$
$$= \begin{pmatrix} \vdots & & \vdots & \vdots & & \vdots \\ \sigma_1\mathbf{u}_1 & \cdots & \sigma_q\mathbf{u}_q & 0\cdot\mathbf{u}_{q+1} & \cdots & 0\cdot\mathbf{u}_M \\ \vdots & & \vdots & \vdots & & \vdots \end{pmatrix} \tag{5.20}$$
$$= \begin{pmatrix} \vdots & & \vdots & \vdots & & \vdots \\ \sigma_1\mathbf{u}_1 & \cdots & \sigma_q\mathbf{u}_q & 0 & \cdots & 0 \\ \vdots & & \vdots & \vdots & & \vdots \end{pmatrix}.$$

Note the irrelevance of $\{\mathbf{u}_i\}_{i=q+1}^{M}$; multiplied by the zero bottom part of $\mathbf{\Sigma}$, they contribute nothing. Consequently, the above system gives rise naturally to the reduced system

$$\mathbf{A}\underbrace{\begin{pmatrix} \vdots & & \vdots \\ \mathbf{v}_1 & \cdots & \mathbf{v}_q \\ \vdots & & \vdots \end{pmatrix}}_{\mathbf{V}_r \in \mathbb{R}^{N\times q}} = \underbrace{\begin{pmatrix} \vdots & & \vdots \\ \mathbf{u}_1 & \cdots & \mathbf{u}_q \\ \vdots & & \vdots \end{pmatrix}}_{\mathbf{U}_r \in \mathbb{R}^{M\times q}} \underbrace{\begin{pmatrix} \sigma_1 & & & \mathbf{O} \\ & \sigma_2 & & \\ \mathbf{O} & & \ddots & \\ & & & \sigma_q \end{pmatrix}}_{\Sigma_r \in \mathbb{R}^{q\times q}} \tag{5.21}$$

where $\mathbf{U}_r$ and $\boldsymbol{\Sigma}_r$ are the reduced $\mathbf{U}$ and $\boldsymbol{\Sigma}$, with only q columns. On the right, each retained $\mathbf{u}_i$ is multiplied by the corresponding nonzero singular value σ_i. This reveals the modal correspondence between $\mathbf{u}_i$ and $\mathbf{v}_i$, with $\mathbf{A}$ operating individually on each $\mathbf{v}_i$, mapping it to $\sigma_i \mathbf{u}_i$,

$$\mathbf{A}\mathbf{v}_i = \sigma_i \mathbf{u}_i, \qquad i = [1, q]. \tag{5.22}$$

A similar simple argument holds for the second biorthogonality condition, $\mathbf{A}^T\mathbf{U} = \mathbf{V}\boldsymbol{\Sigma}^T$, yielding

$$\mathbf{A}^T\mathbf{u}_i = \sigma_i \mathbf{v}_i, \qquad i = [1, q]. \tag{5.23}$$

In the former relation, for each i, combining $\mathbf{A}$'s columns $\mathbf{a}_j$ with weights given by $\mathbf{v}_i$'s N elements v_{ij}, $\sum_{j=1}^{N} a_j v_{ij}$, yields $\sigma_i \mathbf{u}_i$. In the latter relation, for each i, combining $\mathbf{A}$'s rows with $\mathbf{u}_i$'s M elements u_{ij} as weights yields $\sigma_i \mathbf{v}_i$.

5.2 Some Examples

5.2.1 *Example I*

Let's start with the simple numerical example

$$\mathbf{A} = \begin{pmatrix} 3 & 0 \\ 0 & 0 \\ 0 & 2 \end{pmatrix}. \tag{5.24}$$

Since $\mathbf{A}\mathbf{A}^T$ is 3×3 while $\mathbf{A}^T\mathbf{A}$ is 2×2, it seems logical to eigenanalyze the latter for $\mathbf{V}$ and then use the biorthogonality conditions to obtain $\mathbf{U}$. (While $|M - N|$ is in this example entirely trivial, in many applications it may be dramatic, and, in some cases, can mean the difference between a computationally tractable problem to one that is completely out of reach with current or foreseeable hardware.) With

$$\mathbf{A}^T\mathbf{A} = \begin{pmatrix} 9 & 0 \\ 0 & 4 \end{pmatrix}, \qquad \mathbf{A}^T\mathbf{A} - \lambda\mathbf{I} = \begin{pmatrix} 9-\lambda & 0 \\ 0 & 4-\lambda \end{pmatrix}, \tag{5.25}$$

the characteristic equation is

$$(9 - \lambda)(4 - \lambda) = 0 \tag{5.26}$$

with roots $\lambda_1 \equiv \sigma_1^2 = 9$ and $\lambda_2 \equiv \sigma_2^2 = 4$ (recall that it is customary to arrange the singular values in descending order), so that

$$\boldsymbol{\Sigma} = \begin{pmatrix} \sigma_1 & 0 \\ 0 & \sigma_2 \\ 0 & 0 \end{pmatrix} = \begin{pmatrix} 3 & 0 \\ 0 & 2 \\ 0 & 0 \end{pmatrix} \quad \text{and} \quad \boldsymbol{\Sigma}_r = \begin{pmatrix} 3 & 0 \\ 0 & 2 \end{pmatrix}. \tag{5.27}$$

Using $\mathbf{A}^T\mathbf{A}\mathbf{v}_1 = \lambda_1\mathbf{v}_1$ yields

$$\begin{pmatrix} 9 & 0 \\ 0 & 4 \end{pmatrix}\begin{pmatrix} \alpha \\ \beta \end{pmatrix} = 9\begin{pmatrix} \alpha \\ \beta \end{pmatrix}, \tag{5.28}$$

whose first equation means $\alpha = 1$ but whose second equation can be satisfied only with $\beta = 0$, so

$$\mathbf{v}_1 = \begin{pmatrix} 1 \\ 0 \end{pmatrix} \tag{5.29}$$

and

$$\begin{pmatrix} 9 & 0 \\ 0 & 4 \end{pmatrix}\begin{pmatrix} \alpha \\ \beta \end{pmatrix} = 4\begin{pmatrix} \alpha \\ \beta \end{pmatrix} \quad \Longrightarrow \quad \mathbf{v}_2 = \begin{pmatrix} 0 \\ 1 \end{pmatrix}. \tag{5.30}$$

Thus, we have found that

$$\mathbf{V} \equiv \begin{pmatrix} \mathbf{v}_1 & \mathbf{v}_2 \end{pmatrix} = \begin{pmatrix} 1 & 0 \\ 0 & 1 \end{pmatrix}. \tag{5.31}$$

Now to $\mathbf{U}$. We clearly cannot obtain the full $\mathbf{U}$ from our current results, only $\mathbf{u}_1$ and $\mathbf{u}_2$. Since, as was discussed above, $\mathbf{u}_3$ may often not be needed, we simply ignore it for now, getting $\mathbf{u}_1$ and $\mathbf{u}_2$ from

$$\begin{aligned} \mathbf{AV} &= \begin{pmatrix} 3 & 0 \\ 0 & 0 \\ 0 & 2 \end{pmatrix}\begin{pmatrix} 1 & 0 \\ 0 & 1 \end{pmatrix} = \begin{pmatrix} 3 & 0 \\ 0 & 0 \\ 0 & 2 \end{pmatrix} \\ &= \begin{pmatrix} \sigma_1\mathbf{u}_1 & \sigma_2\mathbf{u}_2 \end{pmatrix} \equiv \mathbf{U}_r\boldsymbol{\Sigma}_r, \end{aligned} \tag{5.32}$$

which means—since $\sigma_1 = 3$ and $\sigma_2 = 2$—that

$$\mathbf{u}_1 = \begin{pmatrix} 1 \\ 0 \\ 0 \end{pmatrix} \quad \text{and} \quad \mathbf{u}_2 = \begin{pmatrix} 0 \\ 0 \\ 1 \end{pmatrix}. \tag{5.33}$$

The entire decomposition can be verified by checking that $\mathbf{U}_r\boldsymbol{\Sigma}_r\mathbf{V}^T = \mathbf{A}$.

5.2.2 Example II

Let's consider next

$$\mathbf{A} = \begin{pmatrix} 1 & 0 & -1 & 0 & 1 & 0 & -1 & 0 \\ 0 & 1 & 0 & -1 & 0 & 1 & 0 & -1 \\ -1 & 0 & 1 & 0 & -1 & 0 & 1 & 0 \end{pmatrix} \tag{5.34}$$

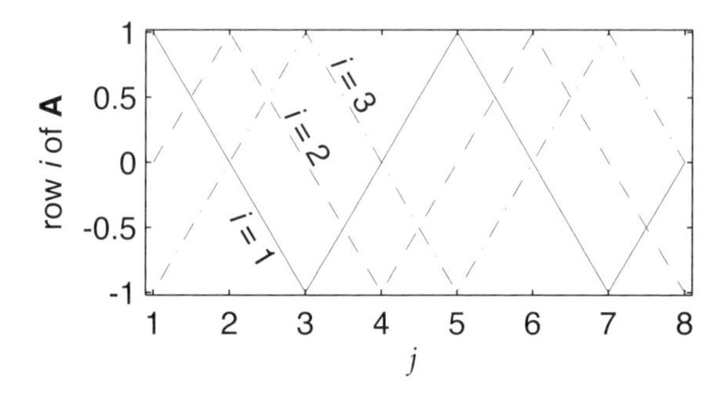

Figure 5.3. The rows of example II's **A**.

shown in fig. 5.3. Note that the mean of each of **A**'s rows is zero, which is useful because with this property $\mathbf{A}^T\mathbf{A}$ and $\mathbf{A}\mathbf{A}^T$ are **A**'s row and column covariance matrices, respectively (the full significance of this will become clearer when we discuss the data analysis operations based on SVD). Because $M = 3$, while $N = 8$, we eigenanalyze the 3×3

$$\mathbf{A}\mathbf{A}^T = \begin{pmatrix} 4 & 0 & -4 \\ 0 & 4 & 0 \\ -4 & 0 & 4 \end{pmatrix}, \tag{5.35}$$

reflecting the fact that **A**'s top row is orthogonal to the second and is the third row's negative. We next solve

$$|\mathbf{A}\mathbf{A}^T - \lambda\mathbf{I}| = \begin{vmatrix} 4-\lambda & 0 & -4 \\ 0 & 4-\lambda & 0 \\ -4 & 0 & 4-\lambda \end{vmatrix} = 0, \tag{5.36}$$

which yields

$$(4-\lambda)\left[(4-\lambda)^2 - 16\right] = 0. \tag{5.37}$$

Clearly, $\lambda_1 = 4$. Alternatively, $(4 - \lambda)^2 = 16 = (\pm 4)^2$ from which

$$4 - \lambda = \pm 4 \quad \Longrightarrow \quad \begin{matrix} \lambda_2 = 8 \\ \lambda_3 = 0 \end{matrix} \tag{5.38}$$

follows. We then rearrange the λs in descending order,

$$\begin{aligned} \lambda_1 &= 8 \\ \lambda_2 &= 4 \\ \lambda_3 &= 0. \end{aligned} \tag{5.39}$$

The mode corresponding to $\lambda_1 = 8$ yields

$$\mathbf{A}\mathbf{A}^T\mathbf{u}_1 = 8\mathbf{u}_1 \tag{5.40}$$

or

$$4\begin{pmatrix} 1 & 0 & -1 \\ 0 & 1 & 0 \\ -1 & 0 & 1 \end{pmatrix}\begin{pmatrix} u_{11} \\ u_{12} \\ u_{13} \end{pmatrix} = 8\begin{pmatrix} u_{11} \\ u_{12} \\ u_{13} \end{pmatrix}. \tag{5.41}$$

From the second row it is clear that $u_{12} = 0$, but the other two rows both yield $u_{13} = -u_{11}$, so one (by no means unique) possibility is

$$\mathbf{u}_1 = \frac{1}{\sqrt{2}}\begin{pmatrix} -1 \\ 0 \\ 1 \end{pmatrix}. \tag{5.42}$$

Similarly, the mode corresponding to $\lambda_1 = 4$ yields

$$\mathbf{A}\mathbf{A}^T\mathbf{u}_2 = 4\mathbf{u}_2 \tag{5.43}$$

or

$$\begin{pmatrix} 1 & 0 & -1 \\ 0 & 1 & 0 \\ -1 & 0 & 1 \end{pmatrix}\begin{pmatrix} u_{21} \\ u_{22} \\ u_{23} \end{pmatrix} = \begin{pmatrix} u_{21} \\ u_{22} \\ u_{23} \end{pmatrix}. \tag{5.44}$$

The second row leaves u_{22} unconstrained, but the top and bottom rows yield $u_{21} = u_{23} = 0$, so (nonuniquely)

$$\mathbf{u}_2 = \begin{pmatrix} 0 \\ 1 \\ 0 \end{pmatrix}. \tag{5.45}$$

Since $q = 2$ here (consistent with $\lambda_3 = 0$, in turn, because **A**'s third row is minus the first row), we can stop here and not evaluate $\mathbf{u}_3$.

Employing the biorthogonality $\mathbf{v}_i = \mathbf{A}^T\mathbf{u}_i/\sigma_i$ yields

$$\mathbf{v}_1 = \frac{1}{2}\begin{pmatrix} -1 \\ 0 \\ 1 \\ 0 \\ -1 \\ 0 \\ 1 \\ 0 \end{pmatrix} \quad \text{and} \quad \mathbf{v}_2 = \frac{1}{2}\begin{pmatrix} 0 \\ 1 \\ 0 \\ -1 \\ 0 \\ 1 \\ 0 \\ -1 \end{pmatrix} \tag{5.46}$$

and our reduced SVD representation of **A** is

$$\begin{aligned} \mathbf{A} &= \mathbf{U}_r\mathbf{\Sigma}_r\mathbf{V}_r^T \\ &= \frac{1}{2\sqrt{2}}\begin{pmatrix} -1 & 0 \\ 0 & \sqrt{2} \\ 1 & 0 \end{pmatrix}\begin{pmatrix} \sqrt{8} & 0 \\ 0 & 2 \end{pmatrix}\begin{pmatrix} -1 & 0 & 1 & 0 & -1 & 0 & 1 & 0 \\ 0 & 1 & 0 & -1 & 0 & 1 & 0 & -1 \end{pmatrix} \end{aligned} \tag{5.47}$$

$$= \frac{1}{\sqrt{8}} \begin{pmatrix} -1 & 0 \\ 0 & \sqrt{2} \\ 1 & 0 \end{pmatrix} \begin{pmatrix} \sqrt{8} & 0 \\ 0 & 2 \end{pmatrix} \begin{pmatrix} -1 & 0 & 1 & 0 & -1 & 0 & 1 & 0 \\ 0 & 1 & 0 & -1 & 0 & 1 & 0 & -1 \end{pmatrix}.$$

5.2.3 *Example III*

Let's consider next an example that is numerically a bit messier:

$$\mathbf{A} = \begin{pmatrix} 1 & 4 \\ 3 & -1 \\ 2 & 6 \end{pmatrix}, \tag{5.48}$$

starting again with the smaller problem of finding $\mathbf{V}$. With

$$\mathbf{A}^T\mathbf{A} = \begin{pmatrix} 14 & 13 \\ 13 & 53 \end{pmatrix}, \quad \mathbf{A}^T\mathbf{A} - \lambda\mathbf{I} = \begin{pmatrix} 14-\lambda & 13 \\ 13 & 53-\lambda \end{pmatrix} \tag{5.49}$$

yields the characteristic equation

$$(14-\lambda)(53-\lambda) - 13^2 = 0 \quad \text{or} \quad \lambda^2 - 67\lambda + 573 = 0, \tag{5.50}$$

with roots $\lambda_1 \equiv \sigma_1^2 \approx 56.93$ and $\lambda_2 \equiv \sigma_2^2 \approx 10.06$, so that

$$\mathbf{\Sigma} \approx \begin{pmatrix} 7.55 & 0 \\ 0 & 3.17 \\ 0 & 0 \end{pmatrix} \quad \text{and} \quad \mathbf{\Sigma}_r \approx \begin{pmatrix} 7.55 & 0 \\ 0 & 3.17 \end{pmatrix}. \tag{5.51}$$

To get $\mathbf{V}$, we solve $\mathbf{A}^T\mathbf{A}\mathbf{v}_i = \lambda_i\mathbf{v}_i$, $i = 1, 2$,

$$\begin{pmatrix} 14 & 13 \\ 13 & 53 \end{pmatrix} \begin{pmatrix} \alpha \\ \beta \end{pmatrix} = 56.93 \begin{pmatrix} \alpha \\ \beta \end{pmatrix} \tag{5.52}$$

and

$$\begin{pmatrix} 14 & 13 \\ 13 & 53 \end{pmatrix} \begin{pmatrix} \gamma \\ \delta \end{pmatrix} = 10.06 \begin{pmatrix} \gamma \\ \delta \end{pmatrix} \tag{5.53}$$

for $(\alpha, \beta, \gamma, \delta)$, yielding

$$\mathbf{v}_1 \approx \begin{pmatrix} 0.29 \\ 0.96 \end{pmatrix}, \quad \mathbf{v}_2 \approx \begin{pmatrix} 0.96 \\ -0.29 \end{pmatrix} \quad \Longrightarrow \quad \mathbf{V} \approx \begin{pmatrix} 0.29 & 0.96 \\ 0.96 & -0.29 \end{pmatrix}. \tag{5.54}$$

To compute $\mathbf{U}_r$, we use

$$\begin{aligned} \mathbf{AV} &\approx \begin{pmatrix} 1 & 4 \\ 3 & -1 \\ 2 & 6 \end{pmatrix} \begin{pmatrix} 0.29 & 0.96 \\ 0.96 & -0.29 \end{pmatrix} \\ &\approx \begin{pmatrix} 4.12 & -0.20 \\ -0.09 & 3.16 \\ 6.32 & 0.18 \end{pmatrix} = \mathbf{U}_r\mathbf{\Sigma}_r \end{aligned} \tag{5.55}$$

and get

$$\begin{pmatrix} 4.12 \\ -0.09 \\ 6.32 \end{pmatrix} = \sigma_1 \mathbf{u}_1 \implies \mathbf{u}_1 \approx \begin{pmatrix} 0.55 \\ -0.01 \\ 0.84 \end{pmatrix} \tag{5.56}$$

$$\begin{pmatrix} -0.20 \\ 3.16 \\ 0.18 \end{pmatrix} = \sigma_2 \mathbf{u}_2 \implies \mathbf{u}_2 \approx \begin{pmatrix} -0.06 \\ 1.00 \\ 0.06 \end{pmatrix}$$

and $\mathbf{U}_r = (\,\mathbf{u}_1 \;\; \mathbf{u}_2\,)$.

All that is left is to check that we can reconstruct $\mathbf{A}$ using the computed matrices:

$$\begin{aligned} \mathbf{U}_r \boldsymbol{\Sigma}_r \mathbf{V}^T &= \begin{pmatrix} 0.55 & -0.06 \\ -0.01 & 1.00 \\ 0.84 & 0.06 \end{pmatrix} \begin{pmatrix} 7.55 & 0 \\ 0 & 3.17 \end{pmatrix} \begin{pmatrix} 0.29 & 0.96 \\ 0.96 & -0.29 \end{pmatrix} \\ &\approx \begin{pmatrix} 1 & 4 \\ 3 & -1 \\ 2 & 6 \end{pmatrix}, \end{aligned} \tag{5.57}$$

i.e., $\mathbf{A}$ is retrieved to within expected round-off error.

5.2.4 *Example IV*

Let's take a quick look at

$$\mathbf{A} = \begin{pmatrix} 2 & 4 & 5 \\ 5 & 4 & 2 \end{pmatrix}. \tag{5.58}$$

Now, we are clearly better off analyzing

$$\mathbf{A}\mathbf{A}^T = \begin{pmatrix} 45 & 36 \\ 36 & 45 \end{pmatrix}, \tag{5.59}$$

and $\det(\mathbf{A}\mathbf{A}^T - \lambda \mathbf{I}) = 0$ means that

$$(45 - \lambda)^2 - 36^2 = 0 \qquad \text{or} \qquad 45 - \lambda = \pm 36, \tag{5.60}$$

with roots $\lambda_1 = 81$ and $\lambda_2 = 9$. These eigenvalues yield

$$\left.\begin{aligned} \mathbf{A}\mathbf{A}^T \mathbf{u}_1 = \lambda_1 \mathbf{u}_1 &\implies \mathbf{u}_1 = \frac{1}{\sqrt{2}} \begin{pmatrix} 1 \\ 1 \end{pmatrix} \\ \mathbf{A}\mathbf{A}^T \mathbf{u}_2 = \lambda_2 \mathbf{u}_2 &\implies \mathbf{u}_2 = \frac{1}{\sqrt{2}} \begin{pmatrix} 1 \\ -1 \end{pmatrix} \end{aligned}\right\} \mathbf{U} = \frac{1}{\sqrt{2}} \begin{pmatrix} 1 & 1 \\ 1 & -1 \end{pmatrix} \tag{5.61}$$

and, since $\sigma_1 = \sqrt{\lambda_1} = 9$ and $\sigma_2 = \sqrt{\lambda_2} = 3$,

$$\boldsymbol{\Sigma}_r = 3 \begin{pmatrix} 3 & 0 \\ 0 & 1 \end{pmatrix} \quad \text{and} \quad \boldsymbol{\Sigma} = 3 \begin{pmatrix} 3 & 0 & 0 \\ 0 & 1 & 0 \end{pmatrix}. \tag{5.62}$$

Finally,

$$\left.\begin{array}{l} \mathbf{v}_1 = \dfrac{1}{\sigma_1}\mathbf{A}^T\mathbf{u}_1 = \dfrac{1}{9\sqrt{2}}\begin{pmatrix}7\\8\\7\end{pmatrix} \\ \mathbf{v}_2 = \dfrac{1}{\sigma_2}\mathbf{A}^T\mathbf{u}_1 = \dfrac{1}{\sqrt{2}}\begin{pmatrix}-1\\0\\1\end{pmatrix} \end{array}\right\} \mathbf{V}_r = \frac{1}{9\sqrt{2}}\begin{pmatrix}7 & -9\\8 & 0\\7 & 9\end{pmatrix}. \tag{5.63}$$

The test

$$\mathbf{U}\Sigma_r\mathbf{V}_r^T = \frac{1}{6}\begin{pmatrix}1 & 1\\1 & -1\end{pmatrix}\begin{pmatrix}3 & 0\\0 & 1\end{pmatrix}\begin{pmatrix}7 & 8 & 7\\-9 & 0 & 9\end{pmatrix} = \mathbf{A} \tag{5.64}$$

reassures us that the decomposition is indeed correct.

5.3 SVD Applications

5.3.1 Data Compression I

Data matrices often contain considerable redundancy. That is, they represent the information in them in a wasteful, suboptimal way. In many real-life situations, this wasteful representation is a real impediment. SVD presents an excellent way to streamline this representation.

Consider the synthetic field

$$\mathbf{F} = f(x, y) = \sin\left(\frac{2\pi x}{100}\right)\sin\left(\frac{2\pi y}{50}\right) \quad \begin{cases} x = 0, 1, \ldots, 100 \\ y = 0, 1, \ldots, 50, \end{cases} \tag{5.65}$$

held in $\mathbf{F} \in \mathbb{R}^{51\times 101}$ and shown in fig. 5.4's upper left panel. Suppose $\mathbf{F}$ was obtained by an orbiting satellite and can only become useful on Earth. Clearly, in such circumstances information transmission, bandwidth, is in great demand; compressing $\mathbf{F}$ to require transmitting substantially less information can be extremely useful. The SVD of $\mathbf{F}$ is just the method we are looking for. Recall the expansion

$$\mathbf{A} = \sum_{i=1}^{q} \sigma_i \mathbf{u}_i \mathbf{v}_i^T = \sum_{i=1}^{q} \mathbf{M}_i \tag{5.66}$$

of $\mathbf{A}$ in terms of its modes. Now examine the upper-right panel of fig. 5.4, which shows the amount of $\mathbf{F}$'s total variance accounted for by each mode. It is obvious that the above series can be truncated,

$$\mathbf{A} = \sigma_1 \mathbf{u}_1 \mathbf{v}_1^T = \mathbf{M}_1, \tag{5.67}$$

without any meaningful loss of information. This means that instead of transmitting 5151 numbers (each of $\mathbf{F}$'s elements), we need to transmit only 153 (1 for σ_1, 51 for $\mathbf{u}_1$, and 101 for $\mathbf{v}_1$), or ~3% of the total amount of data.

Clearly, the example takes the point to an extreme; most data fields, while redundant, represent more than one mode of information. The actual degree of

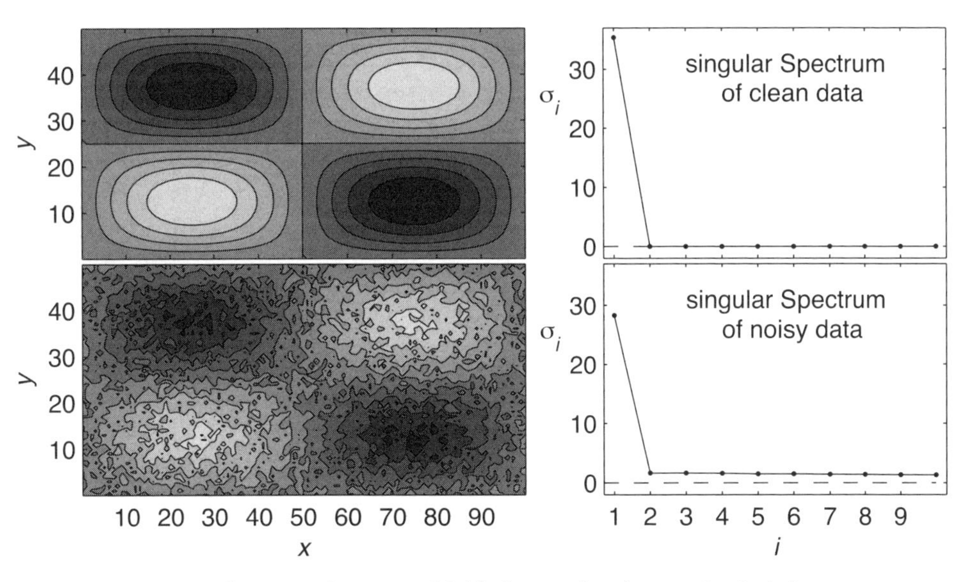

Figure 5.4. Synthetic two-dimensional fields discussed in the text. On the left are shown the clean field (top) and the same field contaminated by noise (bottom). The singular spectra of the fields as a function of mode number are shown on the right.

redundancy (and hence the potential compression saving) is quantified by the falloff rate of the singular spectrum, the fraction of the total variance accounted for by each mode, as a function of mode number (fig. 5.4's right panels).

Note that this is far from merely academic; when your browser downloads an image during high-traffic times, the image seems to gradually appear on the screen as a stepwise increase in focus and clarity. This is a live demonstration of the various modes of the image arriving as individual packets of information, one after the other, each contributing—but less than its predecessors—to the image's ever increasing clarity. The following is a realistic demonstration of the bandwidth saving afforded by SVD in image transmission.

5.3.2 Data Compression II

Consider the image shown in fig. 5.5d. As a color digital image, the visual information can be represented as three 1496×2256 matrices **R**, **G**, and **B**, representing each pixel's intensity (as an integer between 0 and 255) of red, green, and blue, respectively. This visual information can be compressed using

$$\mathbf{D} \equiv \begin{pmatrix} \mathbf{R} \\ \mathbf{G} \\ \mathbf{B} \end{pmatrix} = \mathbf{U}_r \boldsymbol{\Sigma}_r \mathbf{V}^T, \qquad \begin{matrix} \mathbf{D} & \in & \mathbb{R}^{4488 \times 2256} \\ \mathbf{R}, \mathbf{G}, \mathbf{B} & \in & \mathbb{R}^{1496 \times 2256} \end{matrix}. \tag{5.68}$$

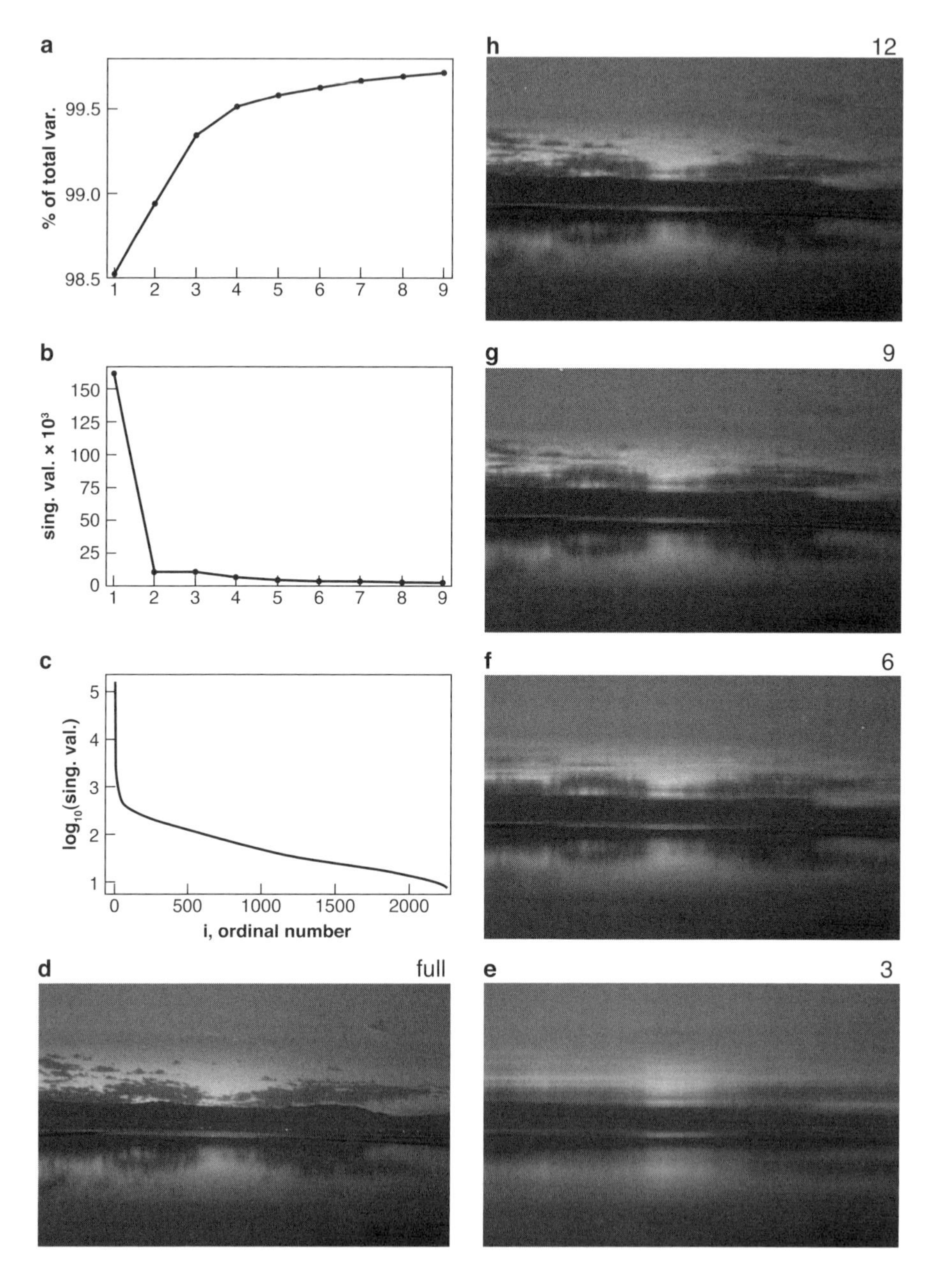

Figure 5.5. SVD use for image data compression. Panel d is a medium-resolution (1496×2256 pixels) digital image of a sunrise over Jordan's Moav mountains I took from the Israeli side of the Dead Sea. Panels e–h are reconstructions of the same image based on the number of SVD modes shown on the panels' upper right. The full singular spectrum of the image is shown (as $\log_{10}$) in panel c, and the leading 9 singular values (σ_i, $1 \leq i \leq 9$) in panel b. Panel a shows the cumulative percentage of variance, $100\sum_{j=1}^{i} \sigma_j^2 / \sum_{i=1}^{2256} \sigma_i^2$. A color version of this figure adorns the book's front cover.

This SVD representation can then be used (fig. 5.5e–h) to devise approximate reconstructions of $\mathbf{D}$ (and thus of $\mathbf{R}$, $\mathbf{G}$, and $\mathbf{B}$),

$$\hat{\mathbf{D}}_c = \sum_{i=1}^{c} \sigma_i \mathbf{u}_i \mathbf{v}_i^T = \begin{pmatrix} \hat{\mathbf{R}}_c \\ \hat{\mathbf{G}}_c \\ \hat{\mathbf{B}}_c \end{pmatrix}, \tag{5.69}$$

where c is some truncation point, and here the overhat denotes "imperfectly reconstructed," not unit norm.

Figure 5.5e–h shows this truncated SVD expansion with $c = 3, 6, 9, 12$. Clearly, in this particular example, the truncated SVD reconstruction affords huge savings. For example, panel h, which comprises >99.76% of the original image's total variance and appears visually to be an extremely good approximation of it (fig. 5.5d), requires only

$$100 \frac{12(4488 + 2256 + 1)}{1496 \cdot 2256 \cdot 3} < 0.8\% \tag{5.70}$$

of the information required by the original image (the denominator; the numerator holds the information requirements of a single mode—$\mathbf{u}_i \in \mathbb{R}^{4488}$, $\mathbf{v}_i \in \mathbb{R}^{2256}$ and the singular value $\sigma_i \in \mathbb{R}$, in parentheses—multiplied by 12 to account for the fact that fig. 5.5h is based on 12 modes). This calculation—in which nearly all the original image's content is reproduced in less than 1% of the expense—demonstrates the incredible hardware or bandwidth economization SVD sometimes permits.

5.3.3 *Filtering or Noise Suppression*

To simulate observed data fields that contain a deterministic portion plus noise, fig. 5.4's lower-left panel shows the original synthetic field $\mathbf{F}$, but contaminated with Gaussian noise:

$$\mathbf{F}_n = f_n(x,y) = \frac{8}{10} \sin\left(\frac{2\pi x}{100}\right) \sin\left(\frac{2\pi y}{50}\right) + \xi, \qquad \xi \sim N(0, 0.1), \tag{5.71}$$

where each noise realization is a random draw from a normal (Gaussian) distribution with zero mean and standard deviation of 0.1. The large-scale structure is still clearly visible, but the noise contamination is substantial. This is quantified by fig. 5.4's lower-right panel, which shows the singular spectrum. While the spectrum of the pristine field $\mathbf{F}$ is strictly a single-mode one, $\sigma_1 \approx 35.4$, the first 3 elements of $\mathbf{F}_n$'s spectrum are $\sigma_1 \approx 28.3$, $\sigma_2 \approx 1.7$, and $\sigma_3 \approx 1.6$. Thus, the effect of the noise is twofold. First, it attenuates the amplitude of the leading mode because of the (futile) attempt to fit noise. Additionally, it redistributes some of the variance of the first mode over the remaining lower modes. SVD can be used to overcome this signal deterioration by filtering out some of the noise, simply by truncation,

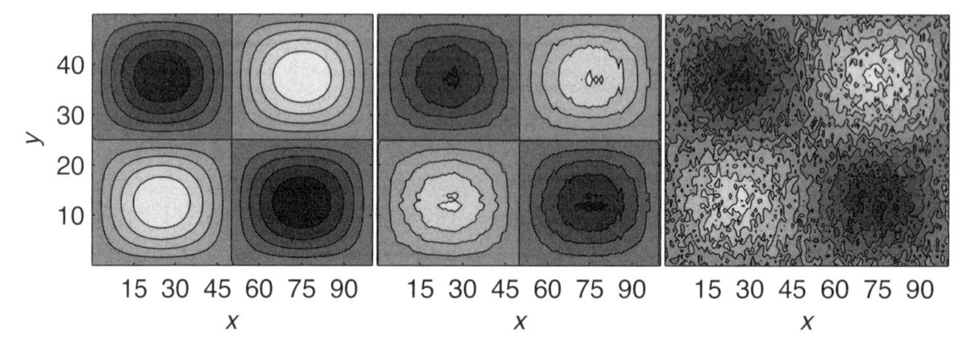

Figure 5.6. (Left) $\mathbf{F}$, the pristine synthetic field; (right) $\mathbf{F}_n$, the noise contaminated $\mathbf{F}$; (middle) the reconstruction of the clean field from the noisy one using only $\mathbf{F}_n$'s leading mode.

$$\hat{\mathbf{F}}_c = \sum_{i=1}^{c} \sigma_i \mathbf{u}_i \mathbf{v}_i^T, \tag{5.72}$$

where again c is some cutoff or truncation point.

Figure 5.6 shows again the pure field (left) and its noisy counterpart (right). The middle panel, obtained with eq. 5.72 and $c = 1$, demonstrates that while some of the noise finds its way to the leading mode, the sharpness of the figure is almost fully restored by the modal decomposition, demonstrating the utility of SVD as an empirical filter.

5.4 Summary

Every $M \times N$ matrix $\mathbf{A}$ can be represented as the product

$$\mathbf{A} = \mathbf{U}\mathbf{\Sigma}\mathbf{V}^T,$$

if $\mathbf{A}$ is real, or

$$\mathbf{A} = \mathbf{U}\mathbf{\Sigma}\mathbf{V}^H,$$

if $\mathbf{A}$ is complex, $\mathbf{A}$'s SVD representation. If $\mathbf{A}$ is real, so are $\mathbf{U}$, $\mathbf{\Sigma}$, and $\mathbf{V}$, the $M \times M$ $\mathbf{U}$ contains the orthonormal eigenvectors of $\mathbf{A}\mathbf{A}^T$ ($\mathbf{A}$'s left singular vectors), the $N \times N$ $\mathbf{V}$ contains the orthonormal eigenvectors of $\mathbf{A}^T\mathbf{A}$ ($\mathbf{A}$'s right singular vectors), and the $M \times N$ $\mathbf{\Sigma}$ has zero off-diagonal elements, diagonal elements 1 through q ($\mathbf{A}$'s singular values, where q is $\mathbf{A}$'s rank) the positive square roots of the nonzero eigenvalues $\mathbf{A}\mathbf{A}^T$ and $\mathbf{A}^T\mathbf{A}$ share, and zero diagonal elements $q + 1$ through $\min(M, N)$ (if this range is nonempty).

The SVD is a spectral representation of $\mathbf{A}$,

$$\mathbf{A} = \underbrace{\sigma_1 \mathbf{u}_1 \mathbf{v}_1^T}_{\text{mode 1}} + \underbrace{\sigma_2 \mathbf{u}_2 \mathbf{v}_2^T}_{\text{mode 2}} + \cdots + \underbrace{\sigma_q r \mathbf{u}_q \mathbf{v}_q^T}_{\text{mode } q}$$

Since the modes (the columns of **U** and **V** and the corresponding diagonal elements of **Σ**) can be arranged in descending order of singular value magnitude, this spectral representation of **A** has the uniquely useful property of being a decreasing monotone, with each mode contributing at least as much as the subsequent one and at most as much as the preceding one.

PART 2

Methods of Data Analysis

SIX

The Gray World of Practical Data Analysis: An Introduction to Part 2

This second part of the book is the crux of the matter: how to analyze actual data. While part 2 builds on part 1, especially on linear algebra fundamentals covered in part 1, the two are not redundant. The main distinguishing characteristic of part 2 is its nuanced grayness.

In the ideal world of algebra (and thus in most of part 1), things are black or white: two vectors are either mutually orthogonal or not, real numbers are either zero or not, a vector either solves a linear system or does not. By contrast, realistic data analysis, the province of part 2, is always gray, always involves subjective decisions. For example, if $\mathbf{a}^T\mathbf{b} \approx 1$ and $\mathbf{a}^T\mathbf{b}/\mathbf{c}^T\mathbf{d} \sim \mathcal{O}(10^{16})$, then in part 1, $\mathbf{c}$ and $\mathbf{d}$ are *not orthogonal*, while in part 2, because $\mathcal{O}(10^{-16})$ is close to machine accuracy (is "almost zero" given the inherent limitations of computers' finite precision arithmetic), for most purposes we will take $\mathbf{c}$ and $\mathbf{d}$ to be *practically* orthogonal, at least relative to $\mathbf{a}$ and $\mathbf{b}$.

Does this spirit of "numerical forgiveness" limit the credibility of data-based science? It can, for poorly posed problems or poorly executed analyses, but it surely doesn't have to, and often does not. First, realistic data never shed perfect light on a question and thus at best provide a statistical answer, which is always relative, as discussed in the next two chapters. Second, models of reality are also never perfect, and issuing predictions based on models always involves asserting, implicitly or explicitly, that the model skill far exceeds its limitations, also a relative and possibly subjective statement.

If you are still uneasy about this grayness, consider medical and nutritional sciences, or economics, to name but three of modern life's central pillars. These fields of knowledge are almost exclusively based on empirical data analysis and on the relative merits of one option over its alternatives. As such, their recommendations are based on observations that always have some—hopefully small but definitely specified—likelihood of having arisen by chance, reflecting no reproducible merit of the recommended option. Yet we all religiously read the *New York Times*' "Well" column or Paul Krugman's op-ed pieces, both addressing questions that almost never have definitive, numerically unambiguous, answers, and that are thus implicitly also based on "numerical forgiveness."

Let's learn some statistics and develop a richer understanding of this introduction.

SEVEN

Statistics in Deterministic Sciences: An Introduction

DATA ANALYSIS IS A BRANCH of statistical science. Perhaps the first question that comes to mind about statistics in the natural sciences, especially those—like physics or engineering—that are based, at least in principle, on well-defined, closed governing equations, is "who needs statistics!?" After all, most physical phenomena of concern to such sciences are governed by fundamental, mostly known, physics. It follows, then, that we should be able to write down the governing equations of the system under investigation and use the dynamical system thus constructed to study the system and predict its future states. The answer to this apparent dilemma is that while the basic physics are known, their application to such complex systems as the ocean, atmosphere, or ecosystems is monumentally difficult. For most such systems, the full problems—the values of all relevant variables at all space and time locations—are essentially intractable, even with the fastest computers. Hence, there is always more to the focus of inquiry that cannot be modeled; we must somehow fill in the gaps. This is where statistics come in.

Until the state of the physical system under investigation is fully quantified (i.e., until the value of every dynamical variable is perfectly known at every point in space and time), there is a certain amount of indeterminacy in every statement we make about the state of the system. For example, when we say "the temperature in New York today is x°F," we fully expect slightly different temperatures, e.g., in Central Park vs. inside an urban canyon. Consequently, despite being governed by deterministic physics (i.e., obeying a specific, unambiguous model with well-defined equations that can be used for prediction and that will yield reproducible predicted values), most variables are best viewed as random, not deterministic, ones. For a particular realization, a random variable can assume any arbitrary value. Some values, however, are less likely than others. This implies the existence of a probability density function (PDF) governing the probability of realizing, in a random draw, a value that falls within a specified range. At some level the idea of the PDF is intuitive. For example, if you visit a weather web site one January night to find out tomorrow's forecast for Boston, you would expect something in the 15°–45°F range, say, or—if you wish to include even incredibly unlikely values—maybe the −5–63°F. If you see 89°F instead, most likely you are looking at the San Diego forecast instead...

To clarify these ideas, let's consider an example. Suppose you want to study a process you believe is periodic in time,

$$y_c = a \cos\left(\frac{2\pi t}{\tau}\right), \tag{7.1}$$

where y_c is the clean (fully deterministic) signal, a is some amplitude, t is time, τ is the period, and, for simplicity, we neglect the possibility of a nonzero phase. Now, as discussed above, realistic signals are never pure or pristine. Rather, they contain the imprint of all the physical processes excluded from the model we have in mind, an imprint we collectively call "noise,"

$$y_n = a \cos\left(\frac{2\pi t}{\tau}\right) + \xi \equiv y_c + \xi, \tag{7.2}$$

where ξ denotes the noise, assumed to have zero mean and unit variance.

Figure 7.1 shows such hypothetical y_c and y_n. This can be, for example, the monthly mean deviation from the climatological annual mean temperatures at a given location for some 10 years. Neglecting all processes other than the seasonal variation of solar radiation, the temperature is expected to behave like the solid curve in fig. 7.1. However, in reality there are always processes that affect the studied signal yet are not represented in the model we have in mind, implicitly or explicitly, of the physical processes governing the signal. Therefore, real signals always look more like fig. 7.1's dots. To characterize the noisy signal, perhaps the first things that come to mind are the signal's mean and variance:

$$\bar{y} = \frac{1}{N}\sum_i y_i, \quad d_y^2 = \frac{1}{N}\sum_i (y_i - \bar{y})^2, \quad \text{and} \quad s_y^2 = \frac{1}{N-1}\sum_i (y_i - \bar{y})^2, \tag{7.3}$$

where the overbar denotes sample mean, our best estimate of the population true mean μ, d_y^2 is the sample mean squared deviation, and s_y^2 is an unbiased estimate of the population's true variance, σ_y^2. Surely, for the processes generating

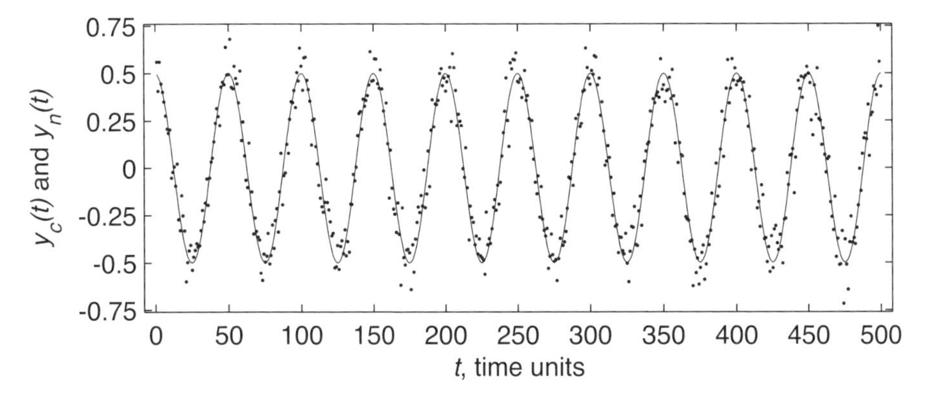

Figure 7.1. Periodic signal (solid), and the same signal, contaminated by Gaussian noise (dots).

y_c or y_n written above, we expect $\bar{y}_c = \bar{y}_n = 0$, as the sinusoid oscillates between $\pm a$ with $\bar{\xi} = 0$. But note two important caveats: if the record is short, any partial cycle included may result in a nonzero mean; and if the resolution of t, δt, is rather coarse (that is, if every cycle is sampled by only a few points), there can be a (spurious) nonzero mean.

In many cases, the mean and variance provide a rough idea about the nature of the signal, but offer relatively little in determining the likelihood of specific deviations from the mean. For that, we need to estimate the process PDF from its short segment we have. To estimate the PDF, we can create a histogram by binning the available values. That is, we divide the range of realized values (say $\pm 2a$ or, in general, $\max(y_n) - \min(y_n)$) into equal bins, (e.g., $\min(y_n) + [i/20, (i+1)/20]$, $i = 0, 1, \cdots, 20[\max(y_n) - \min(y_n)] - 1$), and count the number of readings (realized y_n values) that fall into each one. If we had an infinitely long record (of which fig. 7.1 is a segment), we would have been able to choose arbitrarily narrow bins, and the resultant histogram would have approached the PDF; for actual (finite, often short) signals, we will have to assume that the segment at hand well approximates the full signal. We can also apply various techniques (which all amount to smoothing) to the histogram for a more statistically stable estimate of the PDF. Figure 7.2 demonstrates these points for the data of fig. 7.1.

Many signals arising in the natural sciences contain a robust seasonal component that is not the focus of attention. In climate and other earth sciences, or in agronomy, this often means literally seasonal in the traditional sense of the word, emphasizing the winter–summer cycle. In other cases, this simply means periodic. For example, in the study of traffic congestion, a periodic weekly signal will likely be apparent (because people's weekend driving habits differ from their weekday ones), while solar physics time series are dominated by an 11-year "seasonality." Whatever the salient meaning of "seasonality" is to a particular data set, in many such cases it makes sense to divide the signal into

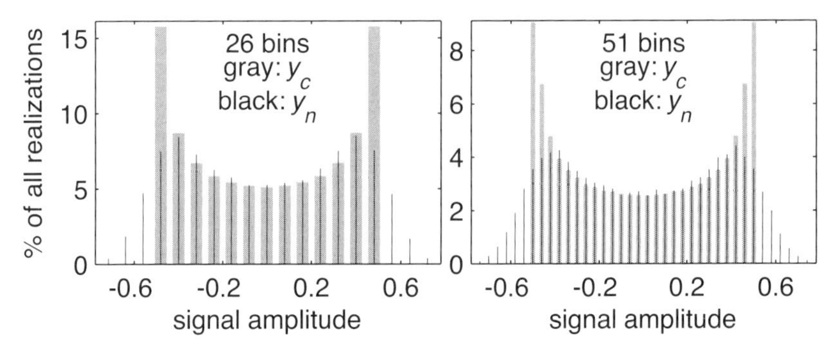

Figure 7.2. Histograms (assumed to reasonably represent the PDFs) of the signals in fig. 7.1. Note the dependence of results on the chosen bin width (and thus bin number).

"summers" and "winters" (or, in general, into times when the signal is reproducibly well above or below the mean) and compute seasonal statistics. More broadly, many signals (like y_n above) contain random and deterministic components, where the deterministic component comprises the part of the signal due to processes whose physics are both well understood and of no interest to the attempted analysis. Examples may include the seasonal or daily cycle contributions to temperature variability. For example, since we know very well why winters are cooler than summers and can predict with good accuracy the onset of future seasons, to many (but certainly not all) questions the ebb and flow of the seasons is merely a distraction from the main questions. When our signal contains such components, with amplitudes high relative to those of the (assumed random) processes we wish to study, it will be reasonable to remove this deterministic part so as to better reveal other, presumably more interesting but subtle, processes. The deterministic parts of commonly encountered data sets may include a linear trend, a regularly periodic portion, an exponential growth or decay, or combinations of those. Figure 7.3 addresses this removal process for a specific example.

What we are left with after removing the deterministic part are anomalies, whose PDF is likely to be completely different from that of the original signal from which they were derived. While the exact meaning of "anomaly" depends on the context, it refers to the deviation of individual readings from some expected value. In climate work, most often the term describes the deviation of a given mean value from the respective climatological mean. For example, the March 1999 temperature anomaly at some specified location is the deviation of the 1999 monthly mean March temperature there from the climatological March mean temperature at that location, where the latter is our best estimate of that location's characteristic March mean temperature, derived by averaging all the available local March means. To be unambiguous, all this information must be articulated when using the term, because anomaly can also mean, for example, the deviation of the 1990–99 mean from the 1945–89 baseline, etc.

Let's examine the PDF of the anomalies derived from fig. 7.1's synthetic signal after the climatological seasonal cycle has been removed. The PDFs (fig. 7.4) are clearly very different from those of the full signal (fig. 7.2). In fact, they seem "normal-like" (if you don't see this, we will give this term precision momentarily). Given that we have contaminated the sinusoidal signal with Gaussian noise, this resemblance of fig. 7.4 to the normal distribution should come as no surprise; it is high time to introduce theoretical distributions in an orderly fashion.

7.1 Probability Distributions

Why are theoretical probability distributions important? Primarily because it turns out that a relatively small number of them can approximately describe a great many physical phenomena. If we have an a priori reason to believe the

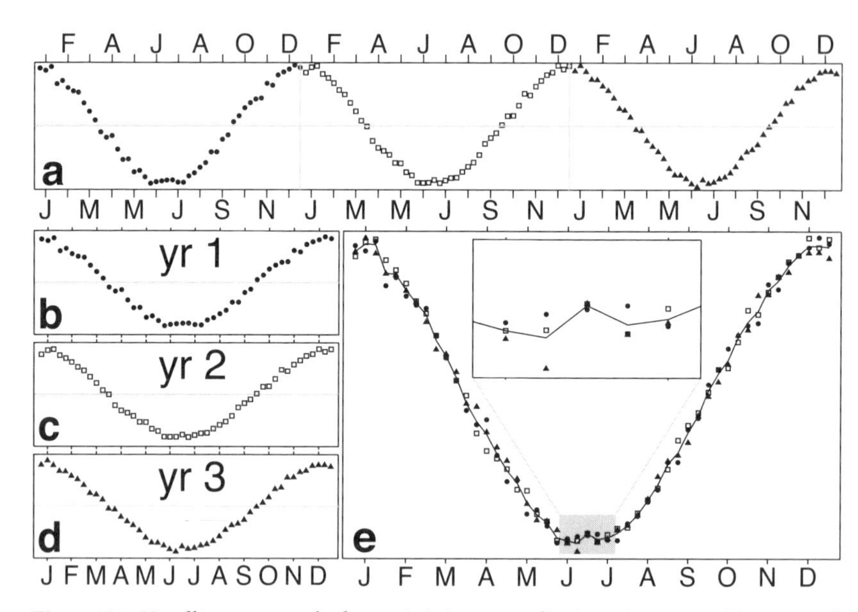

Figure 7.3. Handling apparently deterministic seasonality in a time series. The original 3-year-long series at roughly weekly resolution is shown in panel a with year-specific symbols. While the data exhibit a scatter befitting a random variable, the scatter appears to be about a deterministic seasonal cycle. Panels b–d show the individual years stacked together. In panel e, the climatology—the average of all available cycles, in this case 3 years—is displayed as a solid curve. For each week, the actual data (from each of the 3 available years) are also shown, using the symbol convention of panels a–d. A 5-week subset centered around June (panel e's shaded region) is magnified in panel e's inset. Anomalies (not shown) are the deviation of actual data from their respective mean, e.g., the year 1, week 1 anomaly is the time series actual value of year 1, week 1, minus the overall week 1 climatology (the average of all available week 1, here 3 for each of the available years).

process we are studying follows a certain known theoretical distribution, and if a good fit (in a sense we will define later) is indeed found between the empirical histogram and the anticipated distribution, then the fitted distribution can prove highly useful. It succinctly describes the entire dataset with only a few parameters, thus simplifying calculations and permitting interpolation or extrapolation for probabilities of unrealized events of interest. This is intimately related to significance tests, probably the most widespread use of theoretical distributions (more on that below).

Let's consider the normal (or Gaussian) distribution, a continuous distribution that is very useful for many physical applications, primarily because of the all-important *central limit theorem*. If X is an arbitrarily distributed random variable with mean μ and variance σ^2, the distribution of the estimated sample means $\bar{x}$ using a large number of samples (i.e., repeatedly estimating $\bar{x}$ using

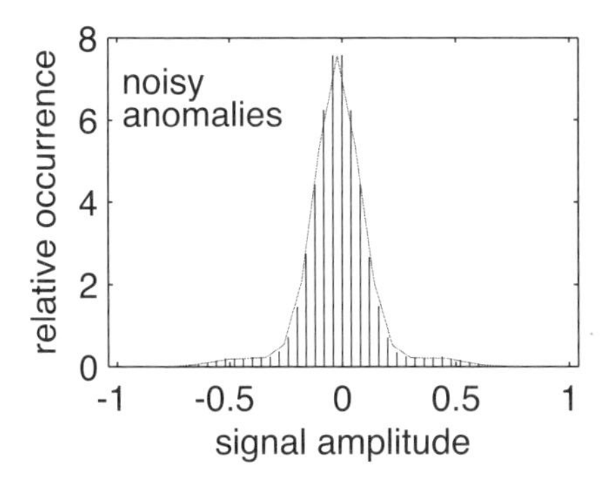

Figure 7.4. Two PDF estimates of the signals in fig. 7.1 after the climatology has been removed. The estimate based on wider, fewer bins is in gray solid, while that based on the finer, more numerous bins is in stems.

a new sample each time) will asymptotically approach normal with $\bar{x} \to \mu$ and $s_{\bar{x}}^2 \to \sigma^2/N$ for large enough N (sample size drawn from X). Amazingly, this result holds with no restrictions on the distribution of X itself!

If X is normally distributed with mean μ and variance σ^2, denoted by either $X \sim N(\mu, \sigma^2)$ or $X = N(\mu, \sigma^2)$, then X's PDF is given by

$$f(x) = \frac{1}{\sigma\sqrt{2\pi}} \exp\left[-\frac{(x-\mu)^2}{2\sigma^2}\right]. \tag{7.4}$$

With any PDF, the probability of obtaining an outcome between a and b is

$$\mathrm{p}(a \le X \le b) = \int_a^b f(x)\,dx \tag{7.5}$$

and thus

$$\int_{-\infty}^{\infty} f(x)\,dx = 1, \tag{7.6}$$

because an outcome *somewhere* along the real line is guaranteed. Note that $f(x)$ is *not* the probability of obtaining *exactly* x; this probability must be identically zero, because

$$\mathrm{p}(a \le X \le a) = \mathrm{p}(X = a) = \int_a^a f(x)\,dx = 0. \tag{7.7}$$

It is therefore meaningless to discuss $\mathrm{p}(X = x)$ which vanishes for any x.

Figure 7.5 presents the PDF of $N(0, 1)$, the zero mean, unit variance (or standard) Gaussian distribution. For applying the normal distribution to significance test, we use the derived cumulative PDF, or CDF, given by

$$F(y) = \int_{-\infty}^{y} f(x)\,dx, \tag{7.8}$$

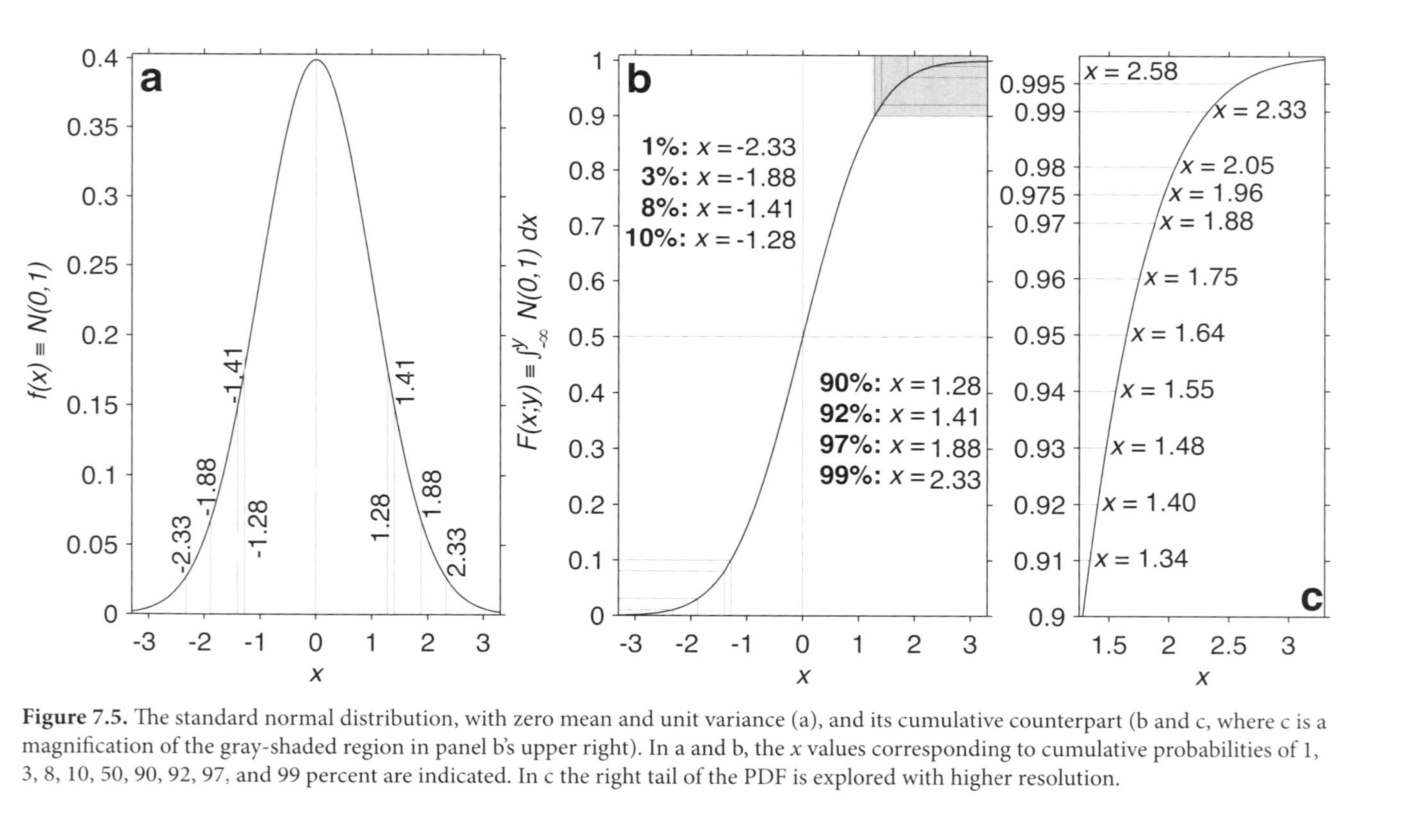

Figure 7.5. The standard normal distribution, with zero mean and unit variance (a), and its cumulative counterpart (b and c, where c is a magnification of the gray-shaded region in panel b's upper right). In a and b, the x values corresponding to cumulative probabilities of 1, 3, 8, 10, 50, 90, 92, 97, and 99 percent are indicated. In c the right tail of the PDF is explored with higher resolution.

shown for a normally distributed X in fig. 7.5's right two panels. The CDF can be used to estimate significance. As significance tests are covered exhaustively by most statistics book, here we only discuss their general spirit briefly and give a simple example that highlights some basic general ideas.

Suppose we seek the population mean μ of a random variable X, of which we have a sample with mean $\bar{x}$. While it is tempting to conclude that $\mu = \bar{x}$, $\bar{x}$ is uncertain because it is derived from a finite sample. It is therefore possible that the calculated $\bar{x}$ is not a robust estimate of the true population mean, which is some other value μ_0, and that the slight $\bar{x} - \mu_0$ deviation arose spuriously and is not unexpected given that $\bar{x}$ is a single realization reflecting no more than the random hand we were dealt. Alternatively, it is also possible that indeed $\mu = \bar{x}$. The spirit of significance testing is to try to indirectly substantiate the $\mu = \bar{x}$ conclusion by identifying the likelihood for obtaining $\bar{x}$ by chance from finite samples drawn from a population whose mean is actually μ_0. Put differently, our significance test answers the question: Can we assert that the likelihood that $\mu = \mu_0$ is smaller than some specified threshold we call significance and denote by p? If so, $\mu = \mu_0$ is invalidated (rejected) at the specified p, and $\mu \neq \mu_0$ is adopted. If, given the circumstances of the test, $\mu = \bar{x}$ is then deemed the only remaining possibility, it is adopted as the truth. This conclusion being based on statistics, it is always only as certain as p is small; we say that "$\mu = \bar{x}$ is significant at significance p." It is no more meaningful to say that something is "significant" without specifying the smallest p tested than it is to say that some destination is "far" without specifying how far or far relative to what.

The test tool is the null hypothesis

$$H_0 : \mu = \mu_0, \tag{7.9}$$

the cautious, skeptical view that the unknown population mean is μ_0, and that the slight deviation of the calculated $\bar{x}$ from μ_0 arose spuriously and not unexpectedly. Because the putative population from which $\bar{x}$ is a realization is the population of random sample-based estimates of the mean, the central limit theorem (CLT) applies and states that $\bar{X} \sim N(\mu, \sigma^2/df)$, where df denotes the number of degrees of freedom. We examine df more closely in the subsequent section. For now, let's assume each x_i is independent of other $x_{j \neq i}$, in which case df is the sample size. The CLT thus states that in this case, the random variable $\bar{X}$, of which $\bar{x}$ is realization, is normally distributed with mean μ and variance σ^2/df). If we test $\mu = \mu_0 = 0$, this becomes $\bar{X} \sim N(0, \sigma^2/df)$.

Using

$$z = \frac{\bar{x} - \mu_0}{\sqrt{s_x^2/df}} = \frac{(\bar{x} - \mu_o)\sqrt{df}}{s_x}, \tag{7.10}$$

we convert $\bar{x}$, presumably $N(0, \sigma^2/N)$-distributed, to $Z \sim N(0, 1)$ (where the numerator is simply $\bar{x}$ because we entertain the possibility that $\mu = 0$) and use tabulated values of $N(0, 1)$ to quantify how likely it is that $\bar{x}$ represents a chance occurrence. This is done by conceptually using fig. 7.5. The shown distribution is what the population of the above-constructed z is expected to look like. Thus, if, for example, the calculated $\bar{x}$ yielded $z = 2.33$, the likelihood of such a sample $\bar{x}$ arising by chance from a population with $\mu = 0$ is 1 in 100. And if $\bar{x}$ yielded $z = 2.58$, the likelihood of such a sample $\bar{x}$ arising by chance from a population with $\mu = 0$ is 5 in 1000.

If the sample is small, say less than 35 elements, the t statistic should be used (as the sample size increases, the t distribution approaches the standard normal one). For the mean, the t statistic is similar to z,

$$t = \frac{\bar{x} - \mu}{s_x / \sqrt{df - 1}}. \tag{7.11}$$

A summary of numerous useful statistics (such as t for various estimated parameters) can be found in appendix I of Philips (1988).[1] For example, suppose the 81-year March mean climatological rainfall in Tel Aviv is 90 mm (wouldn't that be nice . . .), and the standard deviation $s = 18$ mm. What are the 95% confidence intervals around this estimated mean? (In other words: given our current information, what is the δ that is 95% likely to satisfy $90 - \delta \le \mu \le 90 + \delta$, μ being the true mean?) Given the central limit theorem, even though rainfall distribution is wildly different from normal, our estimated mean rainfall is normally distributed around 90, with standard deviation $s_m = s/\sqrt{df} = 18/9 = 2$. With $df = 81$ (because the sample comprises 81 March estimates) and $t_{0.025} \le t \le t_{0.975}$, $-1.99 \le t \le 1.99$. Recalling the above definition of t for the mean, we have

$$s_m t_{0.025} \le \bar{x} - \mu \le s_m t_{0.975} \quad \Longrightarrow \quad \bar{x} - s_m t_{0.975} \le \mu \le \bar{x} - s_m t_{0.025}.$$

Putting in the numerical values, we get

$$90 - 2 \cdot 1.99 \le \mu \le 90 - 2 \cdot (-1.99) \quad \text{or} \quad 86 \le \mu \le 94. \tag{7.12}$$

So we now say that the true mean March rainfall is between 86 and 94 mm, with only 5% chance of being embarrassed. If we repeated the same calculation for only a 5 in 1000 chance to be wrong, we will have to expand the region to span approximately 83–97.

7.2 Degrees of Freedom

Until now we have approximated the number of degrees of freedom (df) as the number of data points. This is a gross, rarely defensible, oversimplification. More accurately, df is the number of independent ways by which the signal can

[1] Philips, John L. (1988) *How to Think About Statistics*, Freeman, New York, ISBN 0-7167-1923-1.

vary; if a time series has M points, it contains *at most M* independent pieces of information (a situation that only exists for pure white noise series).

To make these ideas less abstract, let's consider a sequence for which the order of points matters. We will use the shorthand of "time series" for such a sequence, although the independent variable can also be space or other ordered independent variable. The latter situation arises, e.g., when one measures the temperature as a function of latitude (in which case latitude replaces time as the independent variable, but the order is clearly still important, because the measurement's whole point is to quantify temperature's latitude dependence, and latitude varies smoothly and orderly from pole to pole). Just to be clear, we distinguish this type of data from nonorderly data, such as the relationship (if any) between income and body weight. If such a relationship proves coherent, whether person a's information was measured before, or after, person b's is completely immaterial.

Figure 7.6 (from NCAR's Jim Hurrell[2]) shows the timeseries of observed NAO (North Atlantic Oscillation) Index with various degrees of smoothing. The smoothing uses a *kernel* to transform the *i*th value of the original series, y_i, into the smoother combination

$$\tilde{y}_i = \tfrac{1}{9}\left(y_{i-2} + 2y_{i-1} + 3y_i + 2y_{i+1} + y_{i+2}\right) \tag{7.13}$$

or

$$\begin{aligned}\tilde{y}_i = \tfrac{1}{25}\big(&y_{i-4} + 2y_{i-3} + 3y_{i-2} + 4y_{i-1} + 5y_i \\ &+ 4y_{i+1} + 3y_{i+2} + 2y_{i+3} + y_{i+4}\big),\end{aligned} \tag{7.14}$$

in the middle and bottom panels, respectively. Regardless of the smoothing details, fig. 7.6 clearly shows that while the raw signal appears very noisy, with sequential values appearing very weakly (if at all) related to one another, when rapid variability is suppressed by the running averaging, low-frequency structure emerges, with timescales longer than the smoothing kernel (e.g., the trough to crest shift during 1965–1990). That is, neighboring points are, to some degree, dependent on one another, and not every additional y_i adds a whole new piece of information not previously represented in the data set by earlier y_i's. Put differently, while judging based on the top panel the number of degrees of freedom appears to be the full number of data points, the lower panels suggest otherwise.

While there are various ways to understand the *df* concept, the one I find most illuminating is to view the time series as an N-vector in $\mathbb{R}^N$ (or, more generally, $\mathbb{C}^N$, although here we consider only real-valued fields). To begin with, consider the 3D Cartesian ("physical") space, in which each point is represented

[2] NCAR is the United States National Center for Atmospheric Research, and the data are taken from www.cgd.ucar.edu/cas/jhurrell/indices.html.

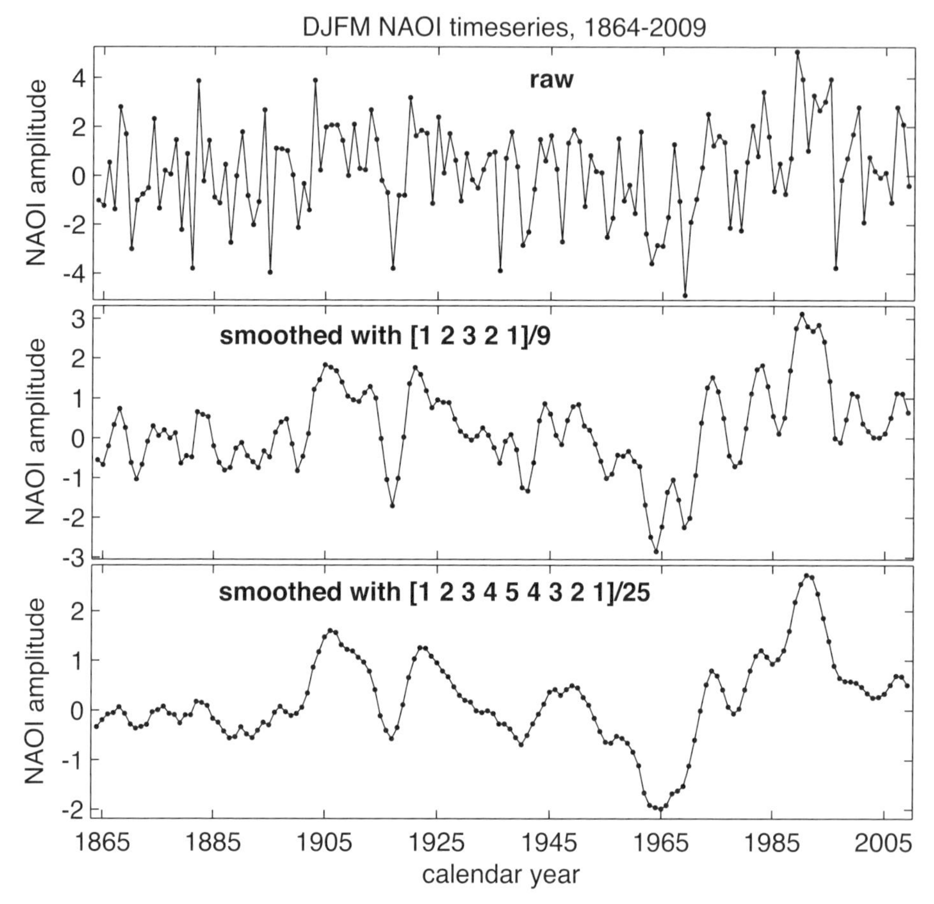

Figure 7.6. The observed annual North Atlantic Oscillation Index, the normalized December–March mean sea level pressure difference between Lisbon, Portugal and Reykjavik, Iceland, assigned to the year of the January–March portion. The panels differ in the applied smoothing; the upper panel shows the raw data, while the other two show the data after the indicated smoothing.

by a 3-vector $\mathbf{v} = (\, v_1\, v_2\, v_3\,)^T$, a 3-element time series in our analogy. This location can be represented as a linear combination of the basis vectors we chose to span $\mathbb{R}^3$, say $\hat{\mathbf{i}}$, $\hat{\mathbf{j}}$, and $\hat{\mathbf{k}}$,

$$\underbrace{\begin{pmatrix} v_1 \\ v_2 \\ v_3 \end{pmatrix}}_{\mathbf{v}} = v_1 \underbrace{\begin{pmatrix} 1 \\ 0 \\ 0 \end{pmatrix}}_{\hat{\mathbf{i}}} + v_2 \underbrace{\begin{pmatrix} 0 \\ 1 \\ 0 \end{pmatrix}}_{\hat{\mathbf{j}}} + v_3 \underbrace{\begin{pmatrix} 0 \\ 0 \\ 1 \end{pmatrix}}_{\hat{\mathbf{k}}}. \tag{7.15}$$

This generalizes straightforwardly to $\mathbb{R}^N$, where

$$\begin{pmatrix} v_1 \\ v_2 \\ \vdots \\ v_{N-1} \\ v_N \end{pmatrix} = v_1 \begin{pmatrix} 1 \\ 0 \\ \vdots \\ 0 \\ 0 \end{pmatrix} + v_2 \begin{pmatrix} 0 \\ 1 \\ \vdots \\ 0 \\ 0 \end{pmatrix} + \cdots + v_{N-1} \begin{pmatrix} 0 \\ 0 \\ \vdots \\ 1 \\ 0 \end{pmatrix} + v_N \begin{pmatrix} 0 \\ 0 \\ \vdots \\ 0 \\ 1 \end{pmatrix}. \tag{7.16}$$

In this view, the N-element time series on the left is represented in $\mathbb{R}^N$ by a set of coefficients scaling the columns of $\mathbf{I}_N$. However, this basis is obviously not unique; there are infinitely many bases available to span $\mathbb{R}^N$, and some do so more parsimoniously than others. To appreciate this, let's look at the simple $\mathbb{R}^5$ example time series $\mathbf{v} = (3\ 3\ 2\ 2\ 1)^T$. First, let's see it when $\mathbb{R}^5$ is spanned by the columns of $\mathbf{I}_5$,

$$\begin{pmatrix} 3 \\ 3 \\ 2 \\ 2 \\ 1 \end{pmatrix} = 3 \begin{pmatrix} 1 \\ 0 \\ 0 \\ 0 \\ 0 \end{pmatrix} + 3 \begin{pmatrix} 0 \\ 1 \\ 0 \\ 0 \\ 0 \end{pmatrix} + 2 \begin{pmatrix} 0 \\ 0 \\ 1 \\ 0 \\ 0 \end{pmatrix} + 2 \begin{pmatrix} 0 \\ 0 \\ 0 \\ 1 \\ 0 \end{pmatrix} + 1 \begin{pmatrix} 0 \\ 0 \\ 0 \\ 0 \\ 1 \end{pmatrix}. \tag{7.17}$$

However, if we are interested in compactness, we can represent $\mathbf{v}$ as

$$\begin{pmatrix} 3 \\ 3 \\ 2 \\ 2 \\ 1 \end{pmatrix} = 3 \begin{pmatrix} 1 \\ 1 \\ 0 \\ 0 \\ 0 \end{pmatrix} + 2 \begin{pmatrix} 0 \\ 0 \\ 1 \\ 1 \\ 0 \end{pmatrix} + 1 \begin{pmatrix} 0 \\ 0 \\ 0 \\ 0 \\ 1 \end{pmatrix}. \tag{7.18}$$

This latter representation is a demonstration that while $\mathbf{v} \in \mathbb{R}^5$, $\mathbf{v}$ has only 3 degrees of freedom, (3, 2, 1); it does not exploit the full scope of possibilities for variability available to it. Now you might think that this is a perverse example with little or no bearing on reality. If you do, think again! Think, for example, of a 100-element time series that is a perfect sine wave with wavelength 10 (such that the time series contains exactly 10 full waves). We can tediously represent the series using the 100 columns of $\mathbf{I}_{100}$ as basis vectors. This will clearly retain the wave information to the exact same extent as the original discrete sine wave. However, we can equally well span it with a single basis vector parallel to $\mathbf{v}$,

$$\mathbf{v} = v\hat{\mathbf{v}} := \|\mathbf{v}\| \left(\frac{\mathbf{v}}{\|\mathbf{v}\|} \right). \tag{7.19}$$

This, again, demonstrates that while our hypothetical sine wave has the entire $\mathbb{R}^{100}$ to vary in, it in fact varies only along a single direction in this space (it satisfies $df = 1$). Thus, if the time series is not pure noise (a state that, believe it or not, is not easy to attain!), we can introduce a coordinate transformation that will reduce the dimensionality of the space in which the time series is a vector.

For example, for the vector in eqs. 7.17 and 7.18, $\mathbb{R}^5$ is simplified because movement is only permitted so as to maintain the same distance from the first and second coordinates, as well as from the third and fourth, with no restrictions on movement along the fifth coordinate; there are only 3, not 5, dimensions along which to move; $df = 3$. In general, the lowest possible space dimension in which the time series can reside is df.

But what about real-world time series? How does one determine their dimensionality? This is not a straightforward question to answer, and the linear view is different from the nonlinear one. In the following chapter we introduce the necessary tools, restricting the discussion to the linear case.

I postpone summarizing this chapter until the end of the following chapter, as the two are intimately related.

EIGHT

Autocorrelation

TO ESTIMATE the degrees of freedom (*df*) of a given sample, one tool at our disposal is the autocorrelation function (acf), ρ. Recall that the common thread of the above examples about *df* was that when neighboring data points are not entirely independent of one another, there is some redundancy in the time series. Linearly quantifying this redundancy is one of the jobs of the acf.

Let's first be sure you remember the correlation coefficient of two random variables X and Y,

$$\rho_{X,Y} = \text{cor}(X,Y) = \frac{\text{cov}(X,Y)}{\sigma_X \sigma_Y} = \frac{E[(X-\mu_X)(Y-\mu_Y)]}{\sigma_Y \sigma_Y}, \tag{8.1}$$

where "cor" and "cov" denote correlation and covariance, σ_X^2 and σ_Y^2 are X's and Y's population variances, E denotes expected value, and $\mu_X = E(X)$ and $\mu_Y = E(Y)$ are X's and Y's expected values (means). We denote specific finite length realizations from these random processes x and y, sets of N element data points each, $\{x_i\}$ and $\{y_i\}$, $i = [1, N]$, also expressed as the vectors $\mathbf{x}, \mathbf{y} \in \mathbb{R}^N$. Then,

$$r_{xy} = \frac{s_{xy}}{\sqrt{s_{xx} s_{yy}}} \equiv \frac{\sum_{i=1}^{N}(x_i - \bar{x})(y_i - \bar{y})}{\sqrt{\sum_{i=1}^{N}(x_i - \bar{x})^2 \sum_{i=1}^{N}(y_i - \bar{y})^2}} = \frac{\mathbf{x}'^T \mathbf{y}'}{\sqrt{\mathbf{x}'^T \mathbf{x}' \mathbf{y}'^T \mathbf{y}'}}, \tag{8.2}$$

where primes denote centering (mean removal).

In a similar vein, the acf of a stochastic process X as a function of lag is defined theoretically by

$$\rho_k = \frac{E[(X_t - \mu_X)(X_{t+k} - \mu_X)]}{\sqrt{E[(X_t - \mu_X)^2 (X_{t+k} - \mu_X)^2]}} \tag{8.3}$$

where E denotes expected value. Note that X is assumed to be stationary, such that μ_X (the true mean, as before) over one segment or another of X is the same. A similar argument applies to the true variance (in the denominator), so the expression simplifies to

$$\rho_k = \frac{E[(X_t - \mu_X)(X_{t+k} - \mu_X)]}{\sigma_X^2}. \tag{8.4}$$

Since for a given finite-length time series we don't know μ_X and σ_X^2, we need to estimate ρ from the sample time series, an estimate we will denote by r. Using the usual estimates of μ_X and σ_X^2 ($\bar{x}$ and s_x^2), we obtain

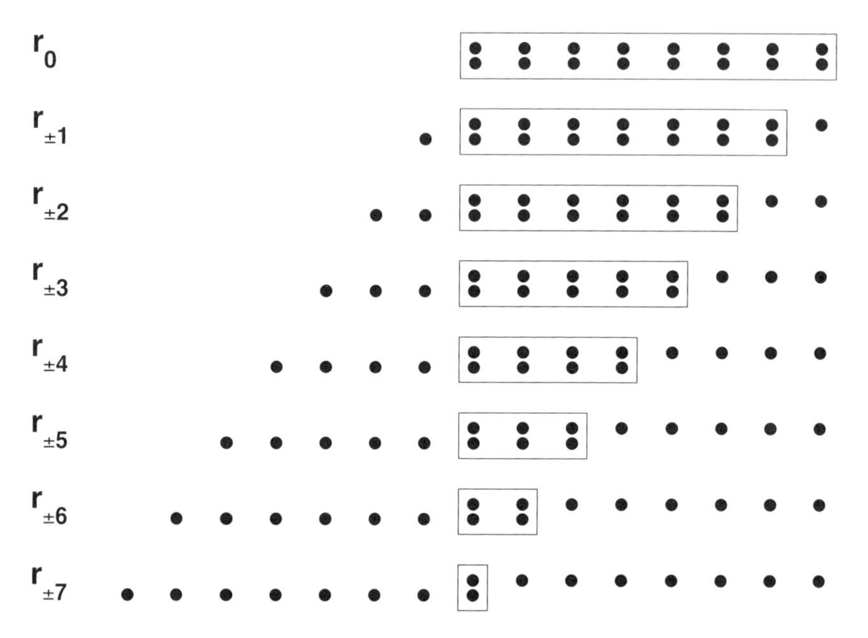

Figure 8.1. A schematic of the autocorrelation function (acf).

$$\mathrm{acf}(k) \equiv r_k = \frac{\sum_{i=1}^{N-k}(x_i - \bar{x})(x_{i+k} - \bar{x})}{\sum_{i=1}^{N}(x_i - \bar{x})^2} \tag{8.5}$$

for an N element time series. If $k_{\max} = K$ is an appreciable fraction of the time series length, say $K > N/20$, it will make sense to take note of the different length of the numerator and denominator time series, replacing eq. 8.5 with

$$r_k = \frac{(N-1)\sum_{i=1}^{N-k}(x_i - \bar{x})(x_{i+k} - \bar{x})}{(N-k-1)\sum_{i=1}^{N}(x_i - \bar{x})^2}. \tag{8.6}$$

Figure 8.2 displays the acf scaling factor $(N-1)/(N-1-k)$ for the challenging (small N, large K), regime, showing that even in the relatively safe $K/N \approx 0.1$ case, eq. 8.5 leads to a >10% error in the large lag part of the acf.

Regardless of the formula used,

$$r_o = \frac{\sum_{i=1}^{N}(x_i - \bar{x})(x_i - \bar{x})}{\sum_{i=1}^{N}(x_i - \bar{x})^2} = 1 \tag{8.7}$$

by construction. Also, $r_k = r_{-k}$ for all k, i.e., the acf is an even function, symmetric about $k = 0$. Note that as k gets larger, we lose more and more information because of the sliding of one segment with respect to the other (see fig. 8.1). In

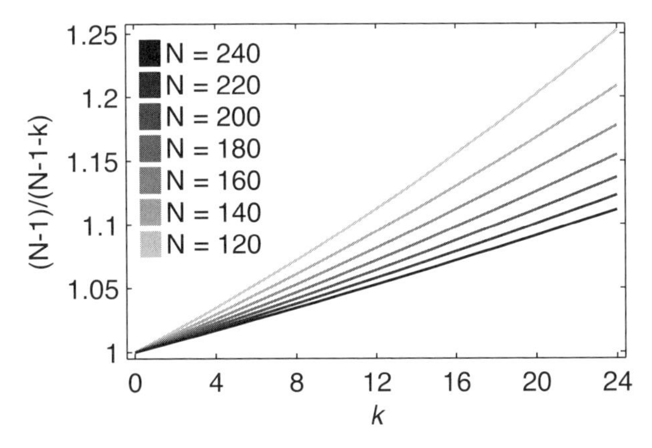

Figure 8.2. Magnitude of the acf scaling factor $(N-1)/(N-1-k)$ for relevant N and k ranges (relatively short series and high maximum lag).

practice, it will make sense to stop at some K for which there are still enough elements in the correlated segments for a reasonably robust r_K; given fig. 8.2, the $K \approx N/20$ choice is in most cases reasonable.

To gain insight into the acf and its behavior, let's examine the acf of a specific signal, the decaying periodic signal shown in fig. 8.3. For how long does a nonzero autocorrelation survive? For the noise-free signal, the time for the autocorrelation to relax to zero is the time it takes the signal's exponential decay to reduce its amplitude to a level comparable to roundoff error. For example, if the analysis is carried out with an expected numerical resolution (the smallest resolved difference between two neighboring real numbers) of $\mathcal{O}(10^{-17})$,

$$e^{-t/120} \approx 10^{-17} \quad \Longrightarrow \quad t \approx -120 \ln(10^{-17}) \approx 4700 \text{ time units.}$$

For the noisy signal, this limit is irrelevant, because the signal's decaying amplitude approaches that of the noise, here approximately 0.1, in as little as $-120 \ln(0.1) \approx 276$ time units. Beyond that point, the acf correlates mostly noise realizations, which are by definition mutually independent. While the structured part of the signal is still coherent, it is dwarfed by incoherent noise, resulting in essentially zero acf. Consistently, panels m—o show that noisy signal acf crests indeed decay with k.

Another practically important aspect of the acf is its dependence on the sampling rate. Figure 8.3 addresses this dependence by showing acf estimates of the same signal sampled at various rates. Since the sampling intervals vary, so do the times to which the fixed maximum lag, $K = 26$, corresponds. For example, when the sampling rate is one t unit (panel a), r_1 correlates time series values at times t with those at times $t \pm 1$ time unit, while for a sampling rate of 8 units

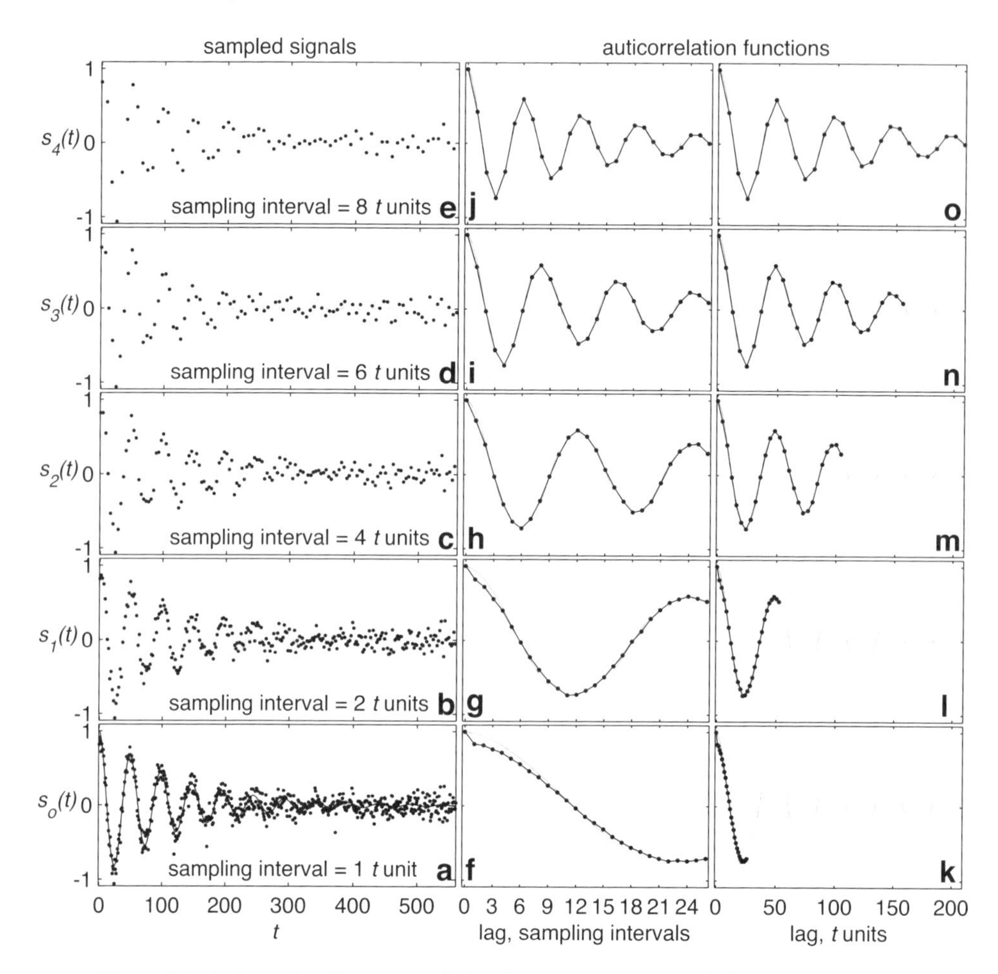

Figure 8.3. A given signal's autocorrelation function estimates at different sampling intervals. The continuous, pure signal, $s(t) = \exp(-t/120)\cos(2\pi t/49)$, is shown in a–e as solid gray curves, while sampled values of its noise-contaminated $[s + N(0, 10^{-2})]$ counterparts are shown by dots. The corresponding estimated acfs are shown in f–o, plotted against lag expressed in increments (k; panels f–j) or time units (k–o). The acfs of the noise-free signal using eqs. 8.5 and 8.6 (eq. 8.6's amplitude decaying more slowly) are shown in continuous thin gray in f–o, while acfs of the noisy signals are shown as dots. The time spans to which the fixed maximum lag, $K = 26$, corresponds at the various sampling rates are shown by the thin vertical bars in a–e.

(panel e), r_1 correlates values at times t and $t \pm 8$ time units. As a result, the longer the sampling interval (the larger δt), the longer the time period a fixed K covers (the gradual rightward shift of the vertical bar from panel a to e) and the coarser the acf sampling over that period (the decreased acf resolution from panels k to o).

The acf's key role is to quantify how mutually related, or independent, neighboring observations are, and—when mutual dependencies are found—the timescales over which they operate. To clarify this, it is useful to represent the centered (zero mean) random anomaly time series $x(t)$ as

$$x(t) = \int_{-\infty}^{t} w(s)x(s)\,ds + \xi(t), \tag{8.8}$$

where ξ is a random instantaneous shock best viewed as externally forced perturbations of the series. An example may be an $x(t)$ representing the time evolution of a given species' abundance at a specific location, with $\xi(t)$ representing perturbations of this abundance induced by processes other than internal ecological dynamics of the species in question. Such forcing processes can be natural (weather, seasonality, or volcanic eruptions) or anthropogenic (hunting, protected competition with livestock, or climate change), but must be external to the species' governing dynamics. Also in (8.8), w is a weighting function that governs the effect of earlier values of the series on its current value. As such, w controls the series' "memory horizon," the longest time earlier values are "remembered." Time series conforming with (8.8) are called autoregressive, because current values are to some extent a function of earlier ones; more on autoregression, AR, after we introduce regression.

The simplest discrete analog of eq. 8.8 is

$$x_i = \alpha\, x_{i-1} + \xi_i, \tag{8.9}$$

the celebrated autoregressive model of order 1, or AR(1), in which the series' current value blends the preceding value x_{i-1} (times α) and the current shock. With eqs. 8.8 or 8.9, the time-series variance (a cumulative function of instantaneous deviations from the zero mean) is the sum of contributions by past values and random shocks. In the physically most likely case of a stable time series (a time series whose variance is bounded from above and roughly independent of time), $\alpha < 1$. Then, the time series variance—by analogy to a steady input–output physical system—is a balance between decay of earlier perturbations (their relaxation with time toward zero, the very definition of stability; the output) and excitation by shocks (the input). The key role of the acf is to quantify how well earlier perturbations are "remembered," and how long this memory can counteract the degrading effect of noise.

The memory plus noise view of time series represented by eqs. 8.8 or 8.9 permits better understanding of the examples of fig. 8.3 and is most pertinent to the signal's *df*. The periodic function plays the role of the deterministic dynamics (e.g., the ecological dynamics governing the internal ebb and flow of the species abundance), while the added noise is $\xi(t)$. Apart from the practical accuracy issues raised earlier, because the deterministic signal (the thin solid curves of fig. 8.3a–e) is periodic, its acf should indefinitely rise to 1 every integer multiple of the period, as shown by the corresponding acfs in panels f–o. That is the deterministic system's unending memory, the generalization of eq. 8.9's α. But how long can it overcome memory degradation by noise contamination?

That is answered by the actual acfs, those of the noise-contaminated signals, the thick solid-dotted curves in fig. 8.3f–o, whose deviations from the thin curves quantify the cumulative degradation by ξ.

Thus, in principle, the acf should yield a characteristic timescale for our random time series, the typical time spacing between two successive mutually independent pieces of information along the time series. Yet this timescale is not unambiguously clear from the acf, and some subjective judgment is needed. The ambiguity arises because, assuming X (the random process of which x is a specific sample) is pure noise, we expect some neighboring time-series values to cluster—and thus yield deceptively high autocorrelations—simply by chance. In an infinite series, these clusters will cancel out one another, and their overall contribution will vanish. Conversely, if the segment of the random process we happened to realize in fact contains such randomly occurring clusters, but not enough of them for mutual cancellation, our estimate of the acf will include some spuriously inflated r_ks. However, given that there are no underlying deterministic physics that govern the emergent autocorrelation (which follows from our assumption that X is pure noise), we must be able to conclude that these nonzero r_ks are spurious; we must be able to tell the wheat (ks with reproducibly and consistently $|r_k| > 0$) from the chaff (ks whose $|r_k| > 0$ occurred by chance). Because there is no perfect method for making this distinction, the approach we will adopt is to compare the estimated r_k to a null expectation, a range of acf values that can be expected from noise series.

Let's suppose the acf is nonzero up to lag K, and zero thereafter ($k > K$). Then it can be shown that for sufficiently large N

$$r_{|k|>0} \approx 0 \quad \text{and} \quad \text{var}(r_k) \approx \frac{1}{N}\left(1 + 2\sum_{k=1}^{K} r_k^2\right). \tag{8.10}$$

Therefore, we can assume a cutoff K and test the computed $r_{k>K}$ against the expected $\{r_k\}$ distribution. In particular, there's nothing to stop us from assuming $K = 0$, in which case we expect $r_0 = 1$ and $r_{|k|>0} \approx 0$, with as many exceptions (nonzero $r_{|k|>0}$) as dictated by the combination of the time-series length and our desired significance level. While the $K = 0$ choice may seem silly (why worry about acf calculations if we expect no significantly nonzero r_ks), remember that this is a null assumption. We set up the conservative, skeptical null hypothesis $r_{|k|>0} = 0$, which we will gladly reject if reality thus dictates. For $K = 0$, $\text{var}(r_k) \approx N^{-1}$ and $\{r_{|k|>0} \sim N(0, N^{-1})$, i.e., the set of nonzero lag acf estimates will be approximately normally distributed with zero mean and standard deviation of $N^{-1/2}$. This means that we expect 95 and 99% of the r_ks to satisfy $|r_k| \leq 1.96 N^{-1/2}$ and $|r_k| \leq 2.58N^{-1/2}$, respectively. (Figure 7.5 clarifies that 1.96 and 2.58 are simply $z_{0.975}$ and $z_{0.995}$, the number of standard deviations to the right of the mean we have to be for the area under the standard normal curve between us and the symmetrical value in the opposite side of the bell to encompass 95 and 99% of the total area.)

To clarify the above ideas, let's reexamine the time series (fig. 7.6) of the North Atlantic Oscillation Index (NAOI) time series, the details of which are described in the caption. Figure 8.4 shows three estimates of the NAOI acf, with the estimated 95 and 99% confidence intervals. The upper panel shows the acf as computed from the full available NAOI time series, while the 2 lower panels present the acf computed from the first and second halves of the time series. If NAOI is stationary, these estimates should converge for large N, and so this kind of plot is a useful simple test. The various independent estimates can also be used together (in what is a called a jackknife estimate) to enhance the robustness of the estimate, but we will not address this here. The acf estimates fall mostly within the bounds expected to bracket 95% of the r_ks (the light shading), and the three estimates are rather different. There is a hint of a 20-year periodicity in panels a and c. But as is (without further corroborating evidence) this is hardly robust. First, the 73-year period considered in panel c contains fewer than 4 realizations of a 20-year cycle, and even the 146-year period of panel a contains only about 7 such realizations, so the likelihood that this apparently significant r_{20} is due to random noise alignment and that it would disappear if longer segments of the same process were available is high. In addition, of the 22 acf estimates $r_{1 \le k \le 22}$, only one exceeds the 95% brackets. Since 1 in 22 is less than 5%, there are fewer significant lags than are expected purely by chance, again casting serious doubt on the validity of the formally significant r_{20}. In summary, based on its acf and accepting no more than 5 chances out of 100 of being wrong, we cannot reject the notion that the NAOI is pure noise.

To conclude this section, let's consider the counterexample of a time series exhibiting memory well in excess of its sampling interval (fig. 8.5). The example is based on time series of hourly mean air temperature anomalies (deviation from seasonal hourly climatologies) representing a semi-rural site 70 miles north of New York City. As we expect summer and winter to behave differently due to their distinct characteristic meteorology, the full time series is split into 3-month seasons (a December–February winter is assigned to the latter of its two calendar years) comprising 90 days in winter and 92 in summer (2160 and 2208 hourly means). Only full seasons, with no missing data, are considered, of which there are 11 winters and 6 summers. For such seasons, the acf is computed individually at 1-hour intervals up to $K = 9$ days (216 hours). In fig. 8.5a,b, these seasonal estimates are shown as solid curves.

Because of the multiple analyzed seasons (and thus acf estimates, curves in fig. 8.5a,b), interpreting and drawing conclusions from the resultant acf estimates is not obvious. The winter estimates, e.g., strongly suggest that air temperature anomalies exhibit small but significant memory as far as 6–7, possibly even 9, days out. But this tentative conclusion is called into question by some of the 11 acfs, which lose significance as early as 2.5 days out; the 6–9 conclusion cannot be considered the final word. Combining discontinuous time-series

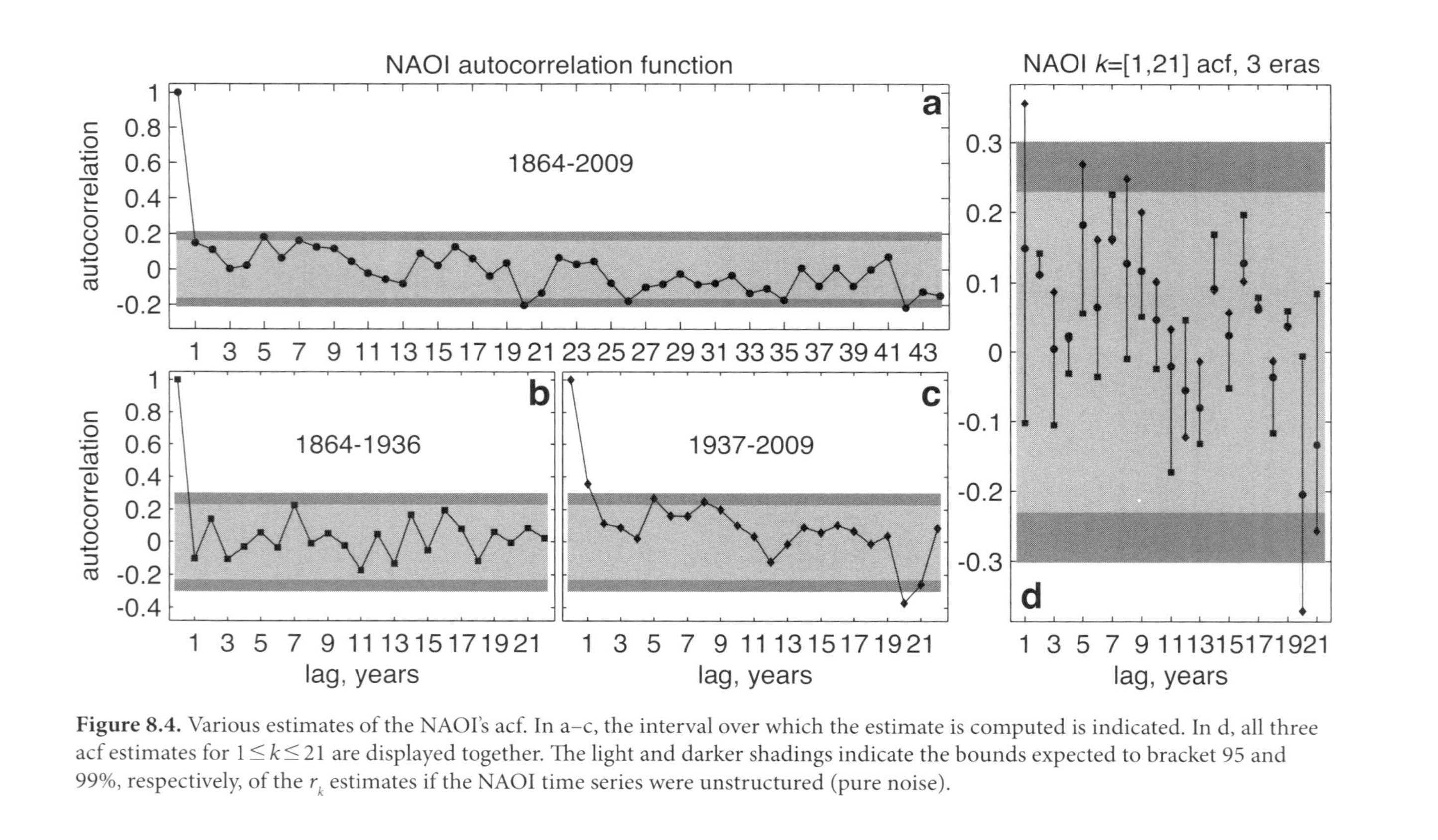

Figure 8.4. Various estimates of the NAOI's acf. In a–c, the interval over which the estimate is computed is indicated. In d, all three acf estimates for $1 \leq k \leq 21$ are displayed together. The light and darker shadings indicate the bounds expected to bracket 95 and 99%, respectively, of the r_k estimates if the NAOI time series were unstructured (pure noise).

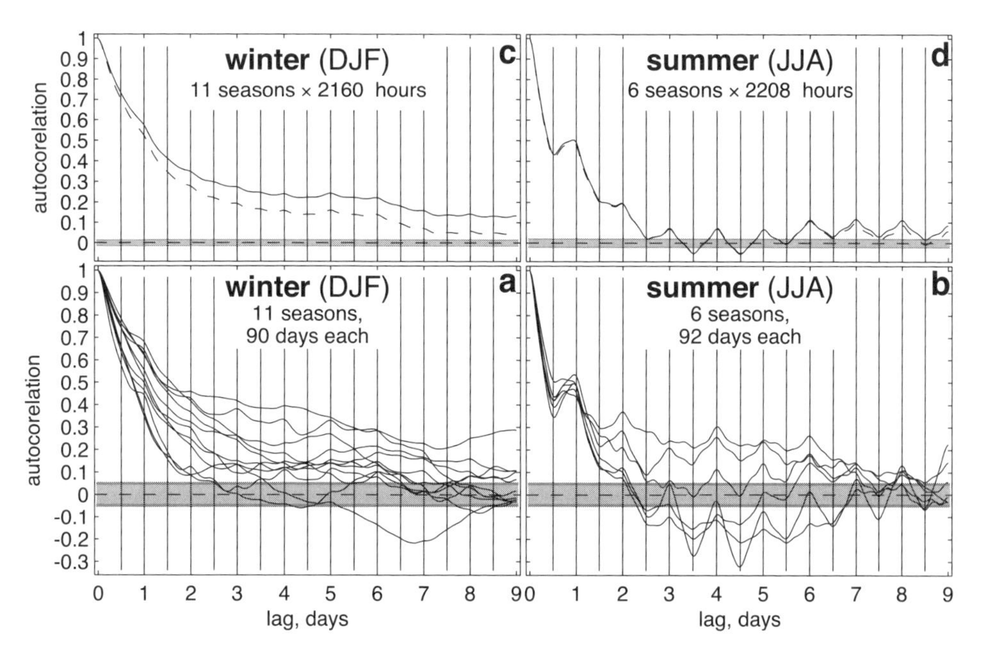

Figure 8.5. Acf estimates derived from 1988 to 2009 hourly mean air temperature anomalies at the Cary Institute of Ecosystem Studies (CIES), in the Hudson Valley, New York, at 41°47.1′N, 73°44.5′W, ~113 km north of New York City, 128 m above sea level. High-frequency (0.25 Hz) measurements are condensed into hourly means that are expressed as anomalies, deviations from the hourly seasonal climatologies. Panels a and b show estimates derived from individual full seasons (solid curves) comprising 90×24 (92×24) hourly means each in winter (summer). Panels c and d show combined global estimates: average of individual seasonal estimates (dashed) and the single estimate given by eq. 8.11 (solid). In all panels, the light and dark shadings show the ranges expected to bracket 95 and 99% of the r_k of a random time series of the same length as the ones actually examined (i.e., the light and dark shadings display the $p \leq 0.05$ and $p \leq 0.01$ significance levels).

segments to form a single, presumably more robust, acf estimate is the second challenge addressed by this example.

The reason this issue arises in this problem is the discontinuity separating successive seasons. If we view individual seasons as random realizations of the same process, then averaging the individual acf estimates may make sense. Alternatively, it is possible to combine all observations by eliminating leading and trailing elements of individual years. Denoting the time-series elements x_i^j, where $i = [1, N_s]$ is the hour index ($N_s = 90 \times 24$ in winter and $N_s = 92 \times 24$ in summer) and $j = [1, J]$ is the season index ($J = 11$ in winter and $J = 6$ in summer), the correlated segments as a function of lag k are

$$k = 0 \begin{cases} \text{segment 1: } (x_1^1 \cdots x_{N_s}^1 \; x_1^2 \cdots x_{N_s}^2 \cdots\cdots x_1^J \cdots x_{N_s}^J) \\ \text{segment 2: } (x_1^1 \cdots x_{N_s}^1 \; x_1^2 \cdots x_{N_s}^2 \cdots\cdots x_1^J \cdots x_{N_s}^J) \end{cases}$$

$$k = 1 \begin{cases} \text{segment 1: } (x_2^1 \cdots x_{N_s}^1 \quad x_2^2 \cdots x_{N_s}^2 \cdots\cdots x_2^J \cdots x_{N_s}^J) \\ \text{segment 2: } (x_1^1 \cdots x_{N_s-1}^1 \; x_1^2 \cdots x_{N_s-1}^2 \cdots\cdots x_1^J \cdots x_{N_s-1}^J) \end{cases}$$

$$k = 2 \begin{cases} \text{segment 1: } (x_3^1 \cdots x_{N_s}^1 \quad x_3^2 \cdots x_{N_s}^2 \cdots\cdots x_3^J \cdots x_{N_s}^J) \\ \text{segment 2: } (x_1^1 \cdots x_{N_s-2}^1 \; x_1^2 \cdots x_{N_s-2}^2 \cdots\cdots x_1^J \cdots x_{N_s-2}^J) \end{cases}$$

where the number of elements correlated at lag k is n_k, above $n_0 = JN_s$, $n_1 = J(N_s - 1)$ and $n_2 = J(N_s - 2)$. For any k, this generalizes to

$$\begin{array}{l} \text{segment 1: } (x_{k+1}^1 \cdots x_{N_s}^1 \quad x_{k+1}^2 \cdots x_{N_s}^2 \quad \cdots\cdots x_{k+1}^J \cdots x_{N_s}^J) \\ \text{segment 2: } (x_1^1 \quad \cdots x_{N_s-k}^1 \; x_1^2 \quad \cdots x_{N_s-k}^2 \cdots\cdots x_1^J \quad \cdots x_{N_s-k}^J) \end{array} \tag{8.11}$$

with length $n_k = J(N_s - k)$.

Either one of the global winter acf estimates (fig. 8.5c) corroborates our earlier suspicion of long-term memory, lending this suspicion the credence it previously lacked. Using the global estimate of eq. 8.11, for example, $r_{5\text{ days}} \approx 0.26$ and $r_{9\text{ days}} \approx 0.15$, small but clearly above the 0.05 and 0.01 significance thresholds. Reassuringly, this conclusion is also supported by the alternative global acf estimate, the mean of individual seasons' estimates (dashed curve in fig. 8.5c). In contrast, in summer air temperature anomalies appear to have shorter memory, fading entirely over ~3 days, with higher memory in integer multiples of a day (see the locally elevated acfs at lags of integer days), reflecting the different meteorologies governing midlatitude night and day.

8.1 Theoretical Autocovariance and Autocorrelation Functions of AR(1) and AR(2)

Since many physical processes lend themselves, at least to some degree, to a low-order AR representation, it will be very valuable to derive the theoretically expected autocovariance and autocorrelation functions of AR(1) and AR(2). The derived theoretical results are most useful as a null expectations against which actual signals' acfs can be analyzed and better understood.

8.1.1 *AR(1)*

Let's consider the AR(1) model $x_i = \alpha x_{i-1} + \xi_i$ (eq. 8.9), assuming $\bar{x} = 0$, i.e., assuming the original signal's mean of has been removed.

8.1.1.1 r_1

- Multiply both sides of (8.9) by x_{i-1},

$$x_i x_{i-1} = \alpha x_{i-1} x_{i-1} + x_{i-1}\xi_i. \tag{8.12}$$

- Apply the expected value operator (denoted by angled brackets, in practice essentially the mean of all such paired occurrences in the available record)

$$\langle x_i x_{i-1} \rangle = \alpha \langle x_i x_i \rangle + \langle x_{i-1}\xi_i \rangle, \tag{8.13}$$

  where $\langle x_{i-1}\xi_i \rangle = 0$ because on average the random shocks are uncorrelated with the signal at any time, and α is taken out of the expectance operator as it is not a statistical quantity. The first right term's modified indices reflect the assumption of stationarity, so that the relations between events depend only on the lag separating them, not on where in the time sequence they occur.
- Carry out the expected value operation

$$c_1 = \alpha c_o, \tag{8.14}$$

  where c_l denotes lag l of the autocovariance function, acvf, of a zero-mean time series

$$c_l \equiv \langle x_{t-l} x_t \rangle = \langle x_t x_{t-l} \rangle = \frac{1}{N-1} \sum_{t=1}^{N-l} x_t x_{t-l}, \tag{8.15}$$

  or, for short time series,

$$c_l = \frac{1}{N-l} \sum_{t=1}^{N-l} x_t x_{t-l}. \tag{8.16}$$

  If the series is long enough to use eq. 8.15, then eq. 8.14 follows because the scaling factor $(N-1)^{-1}$ is common to both terms and thus cancels out. Note that the above alternative short series scaling of the acvf is biased, but approaches the unbiased (8.15) for $l \ll N$.
- Because $r_i = c_i / c_o$ for all i,

$$r_1 = \alpha. \tag{8.17}$$

8.1.1.2 r_2

- Multiply both sides of (8.9) by x_{i-1}, modifying indices to reflect stationarity: $x_{i-1} x_{i+1} = \alpha x_{i-1} x_i + x_{i-1}\xi_{i+1}$.
- Apply expectance, $\langle x_{i-1} x_{i+1} \rangle = \alpha \langle x_{i-1} x_i \rangle + \langle x_{i-1}\xi_{i+1} \rangle$.
- Carry out the expectance $c_2 = \alpha c_1$ or $r_2 = \alpha r_1$.
- Use the previous result $r_1 = \alpha$, $r_2 = \alpha^2$.

8.1.1.3 r_l

- Multiply both sides of (8.9) by x_{i-l+1} and modify indices, $x_{i-l+1} x_{i+1} = \alpha x_{i-l+1} x_i + x_{i-l+1}\xi_{i+1}$.
- Apply the expectance operator, $\langle x_{i-l+1} x_{i+1} \rangle = \alpha \langle x_{i-l+1} x_i \rangle + \langle x_{i-l+1}\xi_{i+1} \rangle$.
- Carry out the expected value operation, $c_l = \alpha^l c_o$ or $r_l = \alpha^l$.

In summary, the theoretical autocorrelation function of an AR(1) process is $r_l = \alpha^l$.

8.1.2 AR(2)

Relying on the previous section derivation, subsequently we develop the machinery in a more abbreviated manner.

The AR(2) model can be written as

$$x_{i+1} = \alpha x_i + \beta x_{i-1} + \xi_{i+1}, \tag{8.18}$$

where we assume again $\{x_i\}$ has a zero mean.

8.1.2.1 r_1

$$\begin{aligned}
\langle x_i x_{i+1} \rangle &= \alpha \langle x_i x_i \rangle + \beta \langle x_i x_{i-1} \rangle + \langle x_i \epsilon_{i+1} \rangle \\
c_1 &= \alpha c_o + \beta c_1 \\
(1-\beta) c_1 &= \alpha c_o \\
c_1 &= \frac{\alpha c_o}{1-\beta} \\
r_1 &= \frac{\alpha}{1-\beta}
\end{aligned}$$

8.1.2.2 r_2

$$\begin{aligned}
\langle x_{i-1} x_{i+1} \rangle &= \alpha \langle x_{i-1} x_i \rangle + \beta \langle x_{i-1} x_{i-1} \rangle + \langle x_{i-1} \epsilon_{i+1} \rangle \\
c_2 &= \alpha c_1 + \beta c_o \\
r_2 &= \alpha r_1 + \beta = \frac{\alpha^2}{1-\beta} + \beta
\end{aligned}$$

8.1.2.3 r_3

$$\begin{aligned}
\langle x_{i-2} x_{i+1} \rangle &= \alpha \langle x_{i-2} x_i \rangle + \beta \langle x_{i-2} x_{i-1} \rangle + \langle x_{i-2} \epsilon_{i+1} \rangle \\
c_3 &= \alpha c_2 + \beta c_1 \\
r_3 &= \alpha r_2 + \beta r_1 = \alpha(\alpha r_1 + \beta) + \beta r_1 \\
r_3 &= (\alpha^2 + \beta) r_1 + \alpha\beta
\end{aligned}$$

8.1.2.4 r_4

$$\begin{aligned}
\langle x_{i-3} x_{i+1} \rangle &= \alpha \langle x_{i-3} x_i \rangle + \beta \langle x_{i-3} x_{i-1} \rangle + \langle x_{i-3} \epsilon_{i+1} \rangle \\
c_4 &= \alpha c_3 + \beta c_2
\end{aligned}$$

$$r_4 = \alpha r_3 + \beta r_2$$

$$r_4 = \alpha\left[\left(\alpha^2 + \beta\right)r_1 + \alpha\beta\right] + \beta\left[\alpha r_1 + \beta\right]$$

$$r_4 = \alpha\left[\alpha^2 r_1 + \beta\left(2r_1 + 1\right)\right] + \beta^2$$

8.1.2.5 r_l

$$\langle x_{i-l+1}x_{i+1}\rangle = \alpha\langle x_{i-l+1}x_i\rangle + \beta\langle x_{i-l+1}x_{i-1}\rangle + \langle x_{i-l+1}\epsilon_{i+1}\rangle$$

$$c_l = \alpha c_{l-1} + \beta c_{l-2}$$

$$r_l = \alpha r_{l-1} + \beta r_{l-2}$$

The use of these results in practice is addressed by figs. 8.6–8.9.

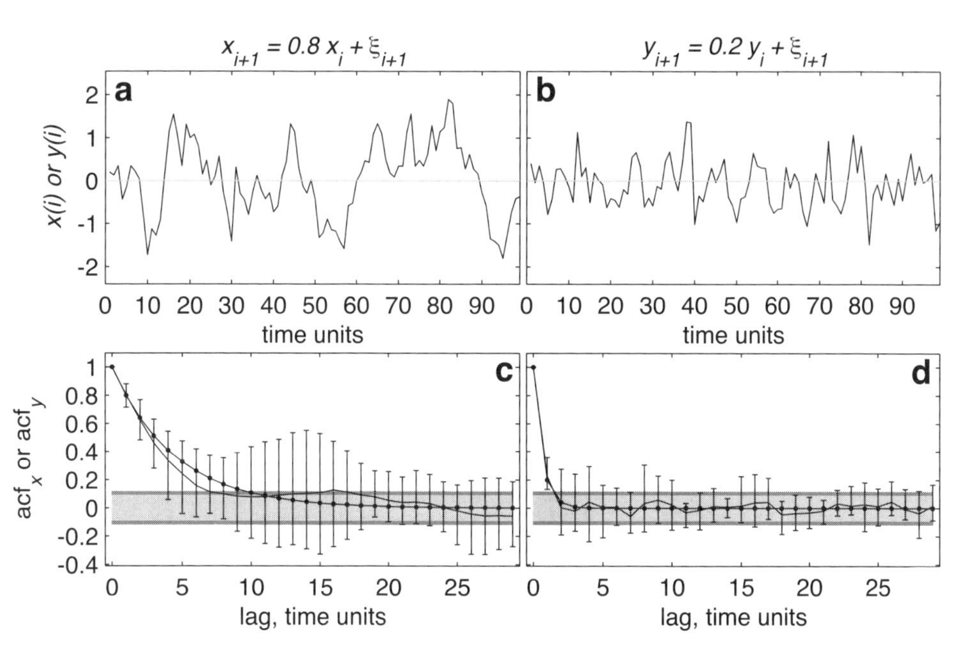

Figure 8.6. Two examples of AR(1) processes with $\alpha = 0.8$ and $\alpha = 0.2$. In both $\xi \sim N(0, 0.25)$. Only the first 100 time units are shown, but the simulations were run through $N = 500$. The lower panels are the autocorrelations. The acf for each of the two time series is estimated 10 times, once using the full time series, and then using partially overlapping 100-element-long continuous segments ($[1, 100]$, $[51, 150]$, ..., $[401, 500]$) of the full 500-element series. Light (dark) shading indicates the range $\{r_l\}$ of noise processes are expected to span. Vertical bars show the full range spanned by the 10 estimates at each lag, with the means shown by dots. Theoretical acfs are shown by thin solid curves.

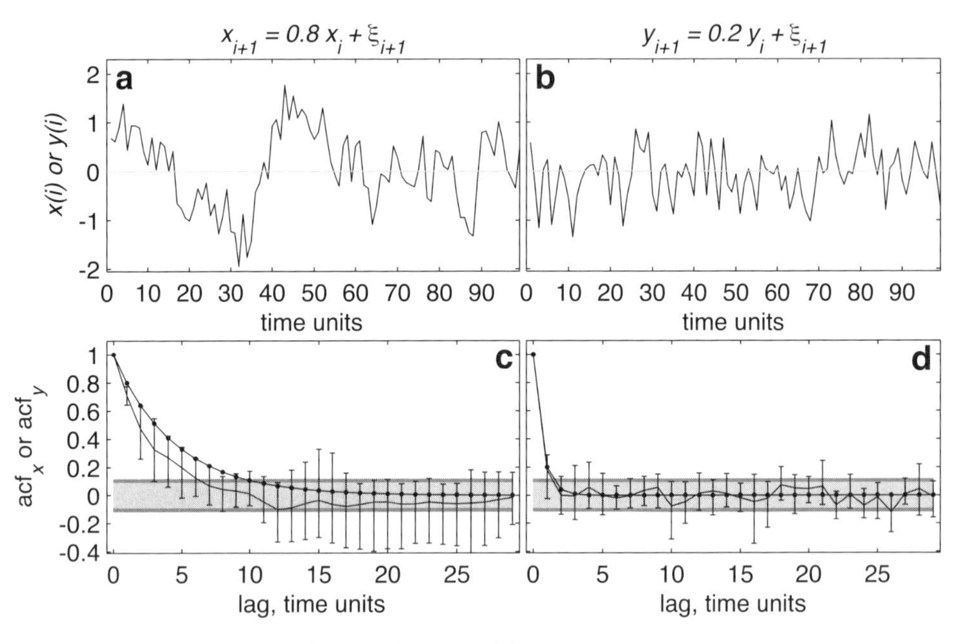

Figure 8.7. Different realizations of the same two AR(1) processes.

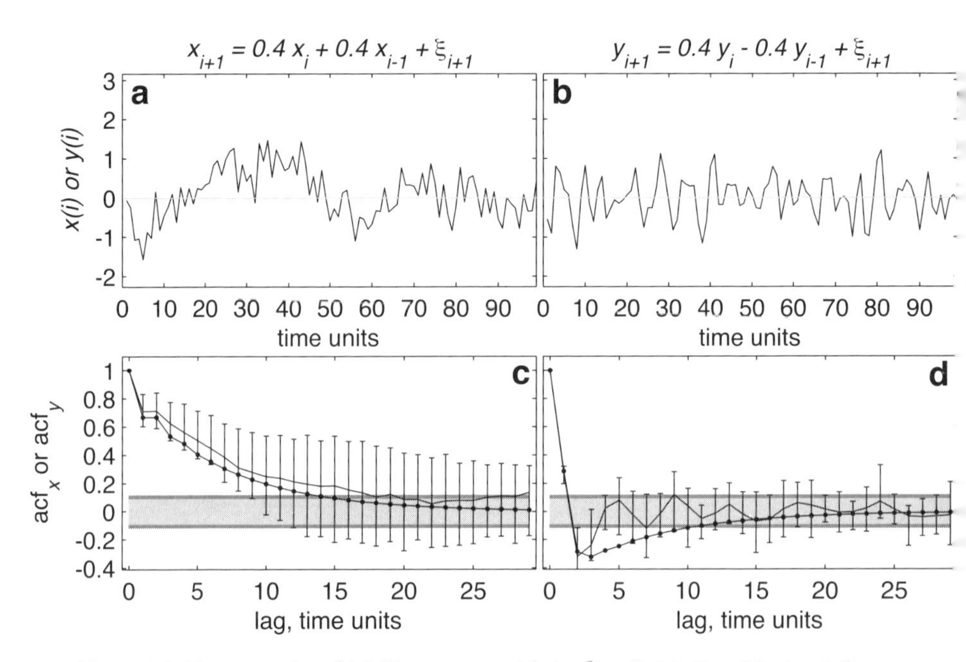

Figure 8.8. Two examples of AR(2) processes with $(\alpha, \beta) = (0.4, 0.4)$ and $(0.4, -0.4)$. All other details and notation are as in fig. 8.6.

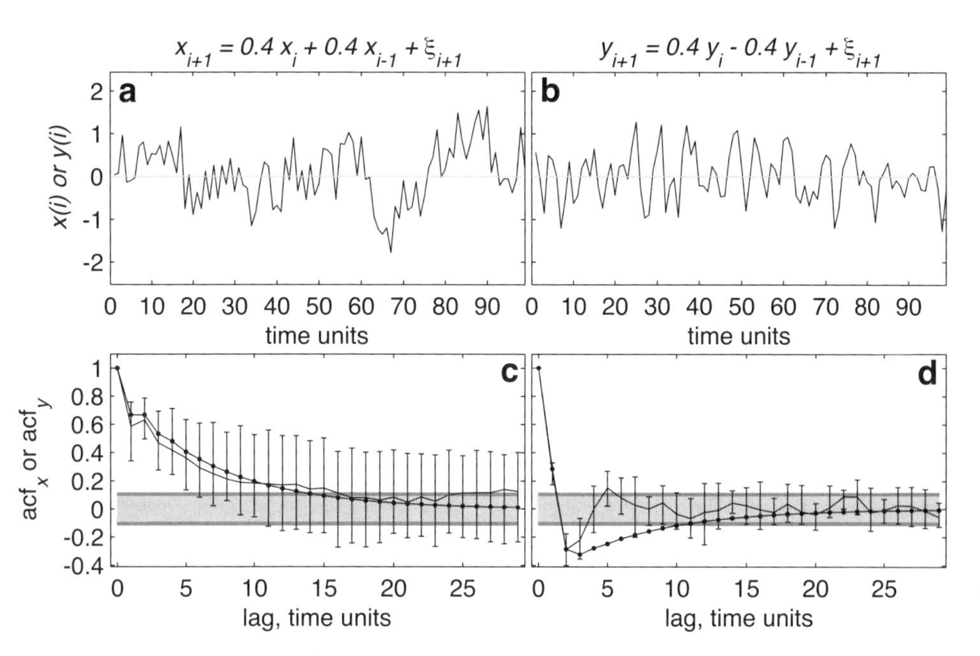

Figure 8.9. Different realizations of the same two AR(2) processes.

8.2 Acf-Derived Timescale

The acf falloff represents the time series' decorrelation scale, the rate of information loss from the time series. It therefore contains, as discussed above, the time series' characteristic timescale, $\tau = T/df = (t_N - t_1)/df$. There are various acf-based df (and thus τ) estimates. Some use regression and will therefore be introduced after general regression has been properly introduced and discussed. However, a few remarks are appropriate here.

It is clear that if the time series is pure noise (satisfying $\tau = 0$), each data point is entirely independent of any of its neighbors, and $df = N$. Even for realistic noise, which must have some $\tau > 0$, $\tau \ll \delta t$, and $df = N$ still. For time series with $\tau > \delta t$, $df < N$. To estimate df, we use some form of integrating[1] or summing[2] the acf,

$$df = \frac{N}{2\int_0^\infty r(l)\,dl} \quad \text{or discretely for large } N \quad df = \frac{N}{1 + 2\sum_{k=1}^{K} r_k}, \qquad (8.19)$$

[1] E.g., Taylor, G. I. (1921) *Proc. London Math. Soc.* **20**(2).

[2] E.g., von Storch, H. and F. W. Zwiers (1999) *Statistical Analysis in Climate Research*, Cambridge University Press, Cambridge, UK, ISBN 0-521-45071-3, eq. 17.5.

assuming $r_{k>K} \approx 0$. Alternatively, if we can reasonably assume the process is an AR(1),

$$df = N\left(\frac{1-\alpha}{1+\alpha}\right), \tag{8.20}$$

where α is the AR(1) coefficient introduced earlier. Table 8.1 gives the τs derived from the acfs of fig. 8.5c,d. The acf-based timescale estimates conform to our expectations: the summer estimates are significantly shorter than the winter ones, reflecting the more rapid falloff of the summer acfs, and the timescale estimates based on the global (solid curves in fig. 8.5c,d) are longer than their respective counterparts based on averaging individual seasonal acfs (dashed lower curves in fig. 8.5c,d). Finally timescale estimates derived from eq. 8.19 are all lower than those derived from eq. 8.20. This is not a general result, as the relative magnitudes of τ estimates derived from eqs. 8.19 and 8.20 depend on the detailed structure of the acf to which the equations are applied.

Because $df = N/\tau$, the τs of table 8.1 readily yield the df estimates reported in table 8.2. Serial correlations result in dramatic reduction of df. To demonstrate

Table 8.1
Estimated air temperature characteristic timescales τ (in days) derived from the acfs shown in fig. 8.5c,d. The table top (bottom) row corresponds to the panels' solid (dashed) curves.

	winter τ		*summer* τ	
	eq. 8.19	*eq. 8.20*	*eq. 8.19*	*eq. 8.20*
combined global acf	5.4	8.0	2.4	2.7
mean of individual seasonal acfs	4.1	7.2	2.2	2.6

Table 8.2
Estimated number of degrees of freedom $df = N/\tau$ in the air temperature time series derived from the acfs shown in fig. 8.5c,d and the timescales τ (in hours) they give rise to (table 8.1). The top (bottom) row corresponds to the panels' solid (dashed) curves.

	winter			*summer*		
		df			*df*	
	N	*eq. 8.19*	*eq. 8.20*		*eq. 8.19*	*eq. 8.20*
combined global acf	23,760	182	123	13,248	232	207
mean of individual seasonal acfs	23,760	242	138	13,248	246	216

a potential effect of this *df* reduction, let's consider the sample mean that justifies rejecting the null hypothesis $\bar{x} = 0$ using eq. 7.10,

$$|\bar{x}| \geq \frac{z s_x}{\sqrt{df}}. \qquad (8.21)$$

To give a concrete example, let's assume $s_x = 10$, and examine a probability of 95% (i.e., $p \leq 0.05$). Because a two-tailed test is appropriate, $z = 1.96$, and

$$|\bar{x}| \geq \frac{19.6}{\sqrt{df}}. \qquad (8.22)$$

For winter, e.g., when $df = N = 23{,}760$, $|\bar{x}| \gtrsim 0.13$, while for $df = 182$ and 123, $|\bar{x}| \gtrsim 1.46$ and 1.77. These are clearly major differences, with the threshold mean growing more than tenfold due to the *df* reductions accompanying the serial correlations.

8.3 Summary of Chapters 7 and 8

The autocorrelation function of a scalar time series is a prime tool for estimating the characteristic timescale separating successive independent realizations in the time series. Any process other than completely random noise has some serial correlations. Even variables as volatile and nondeterministic as measures of stock market performance, when valuated close enough, are unlikely to vary appreciably; the Dow Jones' value at 10:43:45AM is, in all likelihood, extremely similar to what it was at 10:43:44AM, even in a day over which the index lost or gained significantly overall. Thus, a time series of the index at 1-second intervals likely contains significant redundancy; it can be almost as representative and contain almost as much information if degraded to valuation intervals of $T > 1$ second. Identifying an acceptable characteristic timescale T separating successive independent realization in the time series that balances the need to retain maximum information while minimizing storage and transmission burdens is a key role of the autocorrelation function.

In controlling T, the autocorrelation function also controls the time series' degrees of freedom *df*, essentially the series' time length divided by T. As such, the autocorrelation function strongly affects statistical inference, as discussed in chapter 7, because of the dependence of such inferences on known theoretical probability distributions and the dependence of those distributions on *df*.

NINE

Regression and Least Squares

9.1 Prologue

THE FOCUS OF THIS CHAPTER, linear regression, is the process of identifying the unique model that best explains a set of observed data among a specified class of general models. Regression thus occupies a uniquely important position at the very interface of modeling and data analysis. Regression arises very often, in various guises, in handling and analyzing data. Since it is one of the most basic, useful, and frequently employed data analysis tools, and since we will need some understanding of regression in later sections, below we discuss regression in some detail. However, the topic is very broad, and its thorough understanding requires further reading.[1]

9.2 Setting Up the Problem

Regression is the fitting of a certain model (functional form, such as a line, a parabola, a sine wave) to a set of data points. By "fitting," we mean choosing the numerical values of the model parameters so as to render the model as valid as possible, as good as it can be at representing and explaining the data. For example, suppose Newton (as in Sir Isaac) sits one day in his study, only to be confronted with the feeling that $f \propto a$, i.e., that the acceleration of a particle is somehow proportional to the force operating on the particle. Knowing nothing about Newton's laws of motion, he is not entirely sure if the idea is going to work. Worrying that Leibniz is going to beat him to it, he sets out to conduct a simple experiment right away. He is going to exert force (later to be measured in newtons!) on iron canonballs of various sizes and measure, to the best of his ability, the balls' acceleration in response to the applied force.

A few hours later Newton is done, having gathered some measurements, each comprising a ball's weight, the applied force, and the measured acceleration. He plots the results, and they look something like fig. 9.1. Newton is pleased;

[1] Some excellent sources include Strang, G. (1998) *Linear Algebra and its Application*, 3rd ed., Harcourt Brace Jovanovich, San Diego ISBN 0-1555-1005-3; Ryan, T. P. (2008) *Modern Regression Methods*, 2nd ed., Wiley Series in Probability and Statistics, Wiley, New York ISBN 0-4700-8186-4; Draper, N. R. and H. Smith (1998) *Applied Regression Analysis*, 3rd ed., Wiley Series in Probability and Statistics, Wiley, New York ISBN 0-4711-7082-8.

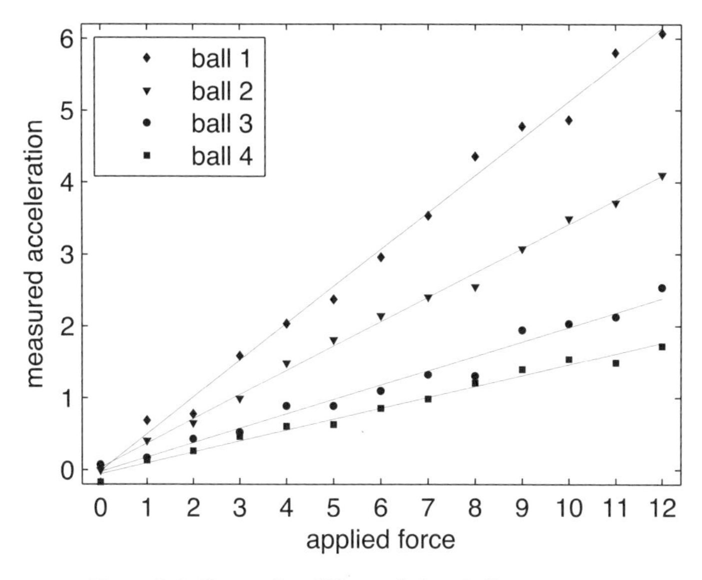

Figure 9.1. The results of Newton's four-ball experiment.

clearly, the idea works! The more force applied to a ball, the larger its acceleration. He is intrigued by the different slopes the various balls display. So he sets out to identify the laws behind the slopes, noting that the measurements are somewhat noisy, exhibiting some scatter around their respective shown straight lines. How to compute an optimal straight line, or, more generally, how to fit a functional form to a set of measurements, is the issue at hand here.

Let's develop the necessary machinery by continuing to follow Newton's work, the next challenge of which is to identify, from the data, the functional relationship between force and acceleration. Denoting by a_i^j the acceleration of ball j during the ith experiment (i.e., while ball j is subject to the ith applied force), Newton writes down the first ball's results (indicated by a superscript index). For the first experiment with the first ball, he writes

$$a_1^1 = \alpha^1 + \beta^1 f_1^1 \tag{9.1}$$

(he wants to be as general as possible, so he allows for both a slope, β^1, and an intercept, α^1; in this particular example that's a bit silly, since we know full well that no acceleration will arise from zero force). Then he continues:

$$\begin{aligned} a_2^1 &= \alpha^1 + \beta^1 f_2^1, \\ a_3^1 &= \alpha^1 + \beta^1 f_3^1, \\ &\vdots \\ a_M^1 &= \alpha^1 + \beta^1 f_M^1, \end{aligned} \tag{9.2}$$

for ball 1's M experiments. There are a number of ways these equations can be solved. Below I favor the matrix approach because of the obvious geometric interpretation it permits.

The first step in the matrix approach is, well, to express the equations as a single matrix equation,

$$\underbrace{\begin{pmatrix} 1 & f_1 \\ 1 & f_2 \\ \vdots & \vdots \\ 1 & f_M \end{pmatrix}}_{A} \underbrace{\begin{pmatrix} \alpha \\ \beta \end{pmatrix}}_{x} = \underbrace{\begin{pmatrix} a_1 \\ a_2 \\ \vdots \\ a_M \end{pmatrix}}_{b}, \tag{9.3}$$

where I dropped the superscripts for neatness, which is possible here because this equation addresses a specific ball only. The first thing we notice is the succinctness; we've reduced the entire set of equations into the deceptively innocent looking $\mathbf{Ax} = \mathbf{b}$, where, in the Newton example, $\mathbf{b}$ holds the M measured accelerations, and $\mathbf{A}$ is the matrix containing the applied forces and a column of 1s allowing for the clearly nonphysical possibility of acceleration in the absence of exerted force. The unknown, sought, vector is $\mathbf{x}$, whose elements are this model's two parameters, the straight line's intercept and slope.

The above $\mathbf{Ax} = \mathbf{b}$ can also be written as

$$\alpha \begin{pmatrix} 1 \\ 1 \\ \vdots \\ 1 \end{pmatrix} + \beta \begin{pmatrix} f_1 \\ f_2 \\ \vdots \\ f_M \end{pmatrix} = \begin{pmatrix} a_1 \\ a_2 \\ \vdots \\ a_M \end{pmatrix}, \tag{9.4}$$

which may make it easier for you to see the condition for the existence of an exact solution: the right-hand-side vector $\mathbf{b}$ must be a linear combination of $\mathbf{A}$'s columns, $\mathbf{b} \in \mathcal{R}(\mathbf{A})$. Here is an example of such a situation (which could not have arisen from the Newton example):

$$\begin{pmatrix} 6 \\ 6 \\ 2 \\ 4 \end{pmatrix} = 6 \begin{pmatrix} 1 \\ 1 \\ 0 \\ 0 \end{pmatrix} + 2 \begin{pmatrix} 0 \\ 0 \\ 1 \\ 2 \end{pmatrix} = \begin{pmatrix} 1 & 0 \\ 1 & 0 \\ 0 & 1 \\ 0 & 2 \end{pmatrix} \begin{pmatrix} 6 \\ 2 \end{pmatrix}. \tag{9.5}$$

In this situation, despite having only two parameters to set (α and β, here 6 and 2), the four observations (the rows of $\mathbf{A}$ and $\mathbf{b}$) happened to make it possible to pick numerical values for the two parameters such that *all* equations are *exactly* satisfied. This is a highly unlikely situation, essentially never realized with real data. It is also not the one Newton encountered (fig. 9.1); there is clearly some scatter of the data points around the shown straight lines $a = \alpha + \beta f$. The scatter does not invalidate Newton's hunch that $a \propto f$. Rather, it is simply due to the imperfect experimental setup: neither the balls' mass nor their acceleration or the applied force are measured perfectly, to say nothing of unconsidered retarding forces (e.g., surface friction, air resistance) that may operate on the balls.

Like all regression problems, Newton's problem is reduced to the single linear vector equation $\mathbf{Ax} = \mathbf{b}$. In the following section, we give a somewhat broad treatment of this general problem.

9.2.1 *The Generality of the Linear* $\mathbf{Ax} = \mathbf{b}$ *Setup*

In the above introduction of the regression problem, the model favored by Newton is very simple, which may give you the sense that the formalism is restricted to simple, linear, models. Note that the regression setup must only be linear *in the parameters*, but need not be itself a linear equation. Let's examine an example that clarifies this point, trying to fit the model

$$y(x) = \alpha\frac{1}{x} + \beta x^3 - \gamma\cos[\ln(x)], \tag{9.6}$$

clearly highly nonlinear in x, to a set of observed (x_i, y_i) pairs. Despite its apparent complexity, this nonlinear model is trivial to set up as a linear regression problem,

$$\mathbf{Ax} \equiv \begin{pmatrix} x_1^{-1} & x_1^3 & -\cos[\ln(x_1)] \\ x_2^{-1} & x_2^3 & -\cos[\ln(x_2)] \\ & \vdots & \\ x_M^{-1} & x_M^3 & -\cos[\ln(x_M)] \end{pmatrix} \begin{pmatrix} \alpha \\ \beta \\ \gamma \end{pmatrix} = \begin{pmatrix} y_1 \\ y_2 \\ \vdots \\ y_M \end{pmatrix} \equiv \mathbf{b}. \tag{9.7}$$

The distinction should be clear: while x^3 or $\cos[\ln(x)]$ are nonlinear in x, any specific x_i used as an input to those nonlinear functions of x yields simple numerical values that become $\mathbf{A}$'s elements. The linearity in the parameters is manifested as a simple multiplicative relationship between the possibly nonlinear in x coefficients ($\mathbf{A}$'s elements) and the parameters, e.g., $-\gamma\cos[\ln(x_1)]$. It is because of this simple multiplicative relationship that we can break down $-\gamma\cos[\ln(x_1)]$ into $-\cos[\ln(x_1)]$ times γ, the former an element of $\mathbf{A}$, the latter an element of the unknown vector $\mathbf{x}$. Contrast this with a term nonlinear in the parameter, e.g., $\sin(\gamma)\cos[\ln(x_1)]$. Unaltered, there is no way to break down this product into the "something times γ" form required for casting the problem as $\mathbf{Ax} = \mathbf{b}$. Since models not linear in their arguments, which frequently arise, are easily reduced to the $\mathbf{Ax} = \mathbf{b}$ form, but models that are nonlinear in their parameters—and thus inconsistent with that form—are most unusual, the linear regression formulation is useful and general.

9.2.2 *The Inverse Problem*

The structure of the system $\mathbf{Ax} = \mathbf{b}$ may be new to some; while as written the equation suggests that $\mathbf{x}$ is our input, with which we evaluate $\mathbf{b}$, the reverse is, in fact, true. That is, in the regression problem, it is **b** that is known, obtained from data, and the solution process amounts to identifying **x** or its nearest compromise $\hat{\mathbf{x}}$ (introduced shortly). That is why this type of problem is often referred to as an "inverse" problem.

If $\mathbf{A} \in \mathbb{R}^{N \times N}$ were square, full rank ($q = N$), and well posed (with all eigenvalues and their reciprocals well above machine accuracy), solving this linear system would be as simple as inverting **A**,

$$\mathbf{x} = \mathbf{A}^{-1}\mathbf{b}, \tag{9.8}$$

which is indeed sometimes the case (as in, e.g., the interpolation problem introduced later in this chapter). But in regression problems, **A** is rectangular, and trying to directly invert it makes no more sense than trying to compute its eigenvalues. That is why solving the inverse problem is more subtle than a simple inversion, as discussed in following sections.

9.3 The Linear System $\mathbf{Ax} = \mathbf{b}$

Linear systems (with $\mathbf{A} \in \mathbb{R}^{M \times N}$) couple $\mathbb{R}^N$ and $\mathbb{R}^M$. Since we envision the system arising from a regression problem, it will be most natural to assume $M > N$, i.e., a system with more equations than unknowns, because if $M \le N$, any sensible researcher would have gathered as much additional data as needed to achieve $M > N$. The system matrix **A** maps $\mathbf{x} \in \mathbb{R}^N$ to $\mathbf{b} \in \mathbb{R}^M$, and both spaces are split by **A** into two complementary, orthogonal subspaces:

$$\left.\begin{matrix} \mathcal{R}(\mathbf{A}) \subseteq \mathbb{R}^M \\ \mathcal{N}(\mathbf{A}^T) \subseteq \mathbb{R}^M \end{matrix}\right\} \text{fully span } \mathbb{R}^M, \qquad \left.\begin{matrix} \mathcal{R}(\mathbf{A}^T) \subseteq \mathbb{R}^N \\ \mathcal{N}(\mathbf{A}) \subseteq \mathbb{R}^N \end{matrix}\right\} \text{fully span } \mathbb{R}^N, \tag{9.9}$$

where $\mathcal{R}$ and $\mathcal{N}$ denote range and null spaces, respectively, with $\mathcal{R}(\mathbf{A}) \perp \mathcal{N}(\mathbf{A}^T)$, $\mathcal{R}(\mathbf{A}) + \mathcal{N}(\mathbf{A}^T) = \mathbb{R}^M$ and $\mathcal{R}(\mathbf{A}^T) \perp \mathcal{N}(\mathbf{A})$, $\mathcal{R}(\mathbf{A}^T) + \mathcal{N}(\mathbf{A}) = \mathbb{R}^N$. The symbol $\perp$ means "orthogonal to," and two vector spaces are mutually orthogonal if every vector from one is orthogonal to every vector from the other (e.g., the (x, y) plane and the z direction in the three-dimensional Euclidean space).

Based on **b** and **A**'s rank q and four fundamental spaces, the system can be over- or underdetermined, or both.

- The system is *underdetermined* when the number of model parameters exceeds the number of meaningful constraints, $q < N$. For example,

$$\underbrace{\begin{pmatrix} 1 & 1 & 0 \\ 1 & 0 & -1 \end{pmatrix}}_{\mathbf{A}} \underbrace{\begin{pmatrix} \alpha \\ \beta \\ \gamma \end{pmatrix}}_{\mathbf{x}} = \underbrace{\begin{pmatrix} a \\ b \end{pmatrix}}_{\mathbf{b}}. \tag{9.10}$$

From the lower equation, $\alpha - \gamma = b$ or $\gamma = \alpha - b$, and from the top, $\alpha + \beta = a$ or $\beta = a - \alpha$, so that

$$\mathbf{x} = \begin{pmatrix} \alpha \\ a-\alpha \\ \alpha-b \end{pmatrix} = \underbrace{\begin{pmatrix} 0 \\ a \\ -b \end{pmatrix}}_{\substack{\text{particular}\\ \text{solution}}} + \alpha \underbrace{\begin{pmatrix} 1 \\ -1 \\ 1 \end{pmatrix}}_{\text{basis}[\mathcal{N}(\mathbf{A})]}, \tag{9.11}$$

with any α, solves the system identically. Generalizing this example, a problem featuring an $\mathbb{R}^N$ null space is underdetermined. Note that this can easily be the case even for systems with nominally more equations than unknowns, provided some of the information is redundant (some rows are very nearly mutually parallel, so that, while $M > N$, $q < N < M$).

- The system is *overdetermined* when there is an inconsistency in the system, $\mathbf{b} \notin \mathcal{R}(\mathbf{A})$. Since this can be true even for $q < \min(M, N)$, a system can be both over- and underdetermined, the two categories are not mutually exclusive.

When we fit a model (a function, such as Newton's $a = \alpha + \beta f$ above) to a set of data, we trust that the proposed physical model is relevant to the data, so that deviations of $\mathbf{b}$'s elements from the model predictions are minor and random. This means that the important, structured, part of $\mathbf{b}$ is indeed from $\mathcal{R}(\mathbf{A})$, or else we would have devised a model that fits the data better, a model with a different $\mathbf{A}$ and thus $\mathcal{R}(\mathbf{A})$ that does contain the important, structured, part of $\mathbf{b}$. Given a model with an appropriate $\mathcal{R}(\mathbf{A})$, the error (or residual),

$$\mathbf{e} := \mathbf{A}\hat{\mathbf{x}} - \mathbf{b} \in \mathbb{R}^M, \tag{9.12}$$

the deviation of the model prediction $\mathbf{A}\hat{\mathbf{x}}$ (where $\hat{\mathbf{x}}$ is the best-fitted parameter vector) from the observed data $\mathbf{b}$, must (1) appear unstructured (be "random looking"); (2) have a characteristic magnitude $\|\mathbf{e}\|$ we can expect from the process at hand; (3) have roughly zero mean, $\langle e_i \rangle = 0$, where angled brackets denote ensemble averaging; (4) have, on average, uncorrelated individual realizations, $\langle e_i e_j \rangle = 0$, $i \neq j$, (or, put differently, have vanishing autocorrelation for all nonzero lags); (5) have a constant variance, roughly independent of the particular subset of the observations used to calculate it; and (6) be uncorrelated with the independent variables ($\mathbf{A}$'s columns).

Having defined a priori what a good fit should look like, we proceed. Because $\mathbf{e}$ is the model error, or model failure, we would naturally like to minimize its magnitude, $e = \|\mathbf{e}\|^2 = \mathbf{e}^T\mathbf{e}$, tuning the free parameters of the model ($\hat{\mathbf{x}}$'s elements) to best reproduce the observations in $\mathbf{b}$. Expanding the scalar error,

$$\begin{aligned} e &= \|\mathbf{e}\|^2 = \mathbf{e}^T\mathbf{e} = \left(\mathbf{A}\hat{\mathbf{x}} - \mathbf{b}\right)^T\left(\mathbf{A}\hat{\mathbf{x}} - \mathbf{b}\right) \\ &= \hat{\mathbf{x}}^T\mathbf{A}^T\mathbf{A}\hat{\mathbf{x}} - \hat{\mathbf{x}}^T\mathbf{A}^T\mathbf{b} - \mathbf{b}^T\mathbf{A}\hat{\mathbf{x}} + \mathbf{b}^T\mathbf{b}. \end{aligned} \tag{9.13}$$

This can be simplified by noting that the second and third terms can be combined, because

$$\hat{\mathbf{x}}^T\mathbf{A}^T\mathbf{b} = \begin{pmatrix} \hat{x}_1 & \cdots & \hat{x}_N \end{pmatrix} \begin{pmatrix} \mathbf{a}_1^T\mathbf{b} \\ \vdots \\ \mathbf{a}_N^T\mathbf{b} \end{pmatrix} = \sum_{i=1}^{N} \hat{x}_i \left(\mathbf{a}_i^T\mathbf{b}\right) \tag{9.14}$$

and

$$\mathbf{b}^T\mathbf{A}\hat{\mathbf{x}} = \begin{pmatrix} \mathbf{b}^T\mathbf{a}_1 & \cdots & \mathbf{b}^T\mathbf{a}_N \end{pmatrix} \begin{pmatrix} \hat{x}_1 \\ \vdots \\ \hat{x}_N \end{pmatrix} = \sum_{i=1}^{N} \hat{x}_i \left(\mathbf{b}^T\mathbf{a}_i\right), \tag{9.15}$$

which is expected as $\mathbf{b}^T\mathbf{A}\hat{\mathbf{x}}$ and $\hat{\mathbf{x}}^T\mathbf{A}^T\mathbf{b}$ are both scalars and $\hat{\mathbf{x}}^T\mathbf{A}^T\mathbf{b} = \left(\mathbf{b}^T\mathbf{A}\hat{\mathbf{x}}\right)^T$. Therefore,

$$e = \underbrace{\hat{x}^T\mathbf{A}^T\mathbf{A}\hat{\mathbf{x}}}_{e_I} - \underbrace{2\hat{x}^T\mathbf{A}^T\mathbf{b}}_{e_{II}} + \underbrace{\mathbf{b}^T\mathbf{b}}_{e_{III}}. \tag{9.16}$$

The expression for e contains an $\hat{\mathbf{x}}$-dependent part $(e_I + e_{II})$ and an $\hat{\mathbf{x}}$-independent part (e_{III}). The magnitude of the former is affected, potentially optimized, by our choice of $\hat{\mathbf{x}}$. The latter, however, is a fixed positive scalar, e's value when $\hat{\mathbf{x}} = \mathbf{0}$. It obviously makes sense to minimize the error magnitude e. The e-minimizing $\hat{\mathbf{x}}$ can be derived in various ways. Because it is such a central result, below I derive it in two ways.

9.3.1 Minimizing e Using Calculus

As the reader is surely well aware, finding extrema of a function entails setting the function's first derivative (slope or gradient) to zero, and, if necessary, determining the nature (maximum, minimum or saddle point) of the extrema using the curvature (second derivative). The only possible novelty for some readers in what follows may be that while e is a scalar (eq. 9.16), its input argument is the N-vector $\hat{\mathbf{x}}$ (i.e., $e : \mathbb{R}^N \mapsto \mathbb{R}$, e is a map from $\mathbb{R}^N$ to the real line). Consequently, e's gradient is N dimensional, $(\partial_{\hat{x}_1}, \partial_{\hat{x}_2}, \ldots, \partial_{\hat{x}_N})$. While obtaining those derivatives is straightforward in principle, it may also be extremely tedious. Consequently, below we break down the differentiation task into several parts, as well as making, and then relaxing, a crucial simplifying assumption, temporarily paying for pedagogical clarity with some loss of generality.

The temporary assumption we now make is that $\mathbf{A}$'s columns are mutually orthonormal. That is, we assume

$$\mathbf{a}_i^T\mathbf{a}_j = \delta_{ij} = \begin{cases} 0 & \text{for} \quad i \neq j \\ 1 & \text{for} \quad i = j, \end{cases} \tag{9.17}$$

where δ_{ij} is the so-called Kronecker delta, so that $\mathbf{A}^T\mathbf{A} = \mathbf{I}_N$, the $\mathbb{R}^N$ identity matrix. While this assumption, which we make strictly for pedagogical expedience, may appear prohibitively restrictive, it is actually not unreasonable and is often encountered in practice. The reason is that in many cases the basis vectors (**A**'s columns) are derived from a trigonometric or other complete orthonormal sets. A clear, relevant, and frequently encountered example of such a situation is the Fourier fit, in which putative periodicities in the data are unearthed by fitting the data to a trigonometric Fourier basis. Here are the details, using a simple specific example.

Suppose you have reason to believe the time-dependent data ($b_i = b(t_i)$, $1 \le i \le M$, **b**'s elements) exhibit a single periodicity, i.e., they vary like $\cos(\omega t + \phi)$, where ω and ϕ are a given set frequency and the sought phase. That is, we wish to entertain the model

$$b_i = \alpha \cos(\omega t_i + \phi) \tag{9.18}$$

with some amplitude α and phase ϕ to be optimized to best fit the data. Here one of the two sought coefficients of the problem, ϕ, appears as an argument of the nonlinear trigonometric function, and linear analysis appears to fall short. Not all is lost, however, because of the trigonometric identity

$$\cos(a + b) = \cos(a)\cos(b) - \sin(a)\sin(b), \tag{9.19}$$

with which eq. 9.18 is recast as

$$b_i = \alpha \cos(\omega t_i)\cos(\phi) - \alpha \sin(\omega t_i)\sin(\phi). \tag{9.20}$$

If we now let

$$f = \alpha \cos(\phi) \quad \text{and} \quad g = -\alpha \sin(\phi), \tag{9.21}$$

eq. 9.18 becomes

$$b_i = f \cos(\omega t_i) + g \sin(\omega t_i) \tag{9.22}$$

and the model fitting problem becomes the familiar linear

$$\mathbf{Ax} = \begin{pmatrix} \cos(\omega t_1) & \sin(\omega t_1) \\ \cos(\omega t_2) & \sin(\omega t_2) \\ \vdots & \vdots \\ \cos(\omega t_M) & \sin(\omega t_M) \end{pmatrix} \begin{pmatrix} f \\ g \end{pmatrix} = \begin{pmatrix} b_1 \\ b_2 \\ \vdots \\ b_M \end{pmatrix} = b, \tag{9.23}$$

with the two sought parameters f and g, from which the original α and ϕ can be obtained.

In the above **A**, and in similar problems in which complete, orthonormal, sets are used to span **A**'s columns, the orthonormality of **A**'s properly normalized columns results in $\mathbf{A}^T\mathbf{A} = \mathbf{I}_N \equiv \mathbf{I} \in \mathbb{R}^{N\times N}$.

9.3.1.1 $e_I = \hat{\mathbf{x}}^T\mathbf{A}^T\mathbf{A}\hat{\mathbf{x}}$

With the orthonormality assumption $\mathbf{A}^T\mathbf{A} = \mathbf{I}_N$, so

$$\hat{\mathbf{x}}^T\mathbf{A}^T\mathbf{A}\hat{\mathbf{x}} = \hat{\mathbf{x}}^T\hat{\mathbf{x}} = \sum_{i=1}^{N} \hat{x}_i^2 \tag{9.24}$$

and

$$\frac{de_I}{d\hat{\mathbf{x}}} = \begin{pmatrix} \partial_{\hat{x}_1} \\ \partial_{\hat{x}_2} \\ \vdots \\ \partial_{\hat{x}_N} \end{pmatrix} e_I = 2\begin{pmatrix} \hat{x}_1 \\ \hat{x}_2 \\ \vdots \\ \hat{x}_N \end{pmatrix} = 2\hat{\mathbf{x}}. \tag{9.25}$$

9.3.1.2 $e_{II} = -2\hat{\mathbf{x}}^T\mathbf{A}^T\mathbf{b}$

$$-2\hat{\mathbf{x}}^T\mathbf{A}^T\mathbf{b} = -2\sum_{i=1}^{N} \hat{x}_i\left(\mathbf{a}_i^T\mathbf{b}\right), \tag{9.26}$$

so

$$\frac{de_{II}}{d\hat{\mathbf{x}}} = \begin{pmatrix} \partial_{\hat{x}_1} \\ \partial_{\hat{x}_2} \\ \vdots \\ \partial_{\hat{x}_N} \end{pmatrix} e_{II} = -2\begin{pmatrix} \mathbf{a}_1^T\mathbf{b} \\ \mathbf{a}_2^T\mathbf{b} \\ \vdots \\ \mathbf{a}_N^T\mathbf{b} \end{pmatrix} = -2\mathbf{A}^T\mathbf{b}. \tag{9.27}$$

9.3.1.3 The overall e

Because e_{III} is $\hat{\mathbf{x}}$ independent, $de_{III}/d\hat{\mathbf{x}} = \mathbf{0}$, and

$$\frac{de}{d\hat{\mathbf{x}}} = \frac{de_I}{d\hat{\mathbf{x}}} + \frac{de_{II}}{d\hat{\mathbf{x}}} = 2\begin{pmatrix} \hat{x}_1 \\ \hat{x}_2 \\ \vdots \\ \hat{x}_N \end{pmatrix} - 2\begin{pmatrix} \mathbf{a}_1^T\mathbf{b} \\ \mathbf{a}_2^T\mathbf{b} \\ \vdots \\ \mathbf{a}_N^T\mathbf{b} \end{pmatrix} = 2\left(\hat{\mathbf{x}} - \mathbf{A}^T\mathbf{b}\right). \tag{9.28}$$

Therefore, the condition for extremum, $de/d\hat{\mathbf{x}} = \mathbf{0}$, requires that

$$\hat{\mathbf{x}} = \mathbf{A}^T\mathbf{b}. \tag{9.29}$$

9.3.1.4 The $\hat{\mathbf{x}}^T\mathbf{A}^T\mathbf{A}\hat{\mathbf{x}}$ term assuming a general $\mathbf{A}$

With a general (not necessarily orthonormal) $\mathbf{A}$,

$$\hat{\mathbf{x}}\mathbf{A}^T\mathbf{A}\hat{\mathbf{x}} = \begin{pmatrix} \hat{x}_1 & \cdots & \hat{x}_N \end{pmatrix}\begin{pmatrix} \cdots & \mathbf{a}_1^T & \cdots \\ & \vdots & \\ \cdots & \mathbf{a}_N^T & \cdots \end{pmatrix}\begin{pmatrix} \vdots & & \vdots \\ \mathbf{a}_1 & \cdots & \mathbf{a}_N \\ \vdots & & \vdots \end{pmatrix}\begin{pmatrix} \hat{x}_1 \\ \vdots \\ \hat{x}_N \end{pmatrix}$$

$$= \begin{pmatrix} \hat{x}_1 & \cdots & \hat{x}_N \end{pmatrix}\begin{pmatrix} \cdots & \mathbf{a}_1^T & \cdots \\ & \vdots & \\ \cdots & \mathbf{a}_N^T & \cdots \end{pmatrix}\sum_{i=1}^{N} \hat{x}_i\mathbf{a}_i$$

$$= \begin{pmatrix} \hat{x}_1 & \cdots & \hat{x}_N \end{pmatrix} \begin{pmatrix} \mathbf{a}_1^T \sum_{i=1}^N \hat{x}_i \mathbf{a}_i \\ \vdots \\ \mathbf{a}_N^T \sum_{i=1}^N \hat{x}_i \mathbf{a}_i \end{pmatrix} \tag{9.30}$$

$$= \begin{pmatrix} \hat{x}_1 & \cdots & \hat{x}_N \end{pmatrix} \begin{pmatrix} \sum_{i=1}^N \hat{x}_i \mathbf{a}_1^T \mathbf{a}_i \\ \vdots \\ \sum_{i=1}^N \hat{x}_i \mathbf{a}_N^T \mathbf{a}_i \end{pmatrix}$$

$$= \hat{x}_1 \left(\hat{x}_1 \mathbf{a}_1^T \mathbf{a}_1 + \cdots + \hat{x}_N \mathbf{a}_1^T \mathbf{a}_N \right) + \hat{x}_2 \left(\hat{x}_1 \mathbf{a}_2^T \mathbf{a}_1 + \cdots + \hat{x}_N \mathbf{a}_2^T \mathbf{a}_N \right)$$
$$+ \cdots + \hat{x}_N \left(\hat{x}_1 \mathbf{a}_N^T \mathbf{a}_1 + \cdots + \hat{x}_N \mathbf{a}_N^T \mathbf{a}_N \right).$$

The final form of eq. 9.30 comprises N additive terms, the ith of which has the general form

$$\hat{x}_i \left(\hat{x}_1 \mathbf{a}_i^T \mathbf{a}_1 + \cdots + \hat{x}_N \mathbf{a}_i^T \mathbf{a}_N \right)$$
$$= \hat{x}_i^2 \mathbf{a}_i^T \mathbf{a}_i + \sum_{j=1}^{i-1} \hat{x}_i \hat{x}_j \mathbf{a}_i^T \mathbf{a}_j + \sum_{j=i+1}^{N} \hat{x}_i \hat{x}_j \mathbf{a}_i^T \mathbf{a}_j. \tag{9.31}$$

Collecting all additive terms, we have

$$\begin{aligned} \hat{\mathbf{x}} \mathbf{A}^T \mathbf{A} \hat{\mathbf{x}} = {} & \hat{x}_1^2 \mathbf{a}_1^T \mathbf{a}_1 + \sum_{j=1}^{0} \hat{x}_1 \hat{x}_j \mathbf{a}_1^T \mathbf{a}_j + \sum_{j=2}^{N} \hat{x}_1 \hat{x}_j \mathbf{a}_1^T \mathbf{a}_j \\ & + \hat{x}_2^2 \mathbf{a}_2^T \mathbf{a}_2 + \sum_{j=1}^{1} \hat{x}_2 \hat{x}_j \mathbf{a}_2^T \mathbf{a}_j + \sum_{j=3}^{N} \hat{x}_2 \hat{x}_j \mathbf{a}_2^T \mathbf{a}_j \\ & \vdots \\ & + \hat{x}_N^2 \mathbf{a}_N^T \mathbf{a}_N + \sum_{j=1}^{N-1} \hat{x}_N \hat{x}_j \mathbf{a}_N^T \mathbf{a}_j + \sum_{j=N+1}^{N} \hat{x}_N \hat{x}_j \mathbf{a}_N^T \mathbf{a}_j, \end{aligned} \tag{9.32}$$

where sums in which the upper summation bound is smaller than the lower bound vanish.

Differentiating with respect to $\hat{x}_i$,

$$\frac{\partial}{\partial \hat{x}_i} \hat{\mathbf{x}} \mathbf{A}^T \mathbf{A} \hat{\mathbf{x}} = \left(2\hat{x}_i \mathbf{a}_i^T \mathbf{a}_i + \sum_{j=1}^{i-1} \hat{x}_j \mathbf{a}_i^T \mathbf{a}_j + \sum_{j=i+1}^{N} \hat{x}_j \mathbf{a}_i^T \mathbf{a}_j \right) \tag{9.33}$$
$$+ \left(\sum_{j=1}^{i-1} \hat{x}_j \mathbf{a}_i^T \mathbf{a}_j + \sum_{j=i+1}^{N} \hat{x}_j \mathbf{a}_i^T \mathbf{a}_j \right).$$

The first group of terms (first row) originates from the ith row in eq. 9.32. The second group of terms (second row) represents the collective contribution from all other ([1, $i-1$], [$i+1, N$]) rows of eq. 9.32, with rows [1, $i-1$] yielding the first of the two second row sums, and rows [$i+1, N$] yielding the second sum. For example, the contribution of row $k \neq i$ is

$$\frac{\partial}{\partial \hat{x}_i}\left(\sum_{j=1}^{k-1} \hat{x}_k \hat{x}_j \mathbf{a}_k^T \mathbf{a}_j + \sum_{j=k+1}^{N} \hat{x}_k \hat{x}_j \mathbf{a}_k^T \mathbf{a}_j\right) = \hat{x}_k \mathbf{a}_k^T \mathbf{a}_i, \tag{9.34}$$

arising from the $j=i$ term, which occurs in the first (second) sum for $k<i$ ($k>i$).

Equation 9.33 can be rearranged. First, the pair of sums disrupted by (i.e., excluding) the $j=i$ term are common to both groups of parenthetical terms (the first and second rows of eq. 9.33), so

$$\frac{\partial}{\partial \hat{x}_i} \hat{\mathbf{x}}\mathbf{A}^T\mathbf{A}\hat{\mathbf{x}} = 2\left(\hat{x}_i \mathbf{a}_i^T \mathbf{a}_i + \sum_{j=1}^{i-1} \hat{x}_j \mathbf{a}_i^T \mathbf{a}_j + \sum_{j=i+1}^{N} \hat{x}_j \mathbf{a}_i^T \mathbf{a}_j\right). \tag{9.35}$$

In addition, the term excluded from the sum by the discontinuity at $j=i$ is $\hat{x}_i \mathbf{a}_i^T \mathbf{a}_i$, is precisely the extra term that precedes the sums in eq. 9.35. As a result,

$$\frac{\partial}{\partial \hat{x}_i} \hat{\mathbf{x}}\mathbf{A}^T\mathbf{A}\hat{\mathbf{x}} = 2\sum_{j=1}^{N} \hat{x}_j \mathbf{a}_i^T \mathbf{a}_j \tag{9.36}$$

and

$$\frac{de_I}{d\hat{\mathbf{x}}} \equiv \frac{d}{d\hat{\mathbf{x}}} \hat{\mathbf{x}}\mathbf{A}^T\mathbf{A}\hat{\mathbf{x}} = 2\begin{pmatrix} \sum_{j=1}^{N} \hat{x}_j \mathbf{a}_1^T \mathbf{a}_j \\ \sum_{j=1}^{N} \hat{x}_j \mathbf{a}_2^T \mathbf{a}_j \\ \vdots \\ \sum_{j=1}^{N} \hat{x}_j \mathbf{a}_N^T \mathbf{a}_j \end{pmatrix} = 2\mathbf{A}^T\mathbf{A}\hat{\mathbf{x}}. \tag{9.37}$$

Putting the two contributions to e together and requiring the slope of the sum to vanish in all directions, as before, yields

$$\frac{de}{d\hat{\mathbf{x}}} = \frac{de_I}{d\hat{\mathbf{x}}} + \frac{de_{II}}{d\hat{\mathbf{x}}} = 2\mathbf{A}^T\mathbf{A}\hat{\mathbf{x}} - 2\mathbf{A}^T\mathbf{b} = \mathbf{0} \tag{9.38}$$

or

$$\mathbf{A}^T\mathbf{A}\hat{\mathbf{x}} = \mathbf{A}^T\mathbf{b}, \tag{9.39}$$

the "normal equations," the solution to the overdetermined regression problem. Note that eq. 9.29 is the special case of eq. 9.39 for the $\mathbf{A}^T\mathbf{A} = \mathbf{I}_N$ case, as required.

9.3.2 *Minimizing e Using Geometry*

The error minimization aspect of overdetermined regression problems can also be addressed geometrically. Because $\mathbf{b} \notin \mathcal{R}(\mathbf{A})$, there is no $\mathbf{x}$ that can solve the problem exactly. As discussed previously, we are neither surprised nor deterred by this observation. Rather, we look for the closest related problem we can solve. The contribution of geometry to this search is in identifying "closest." Whatever compromise solution $\hat{\mathbf{x}}$ we choose, the best it can do is produce the approximate data vector $\hat{\mathbf{b}} \in \mathcal{R}(\mathbf{A})$, thus also producing a nonzero $\mathbf{e} = \mathbf{A}\hat{\mathbf{x}} - \mathbf{b}$. The geometrical interpretation of the problem is to view the choice of $\hat{\mathbf{x}}$ as minimizing $e = \mathbf{e}^T\mathbf{e}$ by ensuring that $\mathbf{e}$ is purely from the $\mathbb{R}^M$ subspace orthogonal to $\mathcal{R}(\mathbf{A})$, $\mathcal{N}(\mathbf{A}^T)$. How this is done is described below.

By way of a refresher, consider the distance $r(x)$ between a point and a line, as shown in fig. 9.2a. We define $x = 0$ as the location along the line from which a normal to the line passes through the point. The distance between the point and the line along that normal is r_0. In one dimension (along x alone),

$$r(x) = \sqrt{r_0^2 + x^2}, \tag{9.40}$$

with extrema when

$$\frac{dr}{dx} = \frac{x}{\sqrt{r_0^2 + x^2}} = 0. \tag{9.41}$$

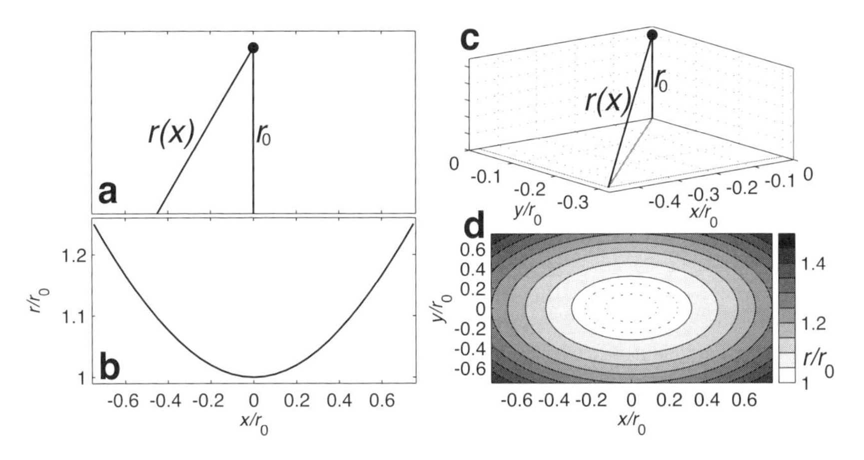

Figure 9.2. The distance between a point and a line (panels a and b) or a plane (panels c and d). Panel a shows the distance $r(x)$ between a point and a line as a function of distance x along the line, where $x = 0$ is the location along the line through which a normal to the line passes through the point. Panel b shows r/r_0 as a function of x/r_0. Panel c shows one quadrant ($x \leq 0$, $y \leq 0$) of the two-dimensional counterpart of the problem shown in panels a and b: the distance $r(x)/r_0$ between a point and the $z = 0$ plane as a function of distance x/r_0 and y/r_0 along the plane, where (0, 0, 0) is the location on the plane from which a normal to the plane passes through the point. Panel d shows r/r_0 as a function of x/r_0 and y/r_0 with solid contours rising from 1.05 at 0.05 interval, and two additional dotted contours at 1.01 and 1.03.

Vindicating intuition, this only holds for $x = 0$. Figure 9.2b shows $r(x)/r_0$, demonstrating how $r(x)$ rises monotonously and symmetrically from its minimum at $x = 0$.

This readily generalizes to 2D, as fig. 9.2c and d shows. Let $\mathbf{x} = (x, y)$ denote a point in the $z = 0$ plane. The distance $r(\mathbf{x})$ between a point and the $z = 0$ plane at (x, y) is

$$r(\mathbf{x}) = \sqrt{r_0^2 + x^2 + y^2} = \sqrt{r_0^2 + \mathbf{x}^T\mathbf{x}}, \tag{9.42}$$

where, like previously, we define (0, 0) as the location in the $z = 0$ plane from which a normal to the plane passes through the off-plane point. The distance between the point and the plane along that normal is still r_0.

Because now r depends on two coordinates, its gradient is two dimensional,

$$\begin{aligned}\frac{dr(\mathbf{x})}{d\mathbf{x}} &= \begin{pmatrix}\partial_x\\ \partial_y\end{pmatrix} r(\mathbf{x}) = \begin{pmatrix}\partial_x\\ \partial_y\end{pmatrix}\sqrt{r_0^2 + x^2 + y^2}\\ &= \frac{1}{2\sqrt{r_0^2 + x^2 + y^2}}\begin{pmatrix}2x\\ 2y\end{pmatrix} = \frac{1}{\sqrt{r_0^2 + x^2 + y^2}}\begin{pmatrix}x\\ y\end{pmatrix}\\ &= \frac{\mathbf{x}}{\sqrt{r_0^2 + x^2 + y^2}} = \frac{\mathbf{x}}{\sqrt{r_0^2 + \mathbf{x}^T\mathbf{x}}}.\end{aligned} \tag{9.43}$$

Consistent with fig. 9.2, $dr/d\mathbf{x} = 0$ only when $x = y = 0$. These results readily generalize to N dimensions, for which $r(\mathbf{x})$ is given by eq. 9.42's rightmost term, and

$$\frac{dr}{d\mathbf{x}} = \begin{pmatrix}\partial_{x_1}\\ \partial_{x_2}\\ \vdots\\ \partial_{x_N}\end{pmatrix} r = \frac{\mathbf{x}}{\sqrt{r_0^2 + \mathbf{x}^T\mathbf{x}}} \tag{9.44}$$

can vanish, and thus attain a minimum, only at the origin.

The message of these simple derivations is that, irrespective of the number of dimensions, the shortest distance between a point and a subspace (the role played by the line and the plane in the above one- and two-dimensional derivations) is always measured along the line that is normal to the subspace while passing through the point.

This lesson is extremely pertinent to the problem of fitting a model to an inconsistent set of measurements, i.e., of solving $\mathbf{A}\mathbf{x} = \mathbf{b}$ with $\mathbf{b} \notin \mathcal{R}(\mathbf{A})$. When we attempt to fit a particularly structured model to a set of data, we trust our model (or, at least, we temporarily do, until proven otherwise by, say, a plot of the residuals). Consequently, we view the failure of $\mathbf{b}$ to reside exclusively in $\mathcal{R}(\mathbf{A})$ not as invalidating the model, but as merely a nuisance due to an imperfect experimental setup. Because we trust the model, we naturally want to choose parameter values for it that will optimize its performance (for a given model structure) by reproducing as much of $\mathbf{b}$ as possible. Since the best any linear model can do is produce $\mathcal{R}(\mathbf{A})$ vectors, we choose an $\hat{\mathbf{x}}$ that yields the $\hat{\mathbf{b}} \in \mathcal{R}(\mathbf{A})$ nearest to $\mathbf{b}$, where distance is, as before, Euclidean. Based on what we have just convinced ourselves of in one and two dimensions, the shortest $\mathbf{b} - \mathbf{A}\hat{\mathbf{x}}$ is the line normal to $\mathcal{R}(\mathbf{A})$ passing through $\mathbf{b}$. Identifying this line is accomplished by setting $\mathbf{A}^T\mathbf{e} = \mathbf{0} \in \mathbb{R}^N$ or $\mathbf{A}^T(\mathbf{A}\hat{\mathbf{x}} - \mathbf{b}) = \mathbf{0}$. Rearranged, this becomes the "normal equations" of statistics:

$$\mathbf{A}^T\mathbf{A}\hat{\mathbf{x}} = \mathbf{A}^T\mathbf{b}. \tag{9.45}$$

It is easiest to visualize the problem with a full rank $\mathbf{A} \in \mathbb{R}^{3\times 2}$. This choice guarantees that the problem can be visualized in a three-dimensional space, the highest dimensional space readily grasped by the human brain and visual system. Because $q = 2$, $\mathcal{R}(\mathbf{A})$ is a plane. This leaves one remaining dimension in $\mathbb{R}^M$ for $\mathcal{N}(\mathbf{A}^T)$, which is therefore a line. For ease of both visualization and the associated algebra, we will also choose an orthonormal $\mathbf{A}$. There is no loss of generality here: from the fact that $\mathbf{A}$ is full rank and $M > N$, we know that $\mathbf{A}$'s columns are linearly independent. Therefore, even if $\mathbf{a}_1^T\mathbf{a}_2 \neq 0$, a modified $\mathbf{a}_2$, call it $\tilde{\mathbf{a}}_2$, satisfying $\mathbf{a}_1^T\tilde{\mathbf{a}}_2 = 0$ can be readily derived by subtracting from $\mathbf{a}_2$ its part in the direction of $\mathbf{a}_1$ (i.e., using the Gram-Schmidt process),

$$\tilde{\mathbf{a}}_2 = \mathbf{a}_2 - \left(\frac{\mathbf{a}_2^T\mathbf{a}_1}{\mathbf{a}_1\mathbf{a}_1}\right)\mathbf{a}_1. \tag{9.46}$$

We are now ready to introduce some examples that will help us develop the idea. For clarity, let's begin with an idealized problem whose $\mathbf{A}$'s columns are orthonormal.

Consider the rank $q = 2$ inconsistent problem

$$\mathbf{A} = \begin{pmatrix} -0.9759 & 0.1659 \\ 0.0976 & 0.9127 \\ 0.1952 & 0.3734 \end{pmatrix} = \begin{pmatrix} \hat{\mathbf{a}}_1 & \hat{\mathbf{a}}_2 \end{pmatrix}, \qquad \mathbf{b} = \begin{pmatrix} -0.2659 \\ 0.8002 \\ 0.6475 \end{pmatrix}, \tag{9.47}$$

with orthonormal columns ($\mathbf{A}^T\mathbf{A} = \mathbf{I}_2$) forming a suitable spanning set for the column space (basis$[\mathcal{R}(\mathbf{A})] = \{\hat{\mathbf{a}}_1, \hat{\mathbf{a}}_2\}$). By solving $\mathbf{A}^T\mathbf{n} = \mathbf{0} \in \mathbb{R}^N$, we also obtain

$$\text{basis}\left[\mathcal{N}(\mathbf{A}^T)\right] = \hat{\mathbf{n}} = \begin{pmatrix} 0.1417 \\ -0.3968 \\ 0.9069 \end{pmatrix}. \tag{9.48}$$

This problem is shown (exactly, not schematically) in fig. 9.3.

The $\mathcal{R}(\mathbf{A})$ plane is slanted with respect to each of the Cartesian planes ($x = x_0$, $y = y_0$, and $z = z_0$). As a result, $\mathbf{b}$'s vertical Cartesian "shadow" ($(\mathbf{b}^T\hat{\mathbf{z}})\hat{\mathbf{z}}$, the vertical dashed gray line in fig. 9.3 between $(b_x, b_y, 0)$ and (b_x, b_y, b_z),

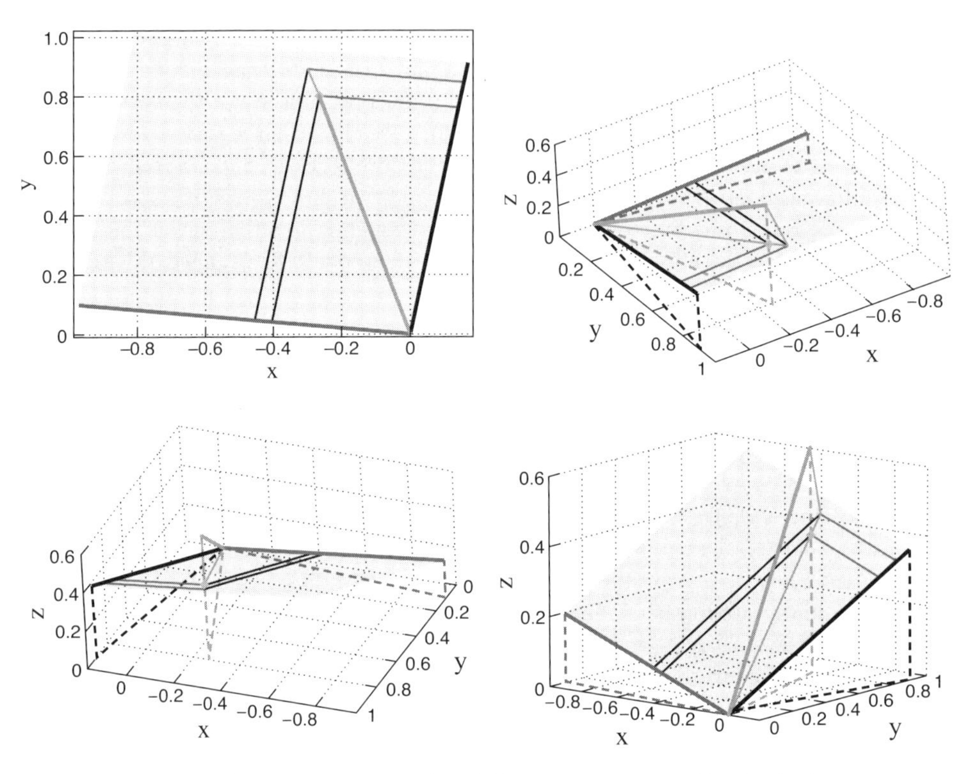

Figure 9.3. Visual representation from four different points of view of the first (orthonormal) 3×2 example best-fit problem discussed in the text. The upper left panel is a plane view from above; all other panels emulate three-dimensional perspectives. Vectors are shown as the line segments connecting the origin, (0, 0, 0), with their "heads" at (x, y, z). Solid thick dark gray and black coordinate lines parallel $\mathbf{a}_1$ ($\mathbf{a}_2$), $\mathbf{A}$'s first (second) column. The right-hand-side vector $\mathbf{b}$ is shown in thick gray. The Cartesian "shadows" (i.e., projections on the $z = 0$ plane, and the orthogonal remainder along $\hat{\mathbf{z}}$) of $\hat{\mathbf{a}}_1$, $\hat{\mathbf{a}}_2$, and $\mathbf{b}$ are shown with dashed lines. The semi-transparent light gray surface is $\mathcal{R}(\mathbf{A})$. The gray diamond shows the intersection of $\mathbf{b}$'s Cartesian vertical component (the vertical dashed gray line) and $\mathcal{R}(\mathbf{A})$. The components of $\mathbf{b}$ parallel, and normal, to $\mathcal{R}(\mathbf{A})$ are given by the thinner solid gray lines. The $\mathbf{a}_1$ and $\mathbf{a}_2$ coordinates of $\mathbf{b}$'s Cartesian and orthogonal projection on $\mathcal{R}(\mathbf{A})$ are given by the two pairs of black and gray right angles contained within $\mathcal{R}(\mathbf{A})$.

intersecting $\mathcal{R}(\mathbf{A})$ at the gray diamond) is not the shortest distance from $\mathbf{b}$ to $\mathcal{R}(\mathbf{A})$. Rather, as discussed above, this shortest distance is measured along the line normal to $\mathcal{R}(\mathbf{A})$ passing through $\mathbf{b}$, shown in fig. 9.3 by the thin solid gray line normal to the $\mathcal{R}(\mathbf{A})$ plane. Because $\mathbf{A}^T\mathbf{A} = \mathbf{I}$, the point at which that line intersects $\mathcal{R}(\mathbf{A})$ is

$$\mathbf{b}_r = \mathbf{A}\mathbf{A}^T\mathbf{b} \equiv \left(\hat{\mathbf{a}}_1^T\mathbf{b}\right)\hat{\mathbf{a}}_1 + \left(\hat{\mathbf{a}}_2^T\mathbf{b}\right)\hat{\mathbf{a}}_2, \tag{9.49}$$

the projection of $\mathbf{b}$ onto $\mathcal{R}(\mathbf{A})$, shown in fig. 9.3 by the thin solid gray line within the $\mathcal{R}(\mathbf{A})$ plane. The $\mathbf{a}_1$ and $\mathbf{a}_2$ coordinates in $\mathcal{R}(\mathbf{A})$ of that point, as well as of the point denoted by the diamond, are shown in fig. 9.3 by solid thin black and dark gray lines, respectively.

Let's next examine a more general problem, by lifting $\mathbf{A}$'s orthonormality requirement, analyzing

$$\mathbf{A} = \begin{pmatrix} -0.2 & -0.8 \\ 0.6 & 4.0 \\ 0.8 & 1.0 \end{pmatrix} = \begin{pmatrix} \mathbf{a}_1 & \mathbf{a}_2 \end{pmatrix}, \qquad \mathbf{b} = \begin{pmatrix} -1.3756 \\ 3.4063 \\ 1.6063 \end{pmatrix}, \tag{9.50}$$

still with $q = 2$, but with

$$\mathbf{A}^T\mathbf{A} = \begin{pmatrix} 5.00 & 4.80 \\ 4.80 & 17.64 \end{pmatrix} \neq \mathbf{I}. \tag{9.51}$$

It will be useful to obtain an orthonormal basis$[\mathcal{R}(\mathbf{A})] = \{\hat{\mathbf{s}}_1, \hat{\mathbf{s}}_2\}$,

$$\hat{\mathbf{s}}_1 = \frac{\mathbf{a}_1}{\|\mathbf{a}_1\|}, \qquad \mathbf{s}_2 = \mathbf{a}_2 - \left(\mathbf{a}_2^T\hat{\mathbf{a}}_1\right)\hat{\mathbf{a}}_1, \qquad \hat{\mathbf{s}}_2 = \frac{\mathbf{s}_2}{\|\mathbf{s}_2\|}, \tag{9.52}$$

and its $\mathbb{R}^3$ complement basis$[\mathcal{N}(\mathbf{A}^T)] = \hat{\mathbf{n}}$ solving $\mathbf{A}^T\hat{\mathbf{n}} = \mathbf{0} \in \mathbb{R}^2$. These three orthonormal vectors are shown in fig. 9.4 by cyan, magenta, and black, respectively.

In this problem, because $\mathbf{A}^T\mathbf{A} \neq \mathbf{I}$, eq. 9.49 is modified to the general expression for the right-hand side reproduced by the least-squares (LS) solution,

$$\mathbf{b}_r \equiv \hat{\mathbf{b}} = \mathbf{A}\hat{\mathbf{x}} = \mathbf{A}\left(\mathbf{A}^T\mathbf{A}\right)^{-1}\mathbf{A}^T\mathbf{b}, \tag{9.53}$$

where the modification of (9.53) relative to (9.49) is necessary to account for $\mathbf{A}$'s nonorthonormality, i.e., for the possibility of $\|\mathbf{a}_1\| \neq 1$, $\|\mathbf{a}_2\| \neq 1$, or $\mathbf{a}_1^T\mathbf{a}_2 \neq 0$. To demonstrate the effect of either of these conditions on the projection of $\mathbf{b}$ onto $\mathcal{R}(\mathbf{A})$, let's consider the general $\mathbf{A} \in \mathbb{R}^{M\times 2}$ case, in which

$$\mathbf{A}^T\mathbf{A} = \begin{pmatrix} \mathbf{a}_1^T\mathbf{a}_1 & \mathbf{a}_1^T\mathbf{a}_2 \\ \mathbf{a}_1^T\mathbf{a}_2 & \mathbf{a}_2^T\mathbf{a}_2 \end{pmatrix} \tag{9.54}$$

so that

$$\left(\mathbf{A}^T\mathbf{A}\right)^{-1} = \frac{1}{\|\mathbf{a}_1\|^2\|\mathbf{a}_2\|^2 - 2\mathbf{a}_1^T\mathbf{a}_2} \begin{pmatrix} \mathbf{a}_2^T\mathbf{a}_2 & -\mathbf{a}_1^T\mathbf{a}_2 \\ -\mathbf{a}_1^T\mathbf{a}_2 & \mathbf{a}_1^T\mathbf{a}_1 \end{pmatrix}. \tag{9.55}$$

In this case, the LS solution is

$$\begin{aligned} \hat{\mathbf{x}} &= \left(\mathbf{A}^T\mathbf{A}\right)^{-1}\mathbf{A}^T\mathbf{b} \\ &= \frac{1}{\|\mathbf{a}_1\|^2\|\mathbf{a}_2\|^2 - 2\mathbf{a}_1^T\mathbf{a}_2} \begin{pmatrix} \mathbf{a}_2^T\mathbf{a}_2 & -\mathbf{a}_1^T\mathbf{a}_2 \\ -\mathbf{a}_1^T\mathbf{a}_2 & \mathbf{a}_1^T\mathbf{a}_1 \end{pmatrix} \begin{pmatrix} \mathbf{a}_1^T\mathbf{b} \\ \mathbf{a}_2^T\mathbf{b} \end{pmatrix} \end{aligned} \tag{9.56}$$

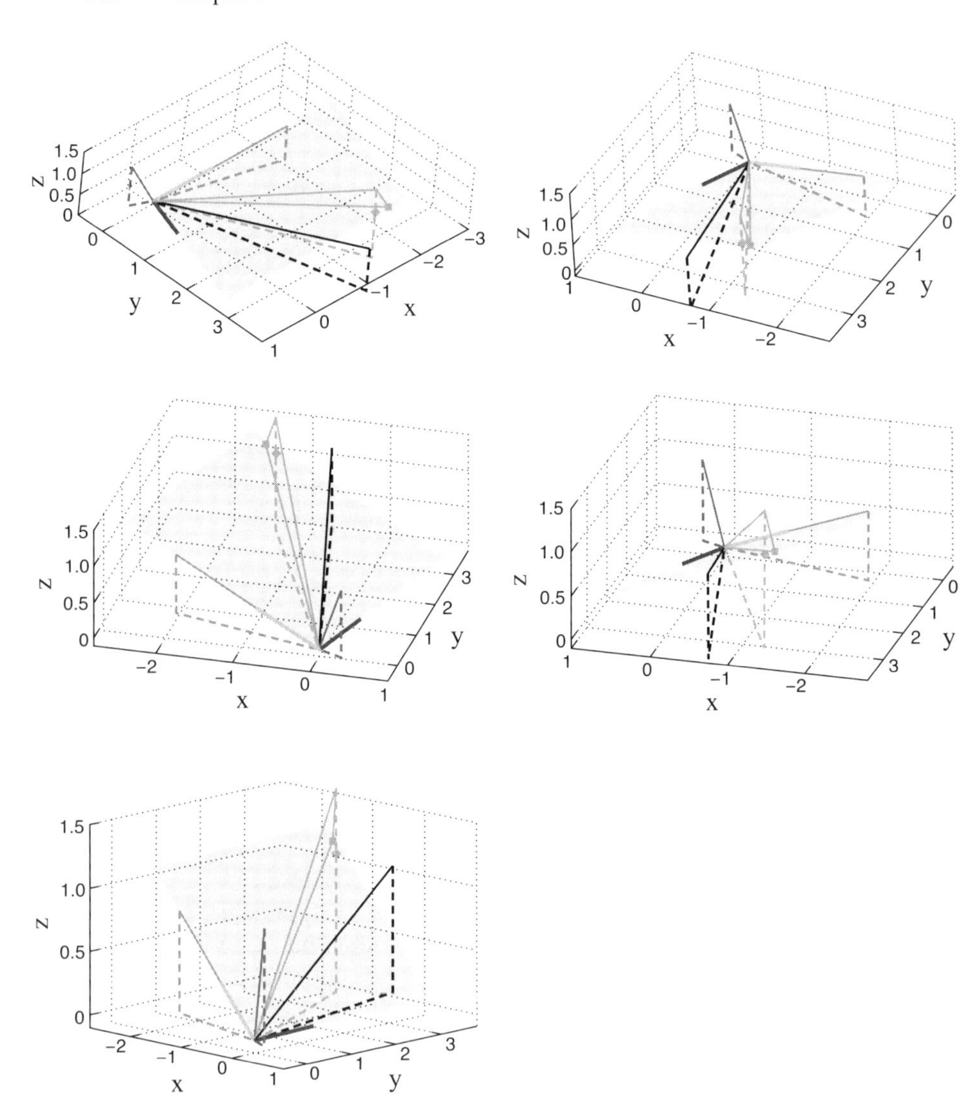

Figure 9.4. Visual representation from various view points of the second (nonorthonormal) 3×2 example best-fit problem discussed in the text. Vectors are again shown as line segments connecting (0, 0, 0) with (x, y, z), and $\mathcal{R}(\mathbf{A})$ is still the semi-transparent gray surface. Solid dark gray (black, lighter gray) lines show $\mathbf{a}_1$ ($\mathbf{a}_2, \mathbf{b}$). Thick light (dark) gray shows the two elements $\hat{\mathbf{s}}_1$ and $\hat{\mathbf{s}}_2$ of an orthonormal spanning set for $\mathcal{R}(\mathbf{A})$, with thin dark gray showing a normalized spanning vector for their $\mathbb{R}^M$ complement, $\mathcal{N}(\mathbf{A}^T)$. Cartesian "shadows" (i.e., projections on the $z = 0$ plane, and the orthogonal remainder along $\hat{\mathbf{z}}$) of $\mathbf{a}_1$, $\mathbf{a}_2$, and $\mathbf{b}$ are again shown with dashed lines. The gray diamond shows the intersection of $\mathbf{b}$'s Cartesian vertical component (the vertical dashed light gray line) and $\mathcal{R}(\mathbf{A})$. The components of $\mathbf{b}$ from $\mathcal{R}(\mathbf{A})$ and $\mathcal{N}(\mathbf{A}^T)$ are given by thinner intermediate gray lines. The least-squares solution, the intersection of $\mathbf{b}$'s $\mathcal{N}(\mathbf{A}^T)$ component with $\mathcal{R}(\mathbf{A})$, is shown by the gray square.

$$= \frac{1}{\|\mathbf{a}_1\|^2\|\mathbf{a}_2\|^2 - 2\mathbf{a}_1^T\mathbf{a}_2}\begin{pmatrix} \|\mathbf{a}_2\|^2\mathbf{a}_1^T\mathbf{b} - (\mathbf{a}_1^T\mathbf{a}_2)(\mathbf{a}_2^T\mathbf{b}) \\ \|\mathbf{a}_1\|^2\mathbf{a}_2^T\mathbf{b} - (\mathbf{a}_1^T\mathbf{a}_2)(\mathbf{a}_1^T\mathbf{b}) \end{pmatrix}. \tag{9.57}$$

If **A**'s columns are orthogonal but not normalized,

$$(\mathbf{A}^T\mathbf{A})^{-1}\mathbf{A}^T\mathbf{b} = \frac{1}{\|\mathbf{a}_1\|^2\|\mathbf{a}_2\|^2}\begin{pmatrix} \|\mathbf{a}_2\|^2\mathbf{a}_1^T\mathbf{b} \\ \|\mathbf{a}_1\|^2\mathbf{a}_2^T\mathbf{b} \end{pmatrix} = \begin{pmatrix} \frac{\mathbf{a}_1^T\mathbf{b}}{\mathbf{a}_1^T\mathbf{a}_1} \\ \frac{\mathbf{a}_2^T\mathbf{b}}{\mathbf{a}_2^T\mathbf{a}_2} \end{pmatrix}. \tag{9.58}$$

If, in addition, the columns are normalized,

$$(\mathbf{A}^T\mathbf{A})^{-1}\mathbf{A}^T\mathbf{b} = \begin{pmatrix} \hat{\mathbf{a}}_1^T\mathbf{b} \\ \hat{\mathbf{a}}_2^T\mathbf{b} \end{pmatrix} = \mathbf{A}^T\mathbf{b}, \tag{9.59}$$

solution (9.49).

9.3.3 *A More General, Weighted, Dual Minimization Formalism*

Above, we treated all errors (**e**'s elements) as equals, which implies that we treat all observations as similarly known. Yet, in some situations this may not be true, as some observations are better constrained than others. Further, even if we have little reason to favor some errors over others, it may be that different variables (to which parts of **b** correspond) exhibit widely disparate physically plausible ranges of variability. In either of the above two situations, it will make sense to weight errors a priori. In addition, if the problem is not fully determined, and thus has an $\mathbb{R}^N$ solution hyperplane, it may make sense to choose a specific solution from this hyperplane that also minimizes the weighted solution norm.

Taking these objectives together, the cost function we wish to minimize, previously $\|\mathbf{A}\hat{\mathbf{x}} - \mathbf{b}\|$, becomes

$$J = \underbrace{(\mathbf{A}\hat{\mathbf{x}} - \mathbf{b})^T\mathbf{R}^{-1}(\mathbf{A}\hat{\mathbf{x}} - \mathbf{b})}_{\substack{\text{minimize weighted} \\ \mathbb{R}^M \text{ error}}} + \underbrace{\hat{\mathbf{x}}^T\mathbf{M}^{-1}\hat{\mathbf{x}}}_{\substack{\text{minimize} \\ \text{weighted } \|\hat{\mathbf{x}}\|}}, \tag{9.60}$$

where the (typically diagonal) **R** quantifies the expected variability of individual elements in **b** or **e**, and **M** does the same for $\hat{\mathbf{x}}$'s individual elements.

Following the logic that guided us in minimizing the uniform error, and using only slightly different derivations, requiring individual element of J's gradient to vanish yields

$$\frac{dJ}{d\hat{\mathbf{x}}} = 2\mathbf{A}^T\mathbf{R}^{-1}\mathbf{A}\hat{\mathbf{x}} - 2\mathbf{A}^T\mathbf{R}^{-1}\mathbf{b} + 2\mathbf{M}^{-1}\hat{\mathbf{x}} = \mathbf{0} \in \mathbb{R}^N \tag{9.61}$$

or, since the assumed independent elements must vanish individually,

$$\mathbf{A}^T\mathbf{R}^{-1}\mathbf{A}\hat{\mathbf{x}} + \mathbf{M}^{-1}\hat{\mathbf{x}} = \mathbf{A}^T\mathbf{R}^{-1}\mathbf{b}. \tag{9.62}$$

Combining the left-hand terms,

$$\left(\mathbf{A}^T\mathbf{R}^{-1}\mathbf{A} + \mathbf{M}^{-1}\right)\hat{\mathbf{x}} = \mathbf{A}^T\mathbf{R}^{-1}\mathbf{b}, \tag{9.63}$$

we obtain the general dual LS solution of the weighted problem,

$$\hat{\mathbf{x}} = \left(\mathbf{A}^T\mathbf{R}^{-1}\mathbf{A} + \mathbf{M}^{-1}\right)^{-1}\mathbf{A}^T\mathbf{R}^{-1}\mathbf{b}. \tag{9.64}$$

Note that if $\mathbf{R} = \mathbf{I} \in \mathbb{R}^M$ and $\mathbf{M} = \mathbf{I} \in \mathbb{R}^N$, this reduces to the nonweighted LS solution, as required. Note also that the rationale for minimizing $\|\hat{\mathbf{x}}\|$—typically in the name of parsimony, or simplicity of the least conjectural solution (as discussed in the next section)—is not as straightforward or logically compelling as minimizing the error and can often be partially waived. Finally, and a bit cynically, adding $\mathbf{M}^{-1}$ to the inverted term can be viewed as no more than sprinkling added power, often with little justification, on the diagonal of an otherwise singular problem, to yield *a* solution. That this solution is robust must be *shown*, not assumed, by the prudent analyst employing (9.64).

9.4 Least Squares: The SVD View

Returning now to the uniform weight problem, for full rank problems, $q = N$, $\mathbf{A}^T\mathbf{A}$ is invertible, and $\hat{\mathbf{x}} = (\mathbf{A}^T\mathbf{A})^{-1}\mathbf{A}^T\mathbf{b}$ is the LS (least-squares) solution. If $\mathbf{A}^T\mathbf{A}$ is singular, this solution does not exist. If $\mathbf{A}^T\mathbf{A}$ is ill-conditioned, $\mathbf{A}^T\mathbf{A}^{-1}$ formally exists, but comprises some poorly determined parts. In the singular or ill-posed cases, the classical LS solution needs modification. This is best achieved using the SVD representation of $\mathbf{A}$, as follows.

Consider the SVD representation $\mathbf{A} = \mathbf{U}\boldsymbol{\Sigma}\mathbf{V}^T$, where $\mathbf{U}$ and $\mathbf{V}$ are orthonormal. Then,

$$\hat{\mathbf{x}} = (\mathbf{A}^T\mathbf{A})^{-1}\mathbf{A}^T\mathbf{b} = \mathbf{V}(\boldsymbol{\Sigma}^T\boldsymbol{\Sigma})^{-1}\boldsymbol{\Sigma}^T\mathbf{U}^T\mathbf{b} \equiv \mathbf{V}\mathbf{D}\mathbf{U}^T\mathbf{b}, \tag{9.65}$$

where $\mathbf{D} \equiv (\boldsymbol{\Sigma}^T\boldsymbol{\Sigma})^{-1}\boldsymbol{\Sigma}^T$, provided $\boldsymbol{\Sigma}^T\boldsymbol{\Sigma}$ is invertible. If $\mathbf{A}$ is rank deficient, some diagonal elements of $\boldsymbol{\Sigma}$ vanish and $\boldsymbol{\Sigma}^T\boldsymbol{\Sigma}$ is singular. The need for N independent columns in $\mathbf{A}$ (assuming $M > N$) for the existence of the LS solution is now better understood. It is necessary for the existence of $(\mathbf{A}^T\mathbf{A})^{-1}$, or the absence of an $\mathbb{R}^N$ null space; $\mathcal{R}(\mathbf{A}^T) = \mathbb{R}^N$. The SVD representation permits broadening the discussion to encompass all possible null spaces and inconsistencies in the problem, distinguishing four cases based on the relative values of q, M and N.

1. *Underdetermined*, $M = q < N$. Since $\mathbf{U}$ has $M = q$ columns, the first q columns, which span $\mathcal{R}(\mathbf{A})$, completely fill $\mathbf{U}$, and no columns are left to span $\mathcal{N}(\mathbf{A}^T)$, which is thus empty; $\mathcal{R}(\mathbf{A}) = \mathbb{R}^M$. On the other hand, $\mathbf{V}$ has N columns, the first q of which span $\mathcal{R}(\mathbf{A}^T)$, and the remaining $N - q$ columns span $\mathcal{N}(\mathbf{A})$. The problem's null space is in $\mathbb{R}^N$.
2. *Overdetermined*, $N = q < M$. The situation is reversed, $\mathbf{V}$'s columns span $\mathcal{R}(\mathbf{A}^T) = \mathbb{R}^N$, leaving $\mathcal{N}(\mathbf{A})$ empty, but $\mathbf{U}$'s $M - q$ last columns span a nontrivial $\mathcal{N}(\mathbf{A}^T)$. The null space of the problem is now in $\mathbb{R}^M$.
3. *Rank deficient*, $q < N < M$. Now both $\mathbf{U}$ and $\mathbf{V}$ have a number of columns that is sufficient to accommodate both a range and a null space.

The problem has null spaces in both $\mathbb{R}^M$ and $\mathbb{R}^N$ and is not invertible. Two possible subcases can be distinguished based on the properties of $\mathbf{b}$.

- *Rank deficient consistent*, $\mathbf{b} \in \mathcal{R}(\mathbf{A})$. Solutions exist but are not unique. The solution is a hyperplane in $\mathbb{R}^N$. All $\mathbf{b}$'s elements corresponding to redundant rows in $\mathbf{A}$ are also zeroed out by the same elementary operations of the Gaussian elimination that zeroed out the rows.
- *Rank deficient inconsistent*, $\mathbf{b} \notin \mathcal{R}(\mathbf{A})$. No exact solution exists, but infinitely many LS ones do.

4. *Well posed*, $q = N = M$. Both $\mathbf{U}$ and $\mathbf{V}$ have only an $\mathcal{R}$-space. The problem has no null space.

9.4.1 "Inversion" of Singular Problems: The Informal View

Case 3 (singular $\boldsymbol{\Sigma}^T\boldsymbol{\Sigma}$) is the most frequently encountered, yet this is just the case for which no LS solution exists. We need to replace $(\boldsymbol{\Sigma}^T\boldsymbol{\Sigma})^{-1}\boldsymbol{\Sigma}^T$ with something so as to extend the conditions for the existence of an inverse to include rank deficient matrices.

To this end, let's consider first the full rank case (assuming $q = N < M$). Then,

$$\hat{\mathbf{x}} = \mathbf{V}\mathbf{D}\mathbf{U}^T\mathbf{b} \tag{9.66}$$

$$= \underbrace{\begin{pmatrix} \vdots & & \vdots \\ \mathbf{v}_1 & \cdots & \mathbf{v}_N \\ \vdots & & \vdots \end{pmatrix}}_{\mathbb{R}^{N\times N}} \underbrace{\begin{pmatrix} \sigma_1^{-1} & & & \vdots \\ & \ddots & & \cdots 0 \cdots \\ & & \sigma_N^{-1} & \vdots \end{pmatrix}}_{\mathbb{R}^{N\times M}} \underbrace{\begin{pmatrix} \mathbf{u}_1^T\mathbf{b} \\ \vdots \\ \mathbf{u}_N^T\mathbf{b} \\ \mathbf{u}_{N+1}^T\mathbf{b} \\ \vdots \\ \mathbf{u}_M^T\mathbf{b} \end{pmatrix}}_{\mathbb{R}^{M\times 1}}$$

$$= \underbrace{\begin{pmatrix} \vdots & & \vdots & \vdots \\ \sigma_1^{-1}\mathbf{v}_1 & \cdots & \sigma_N^{-1}\mathbf{v}_N & \cdots 0 \cdots \\ \vdots & & \vdots & \vdots \end{pmatrix}}_{\mathbb{R}^{N\times M}} \underbrace{\begin{pmatrix} \mathbf{u}_1^T\mathbf{b} \\ \vdots \\ \mathbf{u}_N^T\mathbf{b} \\ \mathbf{u}_{N+1}^T\mathbf{b} \\ \vdots \\ \mathbf{u}_M^T\mathbf{b} \end{pmatrix}}_{\mathbb{R}^{M\times 1}} = \sum_{i=1}^{N} \left(\frac{\mathbf{u}_i^T\mathbf{b}}{\sigma_i} \right) \mathbf{v}_i.$$

This representation demonstrates the problem: when the sum range i includes modes whose $\sigma_i = 0$, the scalar weighting factor for $\mathbf{v}_i$ in the sum is infinite. The more usually encountered case is even worse, with σ_i that correspond to underdetermined modes being not exactly zero, but very small. Then, the weighting factor becomes very large, multiplying essentially undetermined structures ($\mathbf{v}_i$) to completely dominate—and corrupt—the solution. This also suggests the cure: truncate the sum at some safe upper bound $c < q \leq \min(M, N)$, excluding all poorly defined modes. Then

$$\mathbf{x}^+ = \sum_{i=1}^{c<q} \left(\frac{\mathbf{u}_i^T \mathbf{b}}{\sigma_i} \right) \mathbf{v}_i = \mathbf{V}\Sigma^+ \mathbf{U}^T \mathbf{b} \equiv \mathbf{A}^+ \mathbf{b}, \tag{9.67}$$

where $\mathbf{A}^+$ is $\mathbf{A}$'s Moore-Penrose pseudo-inverse defined implicitly above and more rigorously in the following section. Now the summation limit deliberately excludes poorly determined modes. I discuss the choice of c a bit later.

The above expression applies to a general linear system, with arbitrary dimensions and rank. We map $\mathbf{b}$ onto $\mathcal{R}(\mathbf{A})$, by which we overcome the inconsistencies. Next we note that $\mathbf{A}$'s rank deficiency renders the solution nonunique, an $\mathbb{R}^N$ hyperplane. Because $\mathbb{R}^N$ contains a nontrivial null space, any arbitrary combination of null space vectors satisfies the conditions of the problem (chooses a linear combination of $\mathbf{A}$'s columns that reproduces the problem's new right-hand side, $\mathbf{b}$'s projection on $\mathcal{R}(\mathbf{A})$). The way the SVD solution of the general linear problem chooses a unique solution among the infinitely many possible solutions is through minimization of the solution norm by exclusion of null space vectors. The SVD is similar in this respect to the dual weighted minimization and is the solution of the two simultaneous optimization problems:

- minimize $\|\mathbf{A}\hat{\mathbf{x}} - \mathbf{b}\|$, and
- minimize $\|\hat{\mathbf{x}}\|$.

That is, given the inconsistent problem $\mathbf{Ax} = \mathbf{b} \in \mathcal{R}(\mathbf{A})$, we look for the closest possible solvable problem $\mathbf{Ax} = \mathbf{b}_r$, where $\mathbf{b}_r \equiv \mathbf{b}_{\mathcal{R}(\mathbf{A})}$ is the projection of $\mathbf{b}$ onto $\mathcal{R}(\mathbf{A})$. The new problem $\mathbf{Ax} = \mathbf{b}_r$ can be well posed or, more realistically, underdetermined, in which case we minimize the solution norm. Though these two minimizations seem similar, the rationale behind them is quite different. Error $\|\mathbf{e}\|$ minimization is based on our confidence in the validity of the physical model. Since it is adequate, we choose parameters that minimize the minor errors. The choice of the shortest possible $\hat{\mathbf{x}}$, $\mathbf{x}^+$, as the solution of the underdetermined problem $\mathbf{Ax} = \mathbf{b}_r$, is much less physically sound and is only defensible as a means of enforcing parsimony, including in the solution only the minimum required to satisfy the problem, with no $\mathbb{R}^N$ null space component.

9.4.2 *The Choice of c, the Spectral Truncation Point*

In practice, most real-life singular spectra do not have a clear cutoff point separating an obvious range from an obvious null space. Rather, the singular values gradually decline but never quite vanish altogether. Typically, singular spectra exhibit three distinct regions (fig. 9.5): (1) a rapidly declining region, typically spanning the first 5–7% of the total number of modes, $\min(M, N)$; (2) a relatively

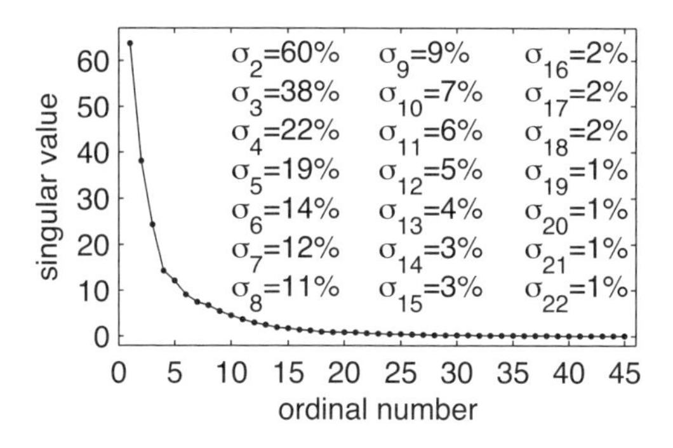

Figure 9.5. A typical singular spectrum, showing σ_i as a function of i. The values of singular values 2 through 22 are reported in the figure as percentages of the leading mode (σ_1; i.e., the shown numbers are $100\sigma_i/\sigma_1$ for $2 \le i \le 22$).

narrow (i.e., comprising relatively small number of modes) "kink" in which the decline of singular values slows considerably (roughly modes 4–12 in fig. 9.5); and (3) a broad plateau of very slow decline of singular values, comprising the remaining right-hand tail of the spectrum. Trailing singular values are rarely outright zero, but are instead much smaller than the leading ones. The ratio σ_1/σ_q of the largest to smallest singular value is the matrix *condition number*, and a span of 2–16 orders of magnitude in singular values is not uncommon.

In the case $\sigma_i > 0 \,\forall i$, there is no formal null space in the smaller of the two spaces (i.e., in $\mathbb{R}^{\min(M,N)}$), but the leading few modes are many orders of magnitude better known than the trailing ones. Thus, depending on the nature of the problem, it would make sense to truncate the singular spectrum at some point. The choice of the cutoff point c is arbitrary and has no objective scientific truth behind it. Often the choice is made based on visual examination of the spectrum, identifying the "kink" and choosing a c near the "kink"'s left end. Others use a criterion like

$$c : 100\frac{\sum_{i=1}^{c} \sigma_i}{\sum_{i=1}^{\min(M,N)} \sigma_i} \ge 85\% \tag{9.68}$$

or similar values that strive to retain most of the useful information while discarding the noise-contaminated right-hand tail.

9.4.3 *"Inversion" of Singular Problems: The Formal View*

For singular cases, define $\boldsymbol{\Sigma}^+ \equiv \lim_{\epsilon \to 0} (\epsilon I_N + \boldsymbol{\Sigma}^T\boldsymbol{\Sigma})^{-1}\boldsymbol{\Sigma}^T$, with the aid of which we generalize the concept of the inverse of a matrix $\mathbf{A}$ (with arbitrary dimensions and rank) to the natural inverse, $\mathbf{A}^+ \equiv \mathbf{V}\boldsymbol{\Sigma}^+\mathbf{U}^T$. With the parameter-dependent diagonal $N \times N$ matrix added to the singular product $\boldsymbol{\Sigma}^T\boldsymbol{\Sigma}$, the LS solution becomes

$$\hat{\mathbf{x}} = \mathbf{V}\left[(\epsilon \mathbf{I}_N + \boldsymbol{\Sigma}^T\boldsymbol{\Sigma})^{-1}\boldsymbol{\Sigma}^T\right]\mathbf{U}^T\mathbf{b}. \tag{9.69}$$

The right-hand side's brackets contain a product of two diagonal matrices. The first is the inverse of an $N \times N$ diagonal matrix, with the first q diagonal elements being $\epsilon + \sigma^2_i$, $1 \leq i \leq q$, and the remaining $N - q$ diagonal elements being ϵ. Since it is a full rank, square diagonal, its inverse exists, with each diagonal element inverted. Thus, $(\epsilon \mathbf{I}_N + \mathbf{\Sigma}^T \mathbf{\Sigma})^{-1}$ has $1/(\epsilon + \sigma^2_i)$, $1 \leq i \leq q$, in the first q diagonal positions, $1/\epsilon$ in the last $N - q$ diagonal positions, and zeros off-diagonally,

$$(\epsilon \mathbf{I}_N + \mathbf{\Sigma}^T \mathbf{\Sigma})^{-1} = \begin{pmatrix} \epsilon + \sigma^2_1 & & & & & \\ & \ddots & & & & \\ & & \epsilon + \sigma^2_q & & & \\ & & & \epsilon & & \\ & & & & \ddots & \\ & & & & & \epsilon \end{pmatrix}^{-1} \tag{9.70}$$

$$= \begin{pmatrix} \frac{1}{\epsilon + \sigma^2_1} & & & & & \\ & \ddots & & & & \\ & & \frac{1}{\epsilon + \sigma^2_q} & & & \\ & & & \frac{1}{\epsilon} & & \\ & & & & \ddots & \\ & & & & & \frac{1}{\epsilon} \end{pmatrix} \in \mathbb{R}^{N \times N}.$$

The second term in eq. 9.69's brackets, $\mathbf{\Sigma}^T$, is $N \times M$ with the first q nonzero diagonal elements being σ_i, $1 \leq i \leq q$, and zeros everywhere else. Pictorially, assuming $q < N < M$, the matrix is

$$\mathbf{\Sigma}^T = \begin{pmatrix} \sigma_1 & & & & \vdots & \\ & \ddots & & & \vdots & \\ & & \sigma_q & \cdots & 0 & \cdots \\ & \vdots & & & \vdots & \\ \cdots & 0 & \cdots & \cdots & 0 & \cdots \\ & \vdots & & & \vdots & \end{pmatrix} \in \mathbb{R}^{N \times M}. \tag{9.71}$$

It is premultiplied by $(\epsilon \mathbf{I}_N + \mathbf{\Sigma}^T \mathbf{\Sigma})^{-1}$ described above. The result is an $N \times M$ matrix with $\sigma_i/(\epsilon + \sigma^2_i)$, $1 \leq i \leq q$, in the first q diagonal positions, and zeros everywhere else (since the last $N - q$ rows of $\mathbf{\Sigma}^T$ map the $1/\epsilon$ $N - q$ diagonal elements of the first matrix to zero). If we now take $\lim_{\epsilon \to 0} (\epsilon \mathbf{I} + \mathbf{\Sigma}^T \mathbf{\Sigma})^{-1} \mathbf{\Sigma}^T$, we get an $N \times M$ matrix whose first q diagonal elements are $\lim_{\epsilon \to 0} \left[\sigma_i/(\epsilon + \sigma^2_i) \right] = 1/\sigma_i$, with zeros elsewhere. We call this matrix $\mathbf{\Sigma}^+$ and use it to define $\mathbf{A}^+$. With this $\mathbf{\Sigma}^+$,

$$\mathbf{x}^+ = \mathbf{A}^+ \mathbf{b} = \mathbf{V} \mathbf{\Sigma}^+ \mathbf{U}^T \mathbf{b}, \tag{9.72}$$

the SVD LS solution of the general linear problem $\mathbf{A}\mathbf{x} = \mathbf{b}$.

9.5 Some Special Problems Giving Rise to Linear Systems

To further develop our understanding of linear systems and to emphasize the importance of the relative magnitudes of M, N and q in the $\mathbf{Ax}=\mathbf{b}$ system, I present here some special cases of the problem.

9.5.1 *Regression vs. Interpolation*

In formulating the LS solution, we assumed our model was sound and sensible, yet imperfect. Consequently, we were willing to settle for a functional fit that does not exactly visit every data point, but is instead reasonably close to all of them. For physically based models, this approach almost universally applies; no model is, or should be considered, perfect. At other times, however, we are motivated not by the desire to better understand the problem's physics, but by the need to deduce from function value observations at discrete independent variable points a continuous function that visits exactly each of the observations yet is continuous throughout the independent variables' domain. This most naturally arises in the context of "table look-up," in which we estimate the value of a function at a specific point, allowing for the possibility that *all* available data points may matter to the function value at the point of interest, not just its immediate neighbors. For the continuous interpolant to visit exactly each of the N data points, the interpolant needs to comprise N independent terms, of which an Nth degree polynomial is a common example.

Let's consider a simple one-dimensional problem of $y=y(x)$, with the $N=7$ discrete $(x_i, y_i)=[x_i, y(x_i)]$ pairs $(0,1)$, $(1,3)$, $(2,4)$, $(3,2)$, $(4,4)$, $(5,5)$, $(6,6)$. Suppose we need to devise an interpolant that fills in the values of y at unobserved x values while passing exactly through the observed points. Let's consider two examples based on the polynomial and cosine bases.

In the general N-point polynomial interpolation problem, we express y_i in terms of a power series in x_i,

$$y_i = \sum_{j=0}^{N-1} c_j x_i^j, \tag{9.73}$$

where i is an index, while the j superscript denotes the jth power. The inversion problem for the coefficients $\{c_j\}$ is

$$\underbrace{\begin{pmatrix} x_1^{N-1} & x_1^{N-2} & x_1^{N-3} & \cdots & x_1^1 & x_1^0 \\ x_2^{N-1} & x_2^{N-2} & x_2^{N-3} & \cdots & x_2^1 & x_2^0 \\ & & \vdots & & & \\ x_{N-1}^{N-1} & x_{N-1}^{N-2} & x_{N-1}^{N-3} & \cdots & x_{N-1}^1 & x_{N-1}^0 \\ x_N^{N-1} & x_N^{N-2} & x_N^{N-3} & \cdots & x_N^1 & x_N^0 \end{pmatrix}}_{\mathbf{A}\in\mathbb{R}^{N\times N}} \underbrace{\begin{pmatrix} c_{N-1} \\ c_{N-2} \\ \vdots \\ c_1 \\ c_0 \end{pmatrix}}_{\mathbf{x}\in\mathbb{R}^N} = \underbrace{\begin{pmatrix} y_1 \\ y_2 \\ \vdots \\ y_{N-1} \\ y_N \end{pmatrix}}_{\mathbf{b}\in\mathbb{R}^N},$$

where the subscripts are indices throughout, while the superscripts of **A**'s elements are powers. For $N = 7$, the above—in which we express y as a power series in x, $y_i = \sum_{j=0}^{6} c_j x_i^j$, $i = [1, 7]$—yields the linear system

$$\underbrace{\begin{pmatrix} 0 & 0 & 0 & 0 & 0 & 0 & 1 \\ 1 & 1 & 1 & 1 & 1 & 1 & 1 \\ 64 & 32 & 16 & 8 & 4 & 2 & 1 \\ 729 & 243 & 81 & 27 & 9 & 3 & 1 \\ 4096 & 1024 & 256 & 64 & 16 & 4 & 1 \\ 15625 & 3125 & 625 & 125 & 25 & 5 & 1 \\ 46656 & 7776 & 1296 & 216 & 36 & 6 & 1 \end{pmatrix}}_{\mathbf{A} \in \mathbb{R}^{7\times 7}} \underbrace{\begin{pmatrix} c_6 \\ c_5 \\ c_4 \\ c_3 \\ c_2 \\ c_1 \\ c_0 \end{pmatrix}}_{\mathbf{x} \in \mathbb{R}^7} = \underbrace{\begin{pmatrix} 1 \\ 3 \\ 4 \\ 2 \\ 4 \\ 5 \\ 6 \end{pmatrix}}_{\mathbf{b} \in \mathbb{R}^7}. \tag{9.74}$$

The unique attribute of the interpolation problem is that its coefficient matrix is square and well posed: $\mathbf{A} \in \mathbb{R}^{N\times N}$, $q = N$. The full rank of the interpolation **A** is a matter of both necessity and common sense. The necessity element arises from the fact that to meet N criteria (devise an interpolant that passes through N points), we need N degrees of freedom, N columns to match the N rows dictated by the number of data points. The common sense element reflects the fact that it will be remarkably foolish to choose (through the analytic description of **A**'s columns, e.g., $a_{ij} = x_i^{N-j}$ in the polynomial interpolation context) an incomplete $\mathbb{R}^N$ spanning set. And the chosen set can only fully span $\mathbb{R}^N$ if **A**'s $q = N$. Because $\mathbf{A} \in \mathbb{R}^{N\times N}$ and $q = N$, the solution of the interpolation problem is simple:

$$\mathbf{x} = \mathbf{A}^{-1}\mathbf{b}, \tag{9.75}$$

used to obtain the thin solid curve of fig. 9.6. Once the $\{c_j\}$ have been identified through inversion (9.75), y can be evaluated at any arbitrary $x_1 \le x \le x_N$ that is not necessarily a node in the inversion (an (x_i, y_i) pair used to construct the inversion that yielded $\{c_j\}$).

The second example shown in fig. 9.6 (by the thick solid curve) uses discretely sampled cosines as **A**'s columns,

$$a_{ij} = \cos\left(\frac{2\pi(j-1)x_i}{x_N - x_1}\right), \qquad i, j = [1, N], \tag{9.76}$$

yielding (to one decimal; you can easily generate it to better accuracy using the above equation) the 7×7 linear system

$$\begin{pmatrix} 1 & 1.0 & 1.0 & 1.0 & 1.0 & 1.0 & 1.0 \\ 1 & 0.6 & -0.3 & -0.9 & -0.9 & -0.1 & 0.7 \\ 1 & -0.3 & -0.9 & 0.7 & 0.5 & -1.0 & -0.0 \\ 1 & -0.9 & 0.7 & -0.4 & -0.0 & 0.4 & -0.7 \\ 1 & -0.9 & 0.5 & -0.0 & -0.5 & 0.9 & -1.0 \\ 1 & -0.1 & -1.0 & 0.4 & 0.9 & -0.6 & -0.7 \\ 1 & 0.7 & -0.0 & -0.7 & -1.0 & -0.7 & -0.0 \end{pmatrix} \begin{pmatrix} c_6 \\ c_5 \\ c_4 \\ c_3 \\ c_2 \\ c_1 \\ c_0 \end{pmatrix} = \begin{pmatrix} 1 \\ 3 \\ 4 \\ 2 \\ 4 \\ 5 \\ 6 \end{pmatrix}.$$

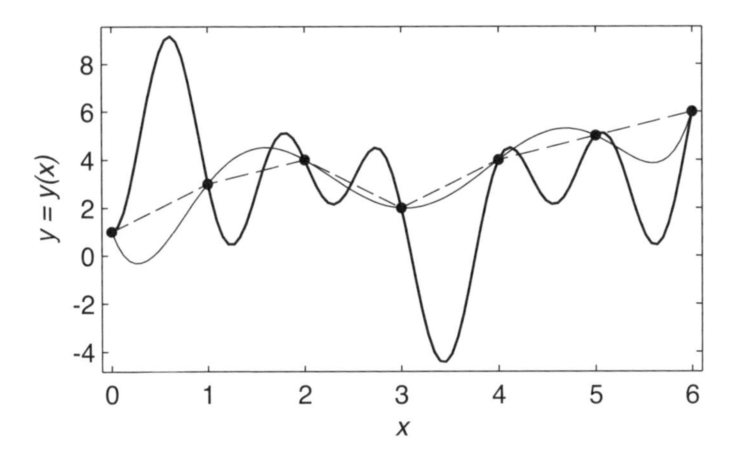

Figure 9.6. The simple 7-point interpolation problem discussed in the text. The observations, at $x_i = i$, $i = [0, 6]$, are shown in filled circles. The dashed line is a piecewise linear interpolant simply connecting the points with straight-line segments. The thin and thick solid curves show the polynomial and cosine fit, respectively.

While interpolation and linear regression are very different in their underlying assumptions and physical motivation, algebraically speaking one can view them on a continuum. Of course, when the fit of a simple model to a data set is poor, we can always add more terms (additional columns in **A**) to improve the fit (whether this makes sense is a different matter and is a statistical rather than algebraic problem). As long as the added terms (columns) are not orthogonal to **b** and are linearly independent of previous ones (so that each added term increases q by 1), the fit is guaranteed to improve until the number of terms reaches N. At that point, regression becomes interpolation, and the fit becomes perfect, with the fitted function passing through every data point. Beyond N, adding terms cannot possibly improve an already perfect fit. In this sense, then, interpolation is the limiting case of regression, when the number of terms becomes N.

9.5.2 Identifying or Removing a Trend

Many signals exhibit a trend. Such trends can be the very focus of inquiry or a distraction interfering with clearly observing and identifying the actual focus. Good examples of the former include the analysis of putative trends in global temperature time series by climate change researchers, or the scrutiny of rising obesity prevalence by public health researchers. As an example of a situation in which the trend is a nuisance, let's examine the atmospheric CO_2 concentration curve at Mauna Loa, Hawaii, shown by the jagged gray line in fig. 9.7. The raw time series clearly exhibits two distinct modes of variability: a seasonal cycle and a rise. The seasonal cycle is mostly due to photosynthetic atmospheric CO_2 drawdown by northern hemisphere vegetation during northern summer.

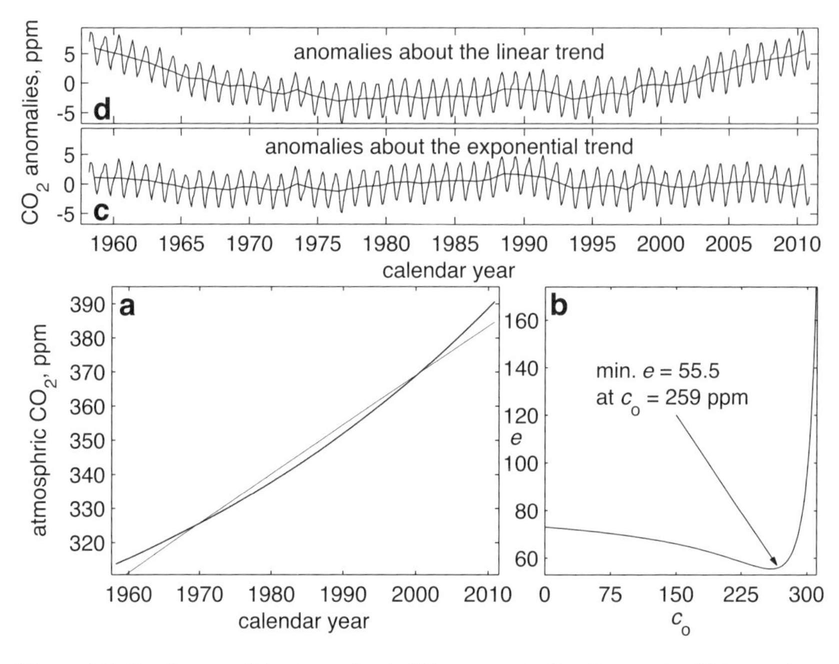

Figure 9.7. Panel a: monthly atmospheric CO_2 concentration time series (thin gray) based on observations taken near the summit of Hawaii's Mauna Loa volcano. In thicker solid black are the straight-line and exponential fits discussed in the text. Panel b: error norm $\|\mathbf{A}\hat{\mathbf{x}} - \mathbf{b}\|$ as a function of c_0 (here the initial CO_2 concentration, not lag 0 covariance). The error norm of the straight-line fit is indicated by the horizontal gray line at $e \approx 83.1$ ppm. Panels c and d: anomalies about the two trend estimates of panel a, with the zero line emphasized. Annual mean anomalies are also shown, in gray.

Consequently, the seasonal cycle's amplitude may be justifiably viewed as an indirect measure of northern hemisphere vegetation vigor and extent. Thus, while a global warming researcher may well zoom in on the rise, a forest ecologist may focus on the seasonal cycle. For that hypothetical analyst, the rise is merely a distraction, which is best eliminated before the real analysis begins.

To identify the trend, the researcher may first try the simplest choice, a linear trend, by solving

$$\begin{pmatrix} 1 & t_1 \\ 1 & t_2 \\ \vdots & \vdots \\ 1 & t_{N-1} \\ 1 & t_N \end{pmatrix} \begin{pmatrix} \alpha \\ \beta \end{pmatrix} = \begin{pmatrix} c_1 \\ c_2 \\ \vdots \\ c_{N-1} \\ c_N \end{pmatrix}, \tag{9.77}$$

where t_i is the ith time point (in decimal calendar year), c_i denotes the corresponding CO_2 measurement, and β is the sought linear trend. Solving this system yields the thin straight line in fig. 9.7a. This is clearly not right; the observations tend to fall above the trend line near the beginning and end of the time series, and below it in the series' midsection. Since this is a systematic, highly structured error, it is emphatically not the random deviations from a fit line characteristic of a good but imperfect fit. This is a structural defect of our model, indicating that a physically better model is needed.

Given the raw data's shape (fig. 9.7a), a sensible candidate is exponential rise,

$$c'(t) := c(t) - c_0 = a \exp(t/\tau), \tag{9.78}$$

where c' is CO_2 deviation from some reference level c_0, a is an amplitude, and τ a timescale. Since this is nonlinear (one of the sought parameters, τ, is an argument into the nonlinear exponential), we take the log,

$$\log(c') = \log(a) + t/\tau, \tag{9.79}$$

define new model unknowns $\alpha \equiv \log(a)$ and $\beta \equiv 1/\tau$, and recast the system as

$$\alpha + \beta t_i = \log(c'_i) \tag{9.80}$$

or

$$\begin{pmatrix} 1 & t_1 \\ 1 & t_2 \\ \vdots & \vdots \\ 1 & t_{N-1} \\ 1 & t_N \end{pmatrix} \begin{pmatrix} \alpha \\ \beta \end{pmatrix} = \begin{pmatrix} \log(c'_1) \\ \log(c'_2) \\ \vdots \\ \log(c'_{N-1}) \\ \log(c'_N) \end{pmatrix}. \tag{9.81}$$

Since the $t = 0$ origin is arbitrary (reflecting no physically significant time point), we cannot objectively and uniquely evaluate $c_0 \equiv c(t = 0)$. We therefore solve the system using a wide range of c_0 values, obtaining a predicted right-hand side for each. The resultant **e** vectors yield a set of e scalars, shown in fig. 7.9b as a function of c_0. The plot makes clear that $c_0 \approx 260$ ppm—not unreasonable since it is very close to our estimated preindustrial atmospheric CO_2—gives the least error. This value is therefore used in fig. 9.7a to obtain the exponential trend curve.

The anomaly curves (fig. 9.7c,d), showing $c'(t)$ deviations from the respective trend estimates, demonstrate the superiority of the exponential estimate for the forest ecologist envisioned in the example. This is best seen by the gray curves showing annual means. While the anomalies about the linear trend (panel d) are highly structured, systematically positive in both ends and negative in between, the anomalies about the exponential trend (panel c) are much less structured, thus better emphasizing the seasonal cycle.

9.5.3 *AR Identification*

If we assume a given zero mean discrete time series $\{x_i\}_1^N$ is an AR process, we would naturally want to estimate the appropriate order p of the AR(p),

$$x_{i+1} = \phi_1 x_i + \phi_2 x_{i-1} + \cdots + \phi_p x_{i-p+1} + \xi_{i+1} \tag{9.82}$$

and its p coefficients $\{\phi_j\}$. There are several methods for obtaining those, two of which are described in this section.

9.5.3.1 Direct Inversion

The first possibility is to form a set of direct inversions,

- $p = 1$: With

  $$x_{i+1} = \phi_1 x_i + \xi_{i+1}, \tag{9.83}$$

  one can form the overdetermined system

  $$\underbrace{\begin{pmatrix} x_2 \\ x_3 \\ \vdots \\ x_N \end{pmatrix}}_{\mathbf{b}} = \underbrace{\begin{pmatrix} x_1 \\ x_2 \\ \vdots \\ x_{N-1} \end{pmatrix}}_{\mathbf{A}} \phi_1, \tag{9.84}$$

  which can be readily solved using the usual least-squares estimator

  $$\hat{\phi}_1 = \left(\mathbf{A}^T\mathbf{A}\right)^{-1}\mathbf{A}^T\mathbf{b} = \frac{\sum_{i=1}^{N-1} x_i x_{i+1}}{\sum_{i=1}^{N-1} x_i^2} = \frac{c_1}{c_0} = r_1, \tag{9.85}$$

  where c_l is again the lth autocovariance, so that r_l is the lth autocorrelation coefficient.
- $p = 2$: With

  $$x_{i+1} = \phi_1 x_i + \phi_2 x_{i-1} + \xi_{i+1}, \tag{9.86}$$

  start by forming the overdetermined system

  $$\underbrace{\begin{pmatrix} x_3 \\ x_4 \\ \vdots \\ x_N \end{pmatrix}}_{\mathbf{b}} = \underbrace{\begin{pmatrix} x_2 & x_1 \\ x_3 & x_2 \\ \vdots & \vdots \\ x_{N-1} & x_{N-2} \end{pmatrix}}_{\mathbf{A}} \underbrace{\begin{pmatrix} \phi_1 \\ \phi_2 \end{pmatrix}}_{\mathbf{\Phi}}. \tag{9.87}$$

  Expressing the solution,

  $$\hat{\mathbf{\Phi}} = \left(\mathbf{A}^T\mathbf{A}\right)^{-1}\mathbf{A}^T\mathbf{b}, \tag{9.88}$$

  analytically is more cumbersome than in the $p = 1$ case. We start with

$$\left(\mathbf{A}^T\mathbf{A}\right)^{-1} = \left[\begin{pmatrix} x_2 & x_3 & \cdots & x_{N-1} \\ x_1 & x_2 & \cdots & x_{N-2} \end{pmatrix}\begin{pmatrix} x_2 & x_1 \\ x_3 & x_2 \\ x_{N-1} & x_{N-2} \end{pmatrix}\right]^{-1}$$

$$= \begin{pmatrix} \sum_{i=2}^{N-1} x_i^2 & \sum_{i=2}^{N-1} x_i x_{i-1} \\ \sum_{i=2}^{N-1} x_i x_{i-1} & \sum_{i=1}^{N-2} x_i^2 \end{pmatrix}^{-1} \tag{9.89}$$

$$= \frac{1}{d}\begin{pmatrix} \sum_{i=1}^{N-2} x_i^2 & -\sum_{i=2}^{N-1} x_i x_{i-1} \\ -\sum_{i=2}^{N-1} x_i x_{i-1} & \sum_{i=2}^{N-1} x_i^2 \end{pmatrix},$$

where d is $\mathbf{A}^T\mathbf{A}$'s determinant,

$$d = \sum_{i=2}^{N-1} x_i^2 \sum_{i=1}^{N-2} x_i^2 - \left(\sum_{i=2}^{N-1} x_i x_{i-1}\right)^2. \tag{9.90}$$

Next, let's use the fact that the time series is stationary, so that autocovariance elements are a function of the lag only, not the particular time series segment on which they are based. In this case,

$$\left(\mathbf{A}^T\mathbf{A}\right)^{-1} = \frac{1}{c_0^2 - c_1^2}\begin{pmatrix} c_0 & -c_1 \\ -c_1 & c_0 \end{pmatrix} \tag{9.91}$$

$$= \frac{1}{c_0^2(1 - r_1^2)}\begin{pmatrix} c_0 & -c_1 \\ -c_1 & c_0 \end{pmatrix}$$

$$= \frac{1}{c_0(1 - r_1^2)}\begin{pmatrix} r_0 & -r_1 \\ -r_1 & r_0 \end{pmatrix}$$

$$= \frac{1}{c_0(1 - r_1^2)}\begin{pmatrix} 1 & -r_1 \\ -r_1 & 1 \end{pmatrix},$$

where c_l is the lag l autocovariance. Similarly,

$$\mathbf{A}^T\mathbf{b} = \begin{pmatrix} x_2 & x_3 & \cdots & x_{N-1} \\ x_1 & x_2 & \cdots & x_{N-2} \end{pmatrix}\begin{pmatrix} x_3 \\ x_4 \\ \vdots \\ x_N \end{pmatrix}$$

$$= \begin{pmatrix} \sum_{i=3}^{N} x_i x_{i-1} \\ \sum_{i=3}^{N} x_i x_{i-2}, \end{pmatrix}, \tag{9.92}$$

which, exploiting again the stationarity of the time series, becomes

$$\mathbf{A}^T\mathbf{b} = \begin{pmatrix} c_1 \\ c_2 \end{pmatrix}. \tag{9.93}$$

Combining the 2 expressions, we have

$$\begin{aligned}(\mathbf{A}^T\mathbf{A})^{-1}\mathbf{A}^T\mathbf{b} &= \frac{1}{c_0(1-r_1^2)}\begin{pmatrix} 1 & -r_1 \\ -r_1 & 1 \end{pmatrix}\begin{pmatrix} c_1 \\ c_2 \end{pmatrix} \\ &= \frac{1}{1-r_1^2}\begin{pmatrix} 1 & -r_1 \\ -r_1 & 1 \end{pmatrix}\begin{pmatrix} r_1 \\ r_2 \end{pmatrix}. \end{aligned} \tag{9.94}$$

Breaking this into individual components, we get

$$\hat{\phi}_1 = \frac{r_1(1-r_2)}{1-r_1^2} \quad \text{and} \quad \hat{\phi}_2 = \frac{r_2 - r_1^2}{1-r_1^2}. \tag{9.95}$$

We can continue exploring $p \geq 3$ cases in this fashion, but the algebra, while not fundamentally different from the $p < 3$ cases, becomes quite prohibitive. For example, for $p = 3$,

$$\mathbf{A}^T\mathbf{A} = \begin{pmatrix} c_0 & c_1 & c_2 \\ c_1 & c_0 & c_1 \\ c_2 & c_1 & c_0 \end{pmatrix}, \tag{9.96}$$

whose determinant, required for the inversion, is the cumbersome

$$\begin{aligned}\det(\mathbf{A}^T\mathbf{A}) &= c_0\left(c_0^2 - 2c_1^2 + 2\frac{c_1^2 c_2}{c_0} - c_2^2\right) \\ &= c_0\left[c_0^2 + 2c_1^2(r_2 - 1) - c_2^2\right]. \end{aligned} \tag{9.97}$$

On premultiplication by the remainder matrix, this becomes a conceptually straightforward yet laborious set of calculations.

Fortunately, there is a better, easier way to obtain the AR coefficients of the arbitrary p, the Yule-Walker (YW) equations.

9.5.3.2 The Yule-Walker Equations

Consider the general AR(p)

$$x_{i+1} = \phi_1 x_i + \phi_2 x_{i-1} + \cdots + \phi_p x_{i-p+1} + \xi_{i+1}. \tag{9.98}$$

1. Lag 1:
   - Multiply both sides of the model by x_i,

$$x_i x_{i+1} = \left(\sum_{j=1}^{p} \phi_j x_i x_{i-j+1}\right) + x_i \xi_{i+1}, \tag{9.99}$$

where i and j are the time and term indices, respectively.

- Take expectation,

$$\langle x_i x_{i+1} \rangle = \left(\sum_{j=1}^{p} \phi_j \langle x_i x_{i-j+1} \rangle \right) + \langle x_i \xi_{i+1} \rangle, \tag{9.100}$$

where $\{\phi_j\}$ are kept outside the expectation operator because they are deterministic, rather than statistical, quantities.

- Note that $\langle x_i \xi_{i+1} \rangle = 0$ because on average the shock (or random perturbation) ξ of the current time is unrelated to—and thus uncorrelated with—previous values of the process, so

$$\langle x_i x_{i+1} \rangle = \sum_{j=1}^{p} \phi_j \langle x_i x_{i-j+1} \rangle. \tag{9.101}$$

- Divide through by $(N-1)$, and use the symmetry of the autocovariance, $c_{-l} = c_l$,

$$c_1 = \sum_{j=1}^{p} \phi_j c_{j-1}. \tag{9.102}$$

- Divide through by c_0,

$$r_1 = \sum_{j=1}^{p} \phi_j r_{j-1}. \tag{9.103}$$

2. Lag 2: Following the same logic and steps (multiplying the model by x_{i-1}, taking expectation, eliminating the zero correlation forcing term, dividing through by $(N-1)$, using $c_{-l} = c_l$ and dividing through by c_0), we get

$$x_{i-1} x_{i+1} = \left(\sum_{j=1}^{p} \phi_j x_{i-1} x_{i-j+1} \right) + x_{i-1} \xi_{i+1}, \tag{9.104}$$

$$\langle x_{i-1} x_{i+1} \rangle = \left(\sum_{j=1}^{p} \phi_j \langle x_{i-1} x_{i-j+1} \rangle \right) + \langle x_{i-1} \xi_{i+1} \rangle \tag{9.105}$$

$$= \sum_{j=1}^{p} \phi_j \langle x_{i-1} x_{i-j+1} \rangle, \tag{9.106}$$

$$c_2 = \sum_{j=1}^{p} \phi_j c_{j-2} \quad \Longrightarrow \quad r_2 = \sum_{j=1}^{p} \phi_j r_{j-2}. \tag{9.107}$$

3. Lags k and p:

- Multiply by x_{i-k-1}:

$$x_{i-k+1} x_{i+1} = \left(\sum_{j=1}^{p} \phi_j x_{i-k+1} x_{i-j+1} \right) + x_{i-k+1} \xi_{i+1}.$$

- Take expectation:

$$\langle x_{i-k+1}x_{i+1}\rangle = \left(\sum_{j=1}^{p}\phi_j\langle x_{i-k+1}x_{i-j+1}\rangle\right) + \langle x_{i-k+1}\xi_{i+1}\rangle.$$

- Eliminate the zero correlation forcing term:

$$\langle x_{i-k+1}x_{i+1}\rangle = \left(\sum_{j=1}^{p}\phi_j\langle x_{i-k+1}x_{i-j+1}\rangle\right).$$

- Divide through by $(N-1)$, and use $c_{-l}=c_l$:

$$c_k = \sum_{j=1}^{p}\phi_j c_{j-k},$$

- Divide through by c_0: $r_k = \sum_{j=1}^{p}\phi_j r_{j-k}$.
- This also leads to the final term, $r_p = \sum_{j=1}^{p}\phi_j r_{j-p}$.

4. Putting it all together: Substituting $r_0 = 1$ gives the equations

$$\begin{array}{lcllllllll} r_1 & = & \phi_1 & + & \phi_2 r_1 & + & \phi_3 r_2 & + & \cdots & + \\ & & \phi_{p-1}r_{p-2} & + & \phi_p r_{p-1} & & & & & \\ r_2 & = & \phi_1 r_1 & + & \phi_2 & + & \phi_3 r_1 & + & \cdots & + \\ & & \phi_{p-1}r_{p-3} & + & \phi_p r_{p-2} & & & & & \\ & & & \vdots & & \vdots & & \vdots & & \vdots \\ r_{p-1} & = & \phi_1 r_{p-2} & + & \phi_2 r_{p-3} & + & \phi_3 r_{p-4} & + & \cdots & + \\ & & \phi_{p-1} & + & \phi_p r_1 & & & & & \\ r_p & = & \phi_1 r_{p-1} & + & \phi_2 r_{p-2} & + & \phi_3 r_{p-3} & + & \cdots & + \\ & & \phi_{p-1}r_1 & + & \phi_p. & & & & & \end{array} \tag{9.108}$$

This can be recast as

$$\mathbf{R\Phi} := \begin{pmatrix} 1 & r_1 & r_2 & \cdots & r_{p-2} & r_{p-1} \\ r_1 & 1 & r_1 & \cdots & r_{p-3} & r_{p-2} \\ & \vdots & & & \vdots & \\ r_{p-2} & r_{p-3} & r_{p-4} & \cdots & 1 & r_1 \\ r_{p-1} & r_{p-2} & r_{p-3} & \cdots & r_1 & 1 \end{pmatrix}\begin{pmatrix} \phi_1 \\ \phi_2 \\ \vdots \\ \phi_{p-1} \\ \phi_p \end{pmatrix}$$

$$= \begin{pmatrix} r_1 \\ r_2 \\ \vdots \\ r_{p-1} \\ r_p \end{pmatrix} := \mathbf{r}, \tag{9.109}$$

the Yule-Walker system. The system (9.109) has as many constraints (equations, $\mathbf{R}$'s rows) as unknowns ($\{\phi_j\}$, $\mathbf{\Phi}$'s elements), and is well posed ($\mathbf{R} \in \mathbb{R}^{p \times p}$ with $q = p$, the latter failing only in the irrelevant case of $r_i = 1 \ \forall i$). Consequently,

$$\mathbf{\Phi} = \mathbf{R}^{-1}\mathbf{r}. \tag{9.110}$$

9.5.3.3 The Partial Autocorrelation Function (pacf)

Equation 9.109 and solution 9.110 offer a simple way of obtaining optimal (least-squares) values of the AR(p) coefficients, provided p is known. But finding the p value appropriate for a given process is challenging. By examining eq. 9.109, one may look for guidance from the acf, which appears to determine $\mathbf{r}$'s length, thus setting p. Reexamining section 8.1, however, dashes this hope, because theoretically expected acfs of AR processes never die out. This is an important shortcoming of the acf. Another, not unrelated, acf limitation is that its kth coefficient simply quantifies how well current observations correlate with past observations k lags back, paying no attention to serial redundancy. For example, the correlation of x_i and x_{i-2} is r_2, but much of this correlation may be an (x_i, x_{i-1}) correlation applied twice. This is clearest with AR(1) processes, in which our best prediction of x_i in terms of x_{i-2} is $\phi_1^2 x_{i-2}$, but this predictive power arises from $x_{i+1} = \phi_1 x_i = \phi_1^2 x_{i-1}$. The pacf—whose kth element is ϕ_p of eq. 9.109 with $p = k$—is the tool designed to overcome these limitation of the acf.

The pacf's kth element quantifies the (x_i, x_{i-k}) correlation after correcting for the contributions of all intermediate lags x_{i-k+j}, $j = [1, k-1]$. While eq. 9.109 can be written for any stationary process, the distinguishing attribute of pacfs of AR(p) processes is that $\phi_{k>p} = 0$. If the magnitude of some process' pacf—originating from 1 at $k = 0$ and descending steadily—fluctuates randomly around the $\phi_k = 0$ line with minuscule amplitudes after some lag p, the underlying process is an AR(p). In other words, the pacf identifies p, the order of the generating AR process, as the lag beyond which the pacf vanishes.

We obtain $\{\phi_i\}$ seriatim, starting with $\phi_0 = r_0 = 1$ (because x_i is perfectly correlated with itself, a truism placed in a broader context below). Progressing to $p = 1$ (and noting that scalars are perfectly fine 1×1 matrices), the system is

$$(1)(\phi_1) = (r_1) \quad \Longrightarrow \quad \phi_1 = r_1, \tag{9.111}$$

which makes sense because to predict x_i using x_{i-1} as our predictor, the least-squares, best predictor is r_1. For $p = 2$ it is

$$\begin{pmatrix} 1 & r_1 \\ r_1 & 1 \end{pmatrix} \begin{pmatrix} \phi_1 \\ \phi_2 \end{pmatrix} = \begin{pmatrix} r_1 \\ r_2 \end{pmatrix}, \tag{9.112}$$

which is solved by

$$\begin{pmatrix} \phi_1 \\ \phi_2 \end{pmatrix} = \frac{1}{1-r_1^2} \begin{pmatrix} 1 & -r_1 \\ -r_1 & 1 \end{pmatrix} \begin{pmatrix} r_1 \\ r_2 \end{pmatrix}, \tag{9.113}$$

yielding

$$\phi_2 = \frac{r_2 - r_1^2}{1-r_1^2}. \tag{9.114}$$

Note again that pacf(k) is defined as ϕ_k when the system (9.109) is constructed with $p = k$. That is, only the kth element of the solution vector at $p = k$ is retained, because $\{\phi_i\}_{i=1}^{k-1}$ are already known (e.g., in the above $p = 2$ case, we need not calculate ϕ_1, because pacf(1) is already known (pacf(1) $= \phi_1$ obtained from solving the $p = 1$ system).

For $p = 3$, the system (9.109) is

$$\begin{pmatrix} 1 & r_1 & r_2 \\ r_1 & 1 & r_1 \\ r_2 & r_1 & 1 \end{pmatrix} \begin{pmatrix} \phi_1 \\ \phi_2 \\ \phi_3 \end{pmatrix} = \begin{pmatrix} r_1 \\ r_2 \\ r_3 \end{pmatrix} \tag{9.115}$$

with

$$\phi_3 = \frac{r_1\left[r_1^2 - r_2(2-r_2) - r_1 r_3\right] + r_3}{(1-r_2)(1+r_2-2r_1^2)}, \tag{9.116}$$

and so on.

While there are efficient methods of computing the pacf exploiting the highly restrictive structure of **R**, in most cases modern personal computers offer more than enough computing power to use the following, simpler but less efficient, well-posed Yule-Walker recursion. First, compute the acf up to a reasonable cutoff, say $N/20$. Then set $\phi_0 = r_0 = 1$ (because again x_i is perfectly "predicted" given x_i) and $\phi_1 = r_1$ (because for lag 1, there is no preceding lag whose predictive contribution must be subtracted). Next, let $\mathbf{r}^{(i)}$ and $\mathbf{R}^{(i)}$ denote eq. 9.109's right-hand side and coefficient matrix when $p = i$, and L some cutoff lag well beyond the reasonably assumed AR order. The following simple algorithm can then be used to obtain the pacf,

```
loop on i between 2 and L
  assign R^(i) and r^(i)  (using Eq. 9.109 with p = i)
  invert for Φ^(i) ≡ (R^(i))^(-1) r^(i)  (using Eq. 9.110)
  set pacf(i) = φ_i^(i),
end of loop on i.
```

The plot of pacf(i) against i over $i = [0, L]$ (e.g., fig. 9.8) can reveal p satisfying $\phi_{|i|>p} \approx 0$, the lag beyond which the pacf approximately vanishes, fluctuating randomly about zero. Since in practice $\{\phi_{|i|>p}\}$ derived from specific finite samples do not identically vanish, we need to quantify the range of values within which the pacf can be considered to "fluctuate randomly about zero." This pacf yardstick is provided by the range of values spanned by the (spuriously nonzero) pacf of pure noise, which is derived from the pacf's assumed variance, $\sim 1/\sqrt{N}$.

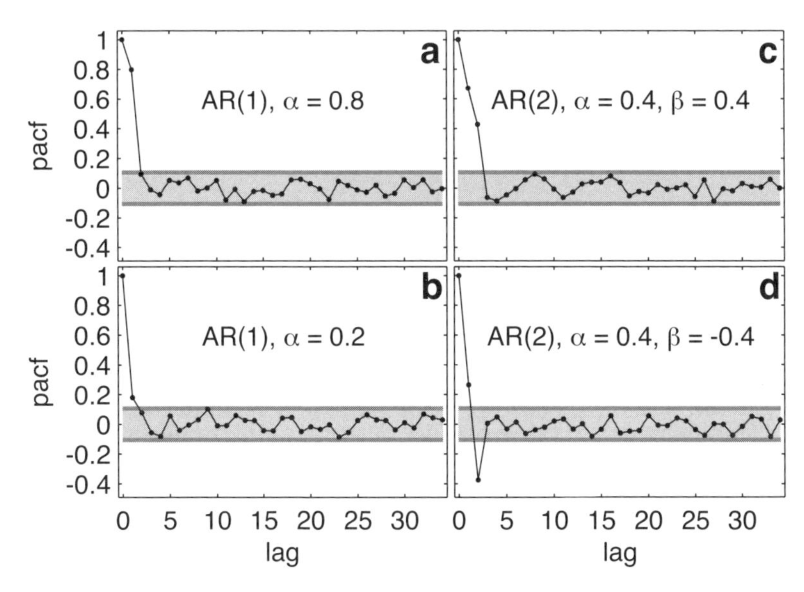

Figure 9.8. Partial autocorrelation estimates for two AR(1) (panels a, b) and two AR(2) (panels c, d) processes. The generating process, with its coefficient choices, is given in each panel. Dark (light) shading shows the area within which 99% (95%) of the estimates derived from a pure noise process are expected to fall.

Note that this well-posed YW-based pacf estimate is entirely equivalent to estimating the pacf—starting again with $\phi_0 = 1$—from a series of direct overdetermined fits over $i = [1, L]$. For example, for $p = 2$, ϕ_2 quantifies the contribution of $\{x_{i-2}\}$ to explaining (hindcasting, "forecasting" values already known) $\{x_i\}$ after accounting for the explanatory contribution of $\{x_{i-1}\}$. Because when $p = 2$ we already know that $\phi_1 = r_1$, we can calculate (using the acf's symmetry) the part of $\{x_{i-2}\}$ explained by $\{x_{i-1}\}$, $\{r_1 x_{i-1}\}$. If $\{x_{i-2}\}$ has a novel contribution, beyond that of $\{x_{i-1}\}$'s, to explaining $\{x_i\}$, $\{x_{i-2} - r_1 x_{i-1}\}$—the part of $\{x_{i-2}\}$ unrelated to $\{x_{i-1}\}$—must exhibit some skill in predicting $\{x_i\}$. The corresponding overdetermined direct fit problem is therefore

$$(\mathbf{x}_1 - r_1\mathbf{x}_2)\phi_2 := \begin{pmatrix} x_1 & - & r_1 x_2 \\ x_2 & - & r_1 x_3 \\ & \vdots & \\ x_{N-2} & - & r_1 x_{N-1} \end{pmatrix} \phi_2 = \begin{pmatrix} x_3 \\ x_4 \\ \vdots \\ x_N \end{pmatrix} := \mathbf{x}_3,$$

which we invert for ϕ_2 with $\mathbf{x}_1 - r_1\mathbf{x}_2$ and $\mathbf{x}_3$ playing the role of $\mathbf{A}$ and $\mathbf{b}$ in our usual linear inversion notation. Because $\mathbf{A}$ here is a vector, the usual least-squares solution, $(\mathbf{A}^T\mathbf{A})^{-1}\mathbf{A}^T\mathbf{b}$, becomes

$$\begin{aligned}
\phi_2 &= \frac{(\mathbf{x}_1 - r_1\mathbf{x}_2)^T\mathbf{x}_3}{(\mathbf{x}_1 - r_1\mathbf{x}_2)^T(\mathbf{x}_1 - r_1\mathbf{x}_2)} \\
&= \frac{\mathbf{x}_1^T\mathbf{x}_3 - r_1\mathbf{x}_2^T\mathbf{x}_3}{\mathbf{x}_1^T\mathbf{x}_1 - 2r_1\mathbf{x}_1^T\mathbf{x}_2 + r_1^2\mathbf{x}_2^T\mathbf{x}_2} \\
&= \frac{c_2 - r_1c_1}{c_0 - 2r_1c_1 + r_1^2c_0} = \frac{r_2 - r_1^2}{1 - 2r_1^2 + r_1^2} = \frac{r_2 - r_1^2}{1 - r_1^2},
\end{aligned} \tag{9.117}$$

identical to (9.114).

9.6.3.4 Estimating the Noise Variance of an AR Process

In most cases, the AR model identification is incomplete without a robust estimate of the noise. Let's therefore derive a method for estimating the noise variance of an AR model.

1. AR(1): We start with the simplest AR model, AR(1),

$$x_{i+1} = \alpha x_i + \xi_{i+1}, \tag{9.118}$$

assuming $\bar{x} = 0$. Rearranging,

$$\xi_{i+1} = x_{i+1} - \alpha x_i, \tag{9.119}$$

which we next use to estimate the noise variance by taking expectation of the above equation's product with itself,

$$\langle \xi_i \xi_i \rangle = \langle x_i x_i \rangle - 2\alpha \langle x_{i+1} x_i \rangle + \alpha^2 \langle x_i x_i \rangle \tag{9.120}$$

where we invoked signal and noise stationarity, $\langle x_j x_j \rangle = \langle x_i x_i \rangle$ and $\langle \xi_j \xi_j \rangle = \langle \xi_i \xi_i \rangle$ for any i and j. Dividing through by $N - 1$, we get

$$\sigma_\xi^2 = c_0 - 2\alpha c_1 + \alpha^2 c_0 = c_0(1 - 2\alpha r_1 + \alpha^2), \tag{9.121}$$

where σ_ξ^2 is the expected noise variance, r_1 is the lag 1 autocorrelation, and c_i is the lag i autocovariance (so that $c_0 = s_x^2$ is x's sample variance). Finally, because for AR(1) $\alpha = r_1$,

$$\sigma_\xi^2 = s_x^2(1 - 2r_1^2 + r_1^2) = s_x^2(1 - r_1^2). \tag{9.122}$$

2. AR(p): For $p > 1$, we take a slightly different strategy. Starting from the AR(p) model

$$x_{i+1} = \phi_1 x_i + \phi_2 x_{i-1} + \cdots + \phi_p x_{i-p+1} + \xi_{i+1}, \tag{9.123}$$

we multiply all terms by x_{i+1} and take expectation, obtaining

$$\begin{aligned}
\langle x_{i+1} x_{i+1} \rangle = \phi_1 \langle x_{i+1} x_i \rangle + \phi_2 \langle x_{i+1} x_{i-1} \rangle + \cdots \\
+ \phi_p \langle x_{i+1} x_{i-p+1} \rangle + \langle x_{i+1} \xi_{i+1} \rangle.
\end{aligned} \tag{9.124}$$

Now the rightmost term combines simultaneous realization of the process and the noise. Because the process value at the current time is not unrelated to the current time shock, the last term does not vanish. Instead, let's use the AR model equation to recast x_{i+1},

$$\langle x_{i+1}\xi_{i+1}\rangle = \langle(\phi_1 x_i + \phi_2 x_{i-1} + \cdots + \phi_p x_{i-p+1} + \xi_{i+1})\xi_{i+1}\rangle.$$

The shock at any given time is clearly uncorrelated with earlier values of the process, so $\langle x_{j\le i}\xi_{i+1}\rangle = 0$. As a result, the only remaining term in the above right-hand side is the last, and

$$\langle x_{i+1}\xi_{i+1}\rangle = \langle \xi_{i+1}\xi_{i+1}\rangle = \sigma_\xi^2, \tag{9.125}$$

where σ_ξ^2 is the AR process' noise variance. Substituting this result into eq. 9.124 and carrying out the expectation operation using the autocovariance function's symmetry yields

$$c_0 = \phi_1 c_1 + \phi_2 c_2 + \cdots + \phi_p c_p + \sigma_\xi^2. \tag{9.126}$$

Rearranging and using $c_0 = s_x^2$ and $c_i/c_0 = r_i$, we get

$$\sigma_\xi^2 = s_x^2(1 - \phi_1 r_1 - \phi_2 r_2 - \cdots - \phi_p r_p) \tag{9.127}$$

or

$$\sigma_\xi^2 = s_x^2\left(1 - \sum_{i=1}^{p} \phi_i r_i\right) \tag{9.128}$$

which reduces to, as it must, our previous result for AR(1).

When analyzing a time series with known p, one can use the above estimate for the noise variance or simply compute the sample noise variance,

$$\operatorname{var}(\{\xi\}) = \operatorname{var}\left(\left\{x_{i+1} - \sum_{j=1}^{p} \phi_i x_{i-j+1}\right\}\right). \tag{9.129}$$

9.5.4 *Fourier Fit*

We have already encountered Fourier-like vectors and a Fourier-like basis in section 4.3.2. There are many reasons Fourier vectors are commonly used. In analyzing apparently periodic time series with unknown frequencies, the main appeal of the Fourier basis is the unambiguous association of each mode (each column of the Fourier matrix, and thus each element of the unknown vector) with a specific frequency. Fourier vectors are also particularly attractive because of their easily achieved orthogonality (and, thus, if desired, orthonormality) and their completeness under simple, readily realized conditions.

For discrete data, there are subtle details regarding the orthogonality and normalization, depending on whether N is even or odd, whether the signal on which the Fourier matrix is to operate is expected to feature even, odd, or no symmetry (e.g., cosine or a quadratic, sine, or a cubic, or most realistic signals, respectively), and other special considerations (e.g., matrix rank and inevitability, overall normalization). For example, a reasonable (neither unique nor ideal) form of an orthonormal Fourier matrix expected to operate on general (not necessarily odd or even) signals of odd length is

$$\mathbf{F} = \begin{pmatrix} \mathbf{c}_0 & \mathbf{c}_1 & \cdots & \mathbf{c}_K & \mathbf{s}_1 & \cdots & \mathbf{s}_K \end{pmatrix}, \tag{9.130}$$

where $K = N/2 - \text{mod}(N, 2)/2$ is the downward rounded signal length, $\mathbf{c}_0$ is the zeroth (constant) cosine vector

$$c_{j0} = \frac{1}{\sqrt{N}}, \qquad 1 \le j \le N, \tag{9.131}$$

$\mathbf{c}_{k\neq 0}$ is the kth cosine vector

$$c_{jk} = \sqrt{\frac{2}{N}} \cos\left(\frac{2\pi k(j-1)}{N}\right), \qquad 1 \le j \le N, \qquad 1 \le k \le K, \tag{9.132}$$

and $\mathbf{s}_{k>0}$ is the kth sine vector

$$s_{jk} = \sqrt{\frac{2}{N}} \sin\left(\frac{2\pi k(j-1)}{N}\right), \qquad 1 \le j \le N, \qquad 1 \le k \le K. \tag{9.133}$$

Figure 9.9 displays an $\mathbf{F}$ of this structure, with $N = 51$ ($x = [0, 50]$). Such $\mathbf{F}$s are square, full rank, and orthonormal, so $\mathbf{F}^T\mathbf{F} = \mathbf{I} \in \mathbb{R}^N$, a computationally

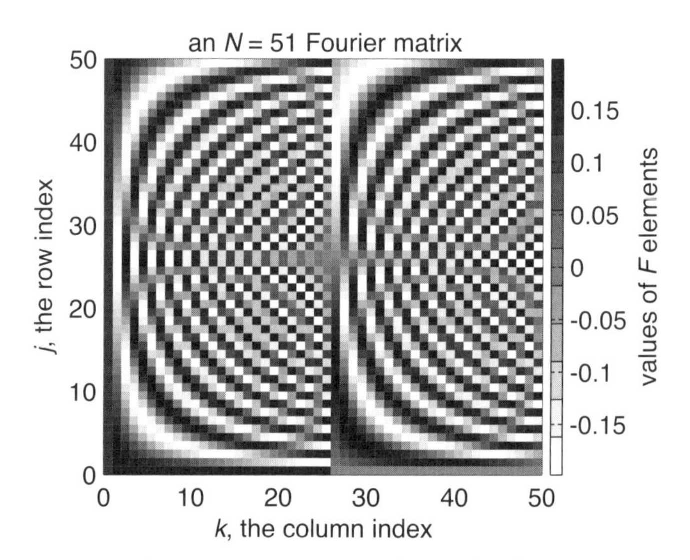

Figure 9.9. The 51 × 51 Fourier matrix discussed in the text (section 9.5.4).

extremely attractive property. On the negative side, while fully spanning $\mathbb{R}^N$ (by virtue of being $N \times N$ with $q = N$) they are not exactly conformal with the traditional discrete Fourier transform.

9.6 Statistical Issues in Regression Analysis

Up to this point, we mostly emphasized algebraic aspects of the regression machinery. But regression analysis of random variables is as much statistical as it is algebraic. This section fills some of the remaining statistical gaps.

Our starting point is the single predictor regression equation $\mathbf{Ap} = \mathbf{y}$, where $y = y(x)$ and x are the predicted and predictor variables, and $\mathbf{p}$ holds the optimized solution. Since previously we studied the regression problem mostly in vector form, let's repeat the derivation here in a scalar form, which will be useful later. With $\mathbf{p} = (\alpha, \beta)$, the regression problem is

$$y_i = \alpha + \beta x_i, \tag{9.134}$$

which we multiply by x_i and sum over $i = [1, N]$,

$$\sum x_i y_i = \alpha N\bar{x} + \beta \sum x_i^2. \tag{9.135}$$

(Below, we will continue to omit summation limits, typically 1 to N, where there is no risk of ambiguity.) Also, recall that

$$\begin{aligned} \sum (x_i - \bar{x})(y_i - \bar{y}) &= \sum x_i y_i - \bar{y} \sum x_i - \bar{x} \sum y_i + \sum \bar{x}\bar{y} \\ &= \sum x_i y_i - N\bar{y}\bar{x} - N\bar{x}\bar{y} + N\bar{x}\bar{y} \\ &= \sum x_i y_i - N\bar{x}\bar{y} \end{aligned} \tag{9.136}$$

(where overbars denote means) and

$$\begin{aligned} \sum (x_i - \bar{x})^2 &= \sum x_i^2 - \bar{x} \sum x_i - \bar{x} \sum x_i + \sum \bar{x}^2 \\ &= \sum x_i^2 - N\bar{x}^2 - N\bar{x}^2 + N\bar{x}^2 \\ &= \sum x_i^2 - N\bar{x}^2 \end{aligned} \tag{9.137}$$

so that

$$\sum x_i y_i = \sum (x_i - \bar{x})(y_i - \bar{y}) + N\bar{x}\bar{y} \tag{9.138}$$

and

$$\sum x_i^2 = \sum (x_i - \bar{x})^2 + N\bar{x}^2. \tag{9.139}$$

Using these relationships, we rewrite

$$\sum x_i y_i = \alpha N\bar{x} + \beta \sum x_i^2 \tag{9.140}$$

as

$$\sum (x_i - \bar{x})(y_i - \bar{y}) + N\bar{x}\bar{y} = \alpha N\bar{x} + \beta \sum (x_i - \bar{x})^2 + \beta N\bar{x}^2. \tag{9.141}$$

To proceed, let's sum the original regression equation,

$$\sum y_i = \sum(\alpha + \beta x_i) \implies N\bar{y} = N\alpha + \beta N\bar{x}$$
$$\implies \bar{y} = \alpha + \beta\bar{x}. \tag{9.142}$$

Rearranging, we get, $\alpha = \bar{y} - \beta\bar{x}$, which we use to eliminate α,

$$\sum(x_i - \bar{x})(y_i - \bar{y}) + N\bar{x}\bar{y} = (\bar{y} - \beta\bar{x})N\bar{x} + \beta\sum(x_i - \bar{x})^2 + \beta N\bar{x}^2,$$
$$\sum(x_i - \bar{x})(y_i - \bar{y}) = N\bar{x}\bar{y} - N\bar{x}\bar{y} - \beta N\bar{x}^2 + \beta N\bar{x}^2 + \beta\sum(x_i - \bar{x})^2$$
$$= \beta\sum(x_i - \bar{x})^2. \tag{9.143}$$

Therefore,

$$\beta = \frac{\sum(x_i - \bar{x})(y_i - \bar{y})}{\sum(x_i - \bar{x})^2} \tag{9.144}$$

and

$$\alpha = \bar{y} - \beta\bar{x} = \bar{y} - \bar{x}\frac{\sum(x_i - \bar{x})(y_i - \bar{y})}{\sum(x_i - \bar{x})^2}. \tag{9.145}$$

Given a regression representation of a process of interest, the first thing we would want is to evaluate how much better off we are (in terms of our information about the predicted variable) *with* the regression than we are *without* it. To answer this question, a useful starting point is the predictand variability $\mathbf{y}^T\mathbf{y}$. By construction, the error e_i in a given observation is the difference between the actual observation y_i and the predicted value $\hat{y}_i$, $e_i = y_i - \hat{y}_i$, so that $y_i = e_i + \hat{y}_i$ or $\mathbf{y} = \mathbf{e} + \hat{\mathbf{y}}$. Therefore,

$$\mathbf{y}^T\mathbf{y} = (\hat{\mathbf{y}} + \mathbf{e})^T(\hat{\mathbf{y}} + \mathbf{e}) = \hat{\mathbf{y}}^T\hat{\mathbf{y}} + \mathbf{e}^T\mathbf{e}. \tag{9.146}$$

If you are puzzled about why the cross-terms above vanish (i.e., why $\hat{\mathbf{y}}^T\mathbf{e} = \mathbf{e}^T\hat{\mathbf{y}} = 0$), note that this is precisely the way we have constructed the least-squares solution $\hat{\mathbf{p}} = (\mathbf{A}^T\mathbf{A})^{-1}\mathbf{A}^T\mathbf{y}$ to the original problem $\mathbf{y} = \mathbf{A}\mathbf{p}$. Recall that because of inconsistencies in the equations, which render the original problem intractable, we solve the closest problem we can solve. That problem is obtained by replacing the observed $\mathbf{y}$ with its part from $\mathbf{A}$'s column space ($\hat{\mathbf{y}}$), discarding the $\mathbf{e}$ remainder from $\mathbf{A}$'s left null space. Consequently, $\hat{\mathbf{y}}^T\mathbf{e} = \mathbf{e}^T\hat{\mathbf{y}} = 0$.

Subtracting $N\bar{y}^2$ from $\mathbf{y}^T\mathbf{y} = \hat{\mathbf{y}}^T\hat{\mathbf{y}} + \mathbf{e}^T\mathbf{e}$ (and recalling that $\bar{\hat{y}} = \bar{y}$ because $y_i = \hat{y}_i + e_i$ and $\sum e_i = 0$), we obtain $\mathbf{y}^T\mathbf{y} - N\bar{y}^2 = \hat{\mathbf{y}}^T\hat{\mathbf{y}} + \mathbf{e}^T\mathbf{e} - N\bar{y}^2$, or

$$\underbrace{\sum_{i=1}^{N}(y_i - \bar{y})^2}_{SS_T} = \underbrace{\sum_{i=1}^{N}(\hat{y}_i - \bar{y})^2}_{SS_R} + \underbrace{\sum_{i=1}^{N}e_i^2}_{SS_E}. \tag{9.147}$$

This splits the total sum of squares about the mean, SS_T, into two contributions.

$$SS_R \equiv \sum_{i=1}^{N} (\hat{y}_i - \bar{y})^2 \quad \text{and} \quad SS_E \equiv \sum_{i=1}^{N} e_i^2. \tag{9.148}$$

The first is the sum of squares of the regression line about its mean (which is *the* mean, because, again, $\bar{\hat{y}} = \bar{y}$). The second is the sum of squared error, or the squared L_2-norm of the error. Figure 9.10 shows the various difference terms for a small sample data set.

Of course, a perfect fit is, by definition, a fit with *no* error; $\hat{y}_i = y_i \; \forall i$, $SS_E = 0$. In such a case,

$$R^2 := 1 - \frac{SS_E}{SS_T} = 1 - 0 = 1. \tag{9.149}$$

In the other extreme, when the fit is utterly useless, $SS_E = SS_T$, and

$$R^2 = 1 - \frac{SS_E}{SS_T} = 1 - 1 = 0. \tag{9.150}$$

You get the picture—R^2 evaluates the goodness of the fit. In fact, R^2 is closely related to something you are very familiar with:

$$\begin{aligned} R^2 &= 1 - \frac{SS_E}{SS_T} = \frac{SS_T - SS_E}{SS_T} = \frac{SS_R}{SS_T} \\ &= \frac{\sum (\hat{y}_i - \bar{y})^2}{\sum (y_i - \bar{y})^2} = \beta^2 \frac{\sum (x_i - \bar{x})^2}{\sum (y_i - \bar{y})^2}, \end{aligned} \tag{9.151}$$

where we have used the regression relations, $y_i = \alpha + \beta x_i$ and $\bar{y} = \alpha + \beta \bar{x}$ (see above), with which

$$(\hat{y}_i - \bar{y}) = [\alpha + \beta x_i - (\alpha + \beta \bar{x})] = \beta(x_i - \bar{x}), \tag{9.152}$$

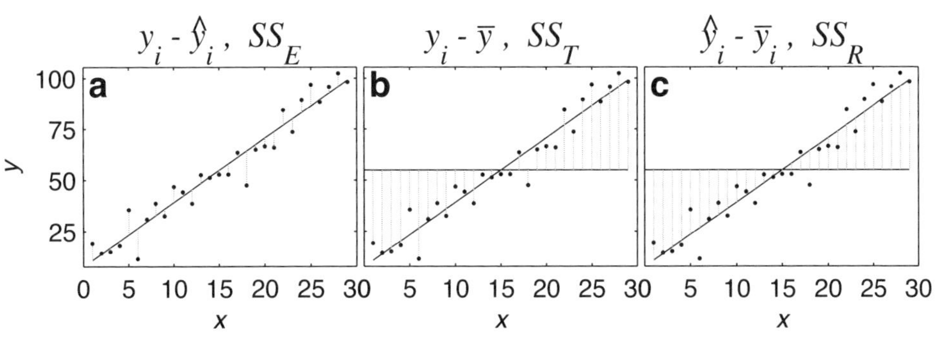

Figure 9.10. The various difference terms for a small sample data set. Panel a shows $y_i - \hat{y}_i$ (whose sum of squares is SS_E), panel b shows $y_i - \bar{y}$ (whose sum of squares is SS_T), and panel c shows $\hat{y}_i - \bar{y}$ (whose sum of squares is SS_R).

yielding the final step of (9.151). Now let's introduce the notation

$$s_{xx} \equiv \sum_{i=1}^{N} (x_i - \bar{x})^2, \qquad s_{yy} \equiv \sum_{i=1}^{N} (y_i - \bar{y})^2$$

and

$$s_{xy} \equiv \sum_{i=1}^{N} (x_i - \bar{x})(y_i - \bar{y}),$$

with the aid of which

$$R^2 = \beta^2 \frac{s_{xx}}{s_{yy}}. \tag{9.153}$$

Next, recall the least-squares solution

$$\beta = \frac{\sum (x_i - \bar{x})(y_i - \bar{y})}{\sum (x_i - \bar{x})^2}, \tag{9.154}$$

which we rewrite, using the above definitions, as $\beta = s_{xy}/s_{xx}$, so that

$$R^2 = \beta^2 \frac{s_{xx}}{s_{yy}} = \frac{s_{xy}^2 \, s_{xx}}{s_{xx}^2 \, s_{yy}} = \frac{s_{xy}^2}{s_{xx} s_{yy}}. \tag{9.155}$$

Recognize this? If not, perhaps taking the square root of the explicit expression will help:

$$\sqrt{R^2} = \frac{\sum (x_i - \bar{x})(y_i - \bar{y})}{\sqrt{\sum [(x_i - \bar{x})^2 (y_i - \bar{y})^2]}}, \tag{9.156}$$

the correlation coefficient. Note that I didn't write $\sqrt{R^2} = R$, although for a single predictor $R = r = r_{xy}$. Thus, in the single-predictor model, the correlation coefficient and the regression slope in the single predictor case are closely related, satisfying the relation

$$\beta = r \sqrt{\frac{s_{yy}}{s_{xx}}}. \tag{9.157}$$

Our next objective is to compare the various sums of squares and further evaluate the fit. However, we must normalize each sum by its number of degrees of freedom *df*. We have already discussed the concept of *df* in a number of cases. As before, in the current context *df* indicates how many independent pieces of information are contained in the various sums. Let's assume, for convenience, that the data are not serially correlated. Remember, though, that this is a highly restrictive requirement, which we cannot expect to be met in general. In the likely case of serial correlations, the appropriate acfs must be used to estimate the *df*.

With the assumed zero serial correlations, the $SS_T = \sum (y_i - \bar{y})^2$ comprises $N - 1$ independent pieces of information, because one of the values is fully determined by the $N-1$ other values and the requirement of zero mean; therefore, the whole problem's $df=N-1$. On the other hand, $SS_R=\sum(\hat{y}_i-\bar{y})^2$ has a single degree of freedom, $df_R=1$, because once β is computed (which corresponds to the single df), all $\hat{y}_i$ values are fully determined by $\hat{y}_i=\bar{y}+\beta(x_i-\bar{x})$, $i=[1,N]$. Since the total problem has $df=N-1$ and the regression accounts for only one of them, $df_R=1$, it follows that the sum of squared residuals, $SS_E=\sum(y_i-\hat{y})^2$, has $N-2$ degrees of freedom; $df_E=N-2$. This is required for conserving the total number of degrees of freedom, $df=df_R+df_E$, but it is also readily understood in terms of the 2 parameters (α and β) the straight-line regression problem has; in the general case of M predictors, the df of SS_E is $df_E=N-M$, corresponding to the total number of parameters.

Given their differing dfs, we now scale the SSs to make them comparable:

$$MS_T=\frac{SS_T}{N-1}=\frac{\sum(y_i-\bar{y})^2}{N-1}, \tag{9.158}$$

$$MS_R=\frac{SS_R}{1}=\frac{\sum(\hat{y}_i-\bar{y})^2}{1} \tag{9.159}$$

(or higher number of predictors, in the case of multiple regression), and

$$MS_E=\frac{SS_E}{N-2}=\frac{\sum(y_i-\hat{y}_i)^2}{N-2}. \tag{9.160}$$

Since the MSs are somewhat redundant, let's focus on the ratio of MS_R and MS_E,

$$\frac{MS_R}{MS_E}=(N-2)\frac{\sum(\hat{y}_i-\bar{y})^2}{\sum(y_i-\hat{y}_i)^2}. \tag{9.161}$$

If we assume the residuals are normally distributed, their squares are chi-squared distributed, and the squares' ratio is F distributed. Therefore, ratio (9.161) is expected to be F distributed,

$$\frac{MS_R}{MS_E}\sim F. \tag{9.162}$$

Figure 9.11 shows F distributions with various df combinations. The likelihood of obtaining the observed rmse reduction by chance can be readily estimated by comparing the computed F to the appropriate tabulated $F^{(df_R,\, df_E)}$.

The essence of the F test of the ratio is shown in fig. 9.12. The SS_T, shown in gray, is a measure of the width of the unconditional distribution (in the absence of any predictor, x, information) of the predictand (y) about its mean $\bar{y}$. Not knowing anything beyond the observed y values, our best estimate of *any* y value is $\bar{y}$, and the error of these estimates will be distributed according to the gray distribution. We now use the x values and the x–y relationship embodied in the regression equation to better predict y values. We assume the error variance of all $(x_i, \hat{y}_i)$ pairs is the same, hence the uniform width of the

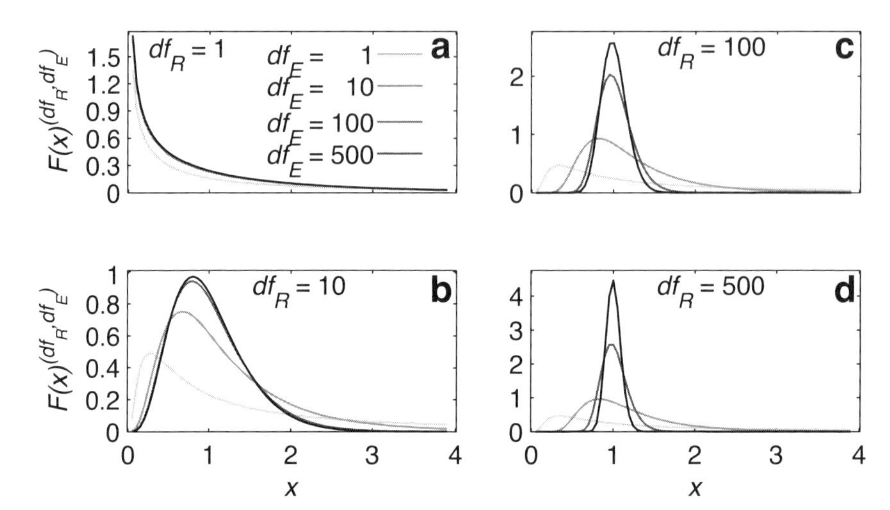

Figure 9.11. The F distribution for various df combinations. Panels a–d present $df_R = (10^0, 10^1, 10^2, 10^3)$, respectively, as indicated. In each panel, the progressively darkening curves present $df_E = (10^0, 10^1, 10^2, 10^3)$, as shown in the legend (shown in panel a but applicable to all panels).

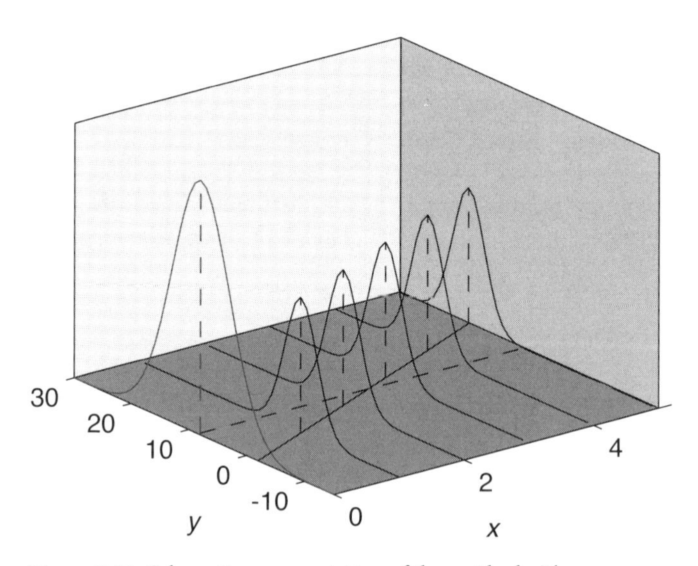

Figure 9.12. Schematic representation of the residuals. The gray distribution represents the expected distribution of the predicted variable y *in the absence of any predictor information*. The black distribution shows the way we imagine the distributions of the residuals about the fit.

black distributions. This width is proportional to $(y_i - \hat{y}_i)$, which is of course the essence of SS_E. The question is: By how much has our uncertainty regarding y values been reduced as a result of incorporating the x (predictor) information? Put differently, the question is: Now that we have introduced predictor (x) information into the problem, how much narrower is the (assumed fixed) distribution width of our y uncertainty (the error; black distributions) than the width of the *unconditional* y uncertainty (gray)? These questions to some extent are addressed by r_{xy} and SS_R-to-SS_E comparison. Most clearly, however, the answer is revealed by their scaled ratio

$$\frac{MS_R}{MS_E} = (N-2)\frac{\sum(\hat{y}_i - \bar{y})^2}{\sum(y_i - \hat{y}_i)^2}, \tag{9.163}$$

i.e., by the analysis of the variance.

Figure 9.13 shows a 20-element example that will, hopefully, clarify things. The upper panels shows an obviously bad fit; but what makes it so bad? Because the predictor–predictand relation is loose at best (it is, in fact, essentially non-existent), knowing x really doesn't help you at all when trying to make statements about y. The fit parameters, shown in table 9.1, tell the tale; $\beta_1 \ll \beta_2$ and $r_{xy,1} \ll r_{xy,2}$, suggesting that perhaps $\beta_1 \approx 0$ is the inevitable conclusion from the computed fit. That $F_1 \ll F_2$ further supports this tentative conclusion.

Figure 9.14 shows the two computed ratios in the context of the $F^{(1,18)}$ cumulative distribution function. The lower panel shows the probability for a random realization from an $F^{(1,18)}$ population being larger than the numbers shown along the horizontal axis. The dashed horizontal lines along $p = .5$ and $p = .05$ show the 50 and 5% likelihoods that a random realization (randomly drawn number) from a population whose distribution follows the $F^{(1,18)}$ one will prove *larger* than the x values (using Octave, `finv(.5,1,18)` = 0.47 and `finv(.95,1,18)` = 4.4 respectively). The interpretation of the shown p values is that the first fit, with $F_1 \approx 0.16$, has approximately 70% chance to have arisen spuriously. The second fit, with $F_2 \approx 3.0$, has about 1-in-10 chance to have arisen spuriously. With this

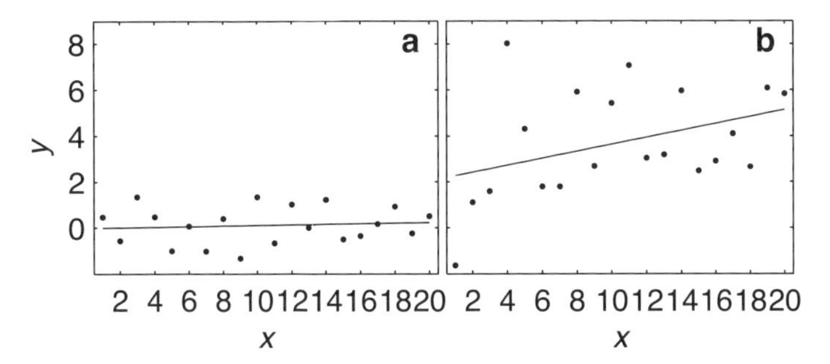

Figure 9.13. A poor fit (a) vs. a somewhat better one (b).

TABLE 9.1
The fit parameters for the shown fits.

	df	*SS*	*MS*	*F*	*p*	β	r_{xy}
1. poor fit:							
Regression	1	0.11	0.11	0.16	0.70	0.01	0.09
Error	18	12.46	0.69				
2. better fit:							
Regression	1	15.12	15.12	3.00	0.10	0.15	0.38
Error	18	90.51	5.03				

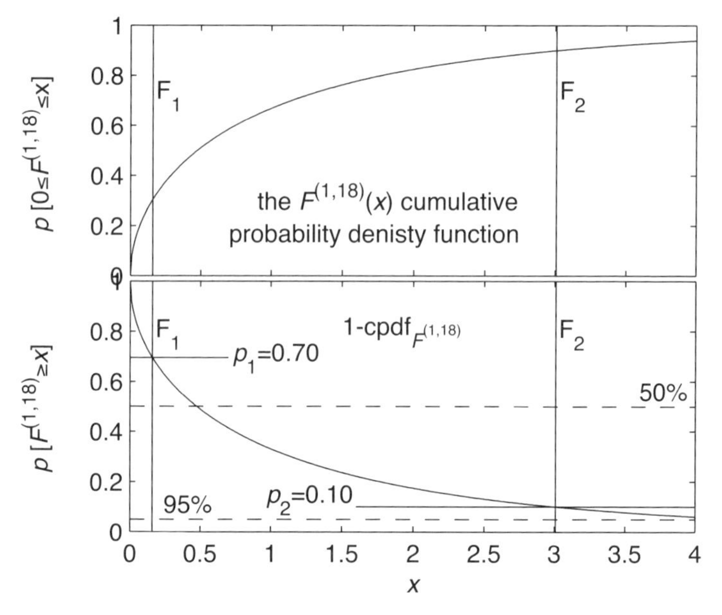

Figure 9.14. The upper panel shows the $F_{1,18}$ cumulative probability density function (i.e., the probability that a random realization from an $F_{1,18}$ population will fall between 0 and x). The two vertical lines are the computed F ratios for the two shown fits. The lower panel shows 1 minus the upper panel's curve, and the probabilities that a random $F_{1,18}$ variable will prove larger than the computed F ratios (70 and 10%, respectively). The dashed lines run along the 50 and 95% probabilities.

information, we have to apply subjective judgment to the question of whether or not the computed levels of uncertainty are acceptable. The usual confidence levels that are considered acceptable are $\geq 95\%$, or $p \leq .05$.

In the context of the regression analysis, the computed F ratios quantify the likelihood that the reduction in mean-squared error by the regression occurred spuriously (by chance). That is, in the absence of predictor information, the best prediction of y is its expected value, $y_i = E(y) = \bar{y}$ for all i. Under these circumstances, our expected error in estimating y values is estimated as simply the observed variability of y about its mean,

$$SS_T = \sum(y_i - \bar{y})^2 \tag{9.164}$$

(which is itself a proxy for the true, but unknown, population SS_T or the related variance).

In the presence of meaningful predictor information, our uncertainty in y values is reduced, because we are provided with information about a variable x that can shed some light on the variable of interest, the predictand y. Under these circumstances, our expected error becomes

$$SS_E = \sum(y_i - \hat{y}_i)^2. \tag{9.165}$$

Thus, the error reduction due to x is $SS_T - SS_E$, which we normalize for convenience to

$$\frac{SS_T - SS_E}{SS_E} = \frac{SS_R}{SS_E}. \tag{9.166}$$

However, as we have already discussed, the numerator and denominator in this expression do not have the same number of degrees of freedom. Thus, for a universal measure of performance we scale these by the appropriate *df* and get the F ratio,

$$F = \frac{MS_R}{MS_E} = (N-2)\frac{\sum(\hat{y}_i - \bar{y})^2}{\sum(y_i - \hat{y}_i)^2}. \tag{9.167}$$

Hence, again, the computed F is a measure of how likely the observed reduction in rmse is to have arisen by chance.

9.6.1 Variance of the Regression Parameters

Considering the simple single-predictor case, let's rewrite

$$\beta = \frac{\sum(x_i - \bar{x})(y_i - \bar{y})}{\sum(x_i - \bar{x})^2} \tag{9.168}$$

as

$$\beta = \frac{\sum(x_i - \bar{x})y_i}{\sum(x_i - \bar{x})^2}, \tag{9.169}$$

which is possible because

$$\sum(x_i - \bar{x})\bar{y} = \bar{y}\sum x_i - \sum\bar{x}\bar{y} = N\bar{x}\bar{y} - N\bar{x}\bar{y} = 0. \tag{9.170}$$

Thus,

$$\beta = \frac{(x_1 - \bar{x})y_1 + (x_2 - \bar{x})y_2 + \cdots + (x_N - \bar{x})y_N}{\sum (x_i - \bar{x})^2}. \tag{9.171}$$

Since the x values are treated here as known, not as realizations from a random process, if we *assume*—as we have previously done (and consistent with fig. 9.12)—that the variance of all y terms is the same, σ^2, then

$$\begin{aligned} \text{var}(\beta) &= \frac{(x_1 - \bar{x})^2\sigma^2 + (x_2 - \bar{x})^2\sigma^2 + \cdots + (x_N - \bar{x})^2\sigma^2}{\sum (x_i - \bar{x})^4} \\ &= \sigma^2 \left[\frac{(x_1 - \bar{x})^2}{\sum (x_i - \bar{x})^4} + \frac{(x_2 - \bar{x})^2}{\sum (x_i - \bar{x})^4} + \cdots + \frac{(x_N - \bar{x})^2}{\sum (x_i - \bar{x})^4} \right] \\ &= \sigma^2 \sum \left[\frac{(x_i - \bar{x})}{\sum (x_i - \bar{x})^2} \right]^2 = \frac{\sigma^2}{s_{xx}^2} \sum (x_i - \bar{x})^2 \\ &= \frac{\sigma^2}{s_{xx}^2} s_{xx} = \frac{\sigma^2}{s_{xx}}. \end{aligned} \tag{9.172}$$

What is left is to define precisely σ and replace it with its sample estimate s. From the look of the above derivation, it would seem that σ^2 is y's variance. There are several problems with this interpretation of σ, however. First, we would like our error estimate to tighten the better the fit. In addition, because of the cancellations above, if we substitute s_y (y's sample standard deviation) for σ, the expression for var(β) becomes oblivious to (x, y) covariability, the essence of what is addressed here. To understand this point, imagine two fits with the same r_{xy}, but one in which y varies over much smaller range compared to case two. Then case 1's variance—and thus also its corresponding var(β)—will be much smaller than case 2's, yet by assumption both fits are equally good.

To overcome both problems, we set

$$s^2 \equiv \frac{\sum_{i=1}^{N} e_i^2}{N-2} = \frac{\sum_{i=1}^{N} (y_i - \hat{y}_i)^2}{N-2},$$

with which

$$\text{var}(\beta) = \frac{\sum_{i=1}^{N} (y_i - \hat{y}_i)^2}{s_{xx}(N-2)},$$

so σ becomes conceptually interpreted as a measure of the width of the black distributions in fig. 9.12.

Therefore, our confidence interval for β is

$$\beta - \frac{t_{\text{sig}}^{N-2} s}{\sqrt{s_{xx}}} \le \beta^{\text{true}} \le \beta + \frac{t_{\text{sig}}^{N-2} s}{\sqrt{s_{xx}}}, \tag{9.173}$$

where t_{sig}^{N-2} is the tabulated value from the t distribution with $df = N - 2$ at the desired level of significance. This can be used to test the formal hypotheses

$$H_0: \ \beta = \gamma, \qquad H_1: \ \beta \neq \gamma, \tag{9.174}$$

where γ is any number, possibly zero, from which we want to be able to distinguish the computed β. If the above brackets exclude γ at the desired significance, H_0 is rejected.

For example, in the two example fits introduced in fig. 9.13, $\sqrt{s_{xx}} \approx 6.08$, $\beta_1 \approx 0.01$ and $\beta_2 \approx 0.15$, $s_1 \approx 0.83$ and $s_2 \approx 2.24$. Taking a desired threshold significance of $p = 0.05$, so that $t_{\text{sig}}^{N-2} = t_{0.975}^{18} \approx 2.1$ (in Matlab or Octave this is obtained using `tinv(1-0.05/2,18) = 2.109`), therefore,

$$\beta_1 - \frac{2.1 \cdot 0.83}{6.08} \leq \beta_1^{\text{true}} \leq \beta_1 + \frac{2.1 \cdot 0.83}{6.08},$$
$$0.01 - 0.29 \leq \beta_1^{\text{true}} \leq 0.01 + 0.29,$$
$$-0.28 \leq \beta_1^{\text{true}} \leq 0.30$$

and

$$\beta_2 - \frac{2.1 \cdot 2.24}{6.08} \leq \beta_2^{\text{true}} \leq \beta_2 + \frac{2.1 \cdot 2.24}{6.08},$$
$$0.15 - 0.77 \leq \beta_2^{\text{true}} \leq 0.15 + 0.77,$$
$$-0.62 \leq \beta_2^{\text{true}} \leq 0.92.$$

Neither fit is particularly useful. If you are surprised by the relatively poor performance of the apparently better fit, note that this is not really a fair comparison, because while $\|\mathbf{e}_1\| / \|\mathbf{y}_1\| \approx 1.0$ and $\|\mathbf{e}_2\| / \|\mathbf{y}_2\| \approx 0.5$, i.e., the scaled error reduction is twice as high in the "better" (second) case, $\|\mathbf{e}_2\| / \|\mathbf{e}_1\| \approx 2.7$, i.e., the second case's error is still twice as large as the first's, because $\|\mathbf{y}_2\| / \|\mathbf{y}_1\| \approx 5.4$, the second case's predictand is still five and a half times larger than that of the "worse" case.

Similar logic leads to a confidence interval for the regression intercept α:

$$\text{var}(\alpha) = \frac{\sigma^2 \sum x_i^2}{N \sum (x_i - \bar{x})^2}, \tag{9.175}$$

where σ is the same one discussed above. Again substituting the sample error variance s for the unknown σ, we obtain confidence intervals for α^{true},

$$\alpha - t_{\text{sig}}^{N-2} \sqrt{\frac{s^2 \sum x_i^2}{N s_{xx}}} \leq \alpha^{\text{true}} \leq \alpha + t_{\text{sig}}^{N-2} \sqrt{\frac{s^2 \sum x_i^2}{N s_{xx}}}, \tag{9.176}$$

which can be used to test $H_0 : \alpha = \gamma$, $H_1 : \alpha \neq \gamma$, hypotheses for any γ, in the same manner similar to the above usage of var(β) to test β-related hypotheses.

9.6.2 *Generalization: Confidence Intervals with Multiple Predictors*

The parameter variance estimates presented in the preceding section are special cases of a broader result addressing the broader context of arbitrary number of predictors. That result unifies the presented var(α) and var(β) estimates and adds covariances as well.

In multiple regression, we explain or predict $\mathbf{y}$ in terms of not one, but many predictors,

$$\mathbf{Ap} = \begin{pmatrix} 1 & x_{11} & x_{12} & \cdots & x_{1P} \\ 1 & x_{21} & x_{22} & \cdots & x_{2P} \\ & & \vdots & & \\ 1 & x_{N-1,1} & x_{N-1,2} & \cdots & x_{N-1,P} \\ 1 & x_{N1} & x_{N2} & \cdots & x_{NP} \end{pmatrix} \begin{pmatrix} \alpha_1 \\ \alpha_2 \\ \vdots \\ \alpha_{P-1} \\ \alpha_P \end{pmatrix}$$

$$= \begin{pmatrix} \mathbf{o} & \mathbf{x}_1 & \mathbf{x}_2 & \cdots & \mathbf{x}_P \end{pmatrix} \mathbf{p} = \begin{pmatrix} y_1 \\ y_2 \\ \vdots \\ y_{N-1} \\ y_N \end{pmatrix} = \mathbf{y}, \tag{9.177}$$

where we allow for a nonzero mean (hence $\mathbf{o}$, the leftmost column of ones), double indices are separated by commas where confusion is possible (e.g., $x_{N-1,1}$), P is the number of predictors, and, as usual, N denotes observation number. Note that this formulation is general, allowing for the possibility that some columns are *functions* of other, unspecified predictors or that some columns are different functions of the same predictor. This can assume the form $x_{ij} \propto \cos(\omega_j t_i)$ (where ω_j is the jth frequency and t_i is the ith time point, as in the Fourier matrix) or, when $x_{ij} = u_i^j$, the ith realization of the predictor variable u raised jth power, as in the polynomial fit.

In this general case the least-squares solution is still

$$\hat{\mathbf{p}} = \left(\mathbf{A}^T\mathbf{A}\right)^{-1}\mathbf{A}^T\mathbf{y}, \tag{9.178}$$

but now the estimated parameters' variance–covariance estimate is

$$\operatorname{var}(\hat{\mathbf{p}}) = \sigma^2\left(\mathbf{A}^T\mathbf{A}\right)^{-1} \tag{9.179}$$

where $\sigma^2 = \Sigma_{i=1}^N e_i^2/(N-2)$ is the misfit variance, as before. Let's first show that this reduces to our individual var(α) and var(β) estimates of the single-predictor case already presented.

With $P = 1$, the regression problem reduces to the familiar form

$$\mathbf{Ap} = \begin{pmatrix} 1 & x_1 \\ 1 & x_2 \\ & \vdots \\ 1 & x_{N-1} \\ 1 & x_N \end{pmatrix} \begin{pmatrix} \alpha_1 \\ \alpha_2 \end{pmatrix} = \begin{pmatrix} \mathbf{o} & \mathbf{x}_1 \end{pmatrix} \mathbf{p} = \begin{pmatrix} y_1 \\ y_2 \\ \vdots \\ y_{N-1} \\ y_N \end{pmatrix} = \mathbf{y}, \tag{9.180}$$

and

$$\begin{aligned}
\text{var}(\hat{\mathbf{p}}) &= \sigma^2 \begin{pmatrix} \mathbf{o}^T\mathbf{o} & \mathbf{o}^T\mathbf{x}_1 \\ \mathbf{x}_1^T\mathbf{o} & \mathbf{x}_1^T\mathbf{x}_1 \end{pmatrix}^{-1} \\
&= \frac{\sigma^2}{N\sum_{i=1}^N x_i^2 - \left(\sum_{i=1}^N x_i\right)^2} \begin{pmatrix} \sum_{i=1}^N x_i^2 & -\sum_{i=1}^N x_i \\ -\sum_{i=1}^N x_i & N \end{pmatrix} \\
&= \frac{\sigma^2}{N\sum_{i=1}^N x_i^2 - N^2\bar{x}^2} \begin{pmatrix} \sum_{i=1}^N x_i^2 & -N\bar{x} \\ -N\bar{x} & N \end{pmatrix} \\
&= \frac{\sigma^2}{\sum_{i=1}^N x_i^2 - N\bar{x}^2} \begin{pmatrix} \frac{\sum_{i=1}^N x_i^2}{N} & -\bar{x} \\ -\bar{x} & 1 \end{pmatrix} \\
&= \frac{\sigma^2}{\sum_{i=1}^N \left(x_i - \bar{x}\right)^2} \begin{pmatrix} \frac{\sum_{i=1}^N x_i^2}{N} & -\bar{x} \\ -\bar{x} & 1 \end{pmatrix}.
\end{aligned} \tag{9.181}$$

Element (1, 1) of this matrix is the variance of $\hat{\mathbf{p}}$'s first element, the constant term coefficient previously denoted α, and here α_1,

$$\text{var}(\alpha_1) = \frac{\sigma^2 \sum_{i=1}^N x_i^2}{N\sum_{i=1}^N \left(x_i - \bar{x}\right)^2}. \tag{9.182}$$

Similarly, element (2, 2) is what we previously denoted var(β), with the more general notation

$$\text{var}(\alpha_2) = \frac{\sigma^2}{\sum_{i=1}^N \left(x_i - \bar{x}\right)^2}, \tag{9.183}$$

as before. And what about the off-diagonal terms? They are the *co*variance of α_1 and α_2,

$$\text{cov}(\alpha_1, \alpha_2) = \frac{-\sigma^2\bar{x}}{\sum_{i=1}^N \left(x_i - \bar{x}\right)^2}. \tag{9.184}$$

In general, the variance estimate of the model's ith parameter ($\hat{\mathbf{p}}$'s ith element) is the ith diagonal element of the var($\hat{\mathbf{p}}$) matrix,

$$\text{var}(\hat{p}_i) = \text{diag}_i\left[\sigma^2\left(\mathbf{A}^T\mathbf{A}\right)^{-1}\right] \equiv \left[\sigma^2\left(\mathbf{A}^T\mathbf{A}\right)^{-1}\right]_{ii}. \tag{9.185}$$

Similarly, the covariance estimate of the model's ith and jth parameters is (cov $\hat{p}_i$, $\hat{p}_j$):

$$\text{cov}(\hat{p}_i, \hat{p}_j) = \left[\sigma^2\left(\mathbf{A}^T\mathbf{A}\right)^{-1}\right]_{ij \text{ or } ji}. \tag{9.186}$$

9.6.3 *Correlation Significance*

While in the business of devising confidence intervals around our regression results, an equally basic and essential result we must estimate confidence intervals for is the correlation coefficient, numerical brackets within which we expect, with specified level of confidence, the correct but unknown correlation coefficient r^{true} to reside.

9.6.3.1 Using the t Test

A commonly employed tool is the t test, stemming from the fact that for normally distributed predictor and predictand data, the correlation coefficient can be transformed into a t-distributed variable with $df = N - 2$ (where, in general, N is not the actual number of data points, but rather the number of *independent* data points, obtained from the acf),

$$t = r\sqrt{\frac{N-2}{1-r^2}}. \tag{9.187}$$

When $t > t_{\text{sig}}^{N-2}$, the calculated correlation is said to be significant at $p = \text{sig}$. For example, if $r = 0.46$ is calculated for an $N = 20$ case, $t = 0.46\sqrt{18/(1-0.46^2)} \approx 2.2$. Consulting tabulated values of inverse t distribution, $t_{0.92}^{18} = 1.5$, $t_{0.96}^{18} = 1.9$, and $t_{0.98}^{18} = 2.2$ (in Octave or Matlab, this is obtained by `tinv([.92 .96 .98],18) = 1.4656 1.8553 2.2137`). Thus, the calculated t fails the significance test at the $p = (1 - 0.98)/2 = 0.01$ ($t < t_{0.98}^{18}$), but passes it at $p = (1 - 0.96)/2 = 0.02$ ($t > t_{0.98}^{18}$). How high should the calculated correlation should be to prove significant at a specified level? To answer this question, we take the inverse (t, r) relation

$$r_{\text{th}} = \frac{t_{\text{sig}}^{N-2}}{\sqrt{N-2+\left(t_{\text{sig}}^{N-2}\right)^2}}, \tag{9.188}$$

where r_{th} is the threshold correlation coefficient, the one required to achieve the specified level of significance. Figure 9.15 displays r_{th} for a wide range of N and five p values collectively spanning the range of commonly used significance levels.

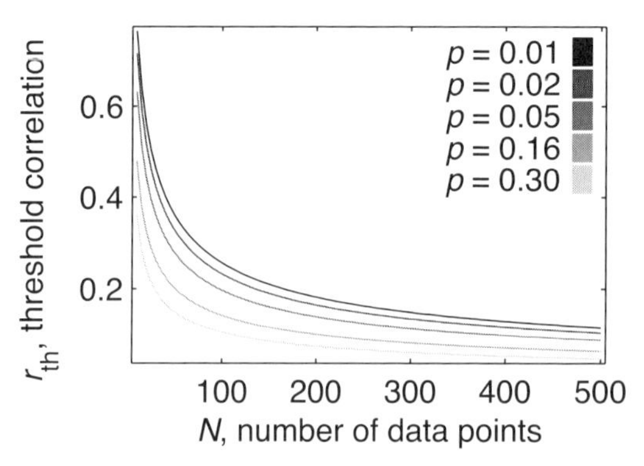

Figure 9.15. The threshold correlation coefficient needed to achieve significance at the levels specified by the gray shades as a function of realization number. In Octave or Matlab, this is obtained by `tinv(1-p/2,N-2)`.

9.6.3.2 USING THE FISHER z TRANSFORM

An alternative available too is the Fisher z-transform. Using the Fisher transform, these brackets are obtained from

$$\frac{1}{2}\ln\left(\frac{1+r}{1-r}\right)-\frac{z_{\text{sig}}}{\sqrt{N-3}}\leq\frac{1}{2}\ln\left(\frac{1+r^{\text{true}}}{1-r^{\text{true}}}\right)\leq\frac{1}{2}\ln\left(\frac{1+r}{1-r}\right)+\frac{z_{\text{sig}}}{\sqrt{N-3}}, \qquad (9.189)$$

where z_{sig} is the $N(0, 1)$ tail area at the specified significance level. (Because the expected distribution of the Fisher transformed correlation coefficient r is normally distributed with variance equal to $1/(N-3)$, $z_{\text{sig}}(N-3)^{-1/2}$ is z_{sig} expressed in units of the expected distribution width.)

For example, let's test the correlation for the better fit above, $r = 0.38$, using the $p = 0.10$ threshold. The necessary elements of our estimate are

$$\frac{z_{1-\frac{0.1}{2}}}{\sqrt{N-3}}=\frac{1.64}{\sqrt{20-3}}=0.40$$

and

$$\frac{1}{2}\ln\left(\frac{1+r}{1-r}\right)=\frac{1}{2}\ln\left(\frac{1+0.38}{1-0.38}\right)=0.40,$$

with which

$$0.40-0.40\leq\frac{1}{2}\ln\left(\frac{1+r^{\text{true}}}{1-r^{\text{true}}}\right)\leq 0.40+0.40,$$

$$0.00\leq\frac{1}{2}\ln\left(\frac{1+r^{\text{true}}}{1-r^{\text{true}}}\right)\leq 0.80,$$

$$\exp(0.00)=1.00\leq\frac{1+r^{\text{true}}}{1-r^{\text{true}}}\leq 4.95=\exp(0.80\cdot 2),$$

$$0.00\leq r^{\text{true}}\leq 0.66.$$

The range is broad and flirts with zero; the results are ambiguous. The correlation coefficient appears potentially, but unpersuasively, different from zero. If our desired confidence level is to have no more than 1-in-10 chance of being wrong, we may just barely be forced reject the null hypothesis

$$H_0:\ r^{\text{true}}=0, \qquad H_1:\ r^{\text{true}}\neq 0.$$

To show what an emphatic failure to reject looks like, let's now insist on $p < 0.01$,

$$\frac{z_{1-\frac{0.01}{2}}}{\sqrt{N-3}}=\frac{2.58}{\sqrt{20-3}}=0.62$$

and

$$0.40-0.62\leq\frac{1}{2}\ln\left(\frac{1+r^{\text{true}}}{1-r^{\text{true}}}\right)\leq 0.40+0.62,$$

$$0.64 \le \frac{1 + r^{\text{true}}}{1 - r^{\text{true}}} \le 7.69,$$

$$-0.22 \le r^{\text{true}} \le 0.77.$$

Now zero is unequivocally well within our estimated range, and we cannot reject

$$H_0: \ r^{\text{true}} = 0$$

at $p < 0.01$. This is the most a significance test can do; it will never prove anything. What *can* happen is a failure to reject a given tested hypothesis. In the final example, we most certainly did not prove that $r^{\text{true}} = 0$. Instead, we failed to conclude that $|r^{\text{true}}| > 0$, which left the least conjectural, default alternative, $r^{\text{true}} \approx 0$, standing.

9.6.4 *Multiple Regression: The Succinct Protocol*

In addition to prediction, in many cases, like in the Newton story used to introduce regression early in this chapter, regression is used to enhance mechanistic understanding of the studied process. This often involves weighing various alternative explanations of the studied phenomenon, eventually favoring one. Algebraically, each of the explanations yields a column or a group of columns that represent one explanation, followed by one or more columns representing the alternative process, and so on. Thus, in principle, we may begin with many potential predictors, many columns vying for a spot in the eventual **A**. Our job as researchers, in this case, is to sort out the wheat from the chaff, ending up with a carefully selected subset of predictors after discarding ones that failed to live up to their potential promise. This process—which implicitly invokes the principle of parsimony, or Occam's razor—is often the key step toward mechanistic, explanatory understanding, because it identifies the simplest, least conjectural, and thus most likely real underlying dynamics of the studied process.

How do we apply rigorous reasoning to choosing which predictors, **A**'s columns, are in and which are out? This is an elaborate procedure, discussed in much more details in classical regression text.[2] Let's review here the basic, central elements.

9.6.4.1 Step 1

A reasonable first step in the selection process would be to compute (using eq. 8.2) the correlation coefficient between the predictand, $\mathbf{y} \equiv \{y_i\}_{i=1}^{N}$, and each of the candidate predictors, $\{r_{\mathbf{x}_j\mathbf{y}}\}$, and the significance to which each $r_{\mathbf{x}_j\mathbf{y}}$ corresponds (using the methods of section 9.6.3). While all considered predictors are of the same length, their acf-based effective *df*s (see section 8.2)—and thus

[2] Two excellent introductory yet comprehensive personal favorite sources are Ryan, T. P. (1996) *Modern Regression Methods*, Wiley, New York, ISBN 0-471-52912-5; Draper, N. R. and H. Smith (1998) *Applied Regression Analysis*, 3rd ed., Wiley, New York, ISBN 0-471-17082-8.

their significance—can vary. Following these calculations, we will discard all considered predictors that are individually insignificantly correlated with the predictand. It would also make sense to choose the one with the highest $r_{\mathbf{x}_j\mathbf{y}}$ (or the most significant one, if there is a conflict) as the one sure to make it into the final model. This rule may need to be overridden, however, when there exists an alternative predictor with a similar (or similarly significant) correlation but a more compelling mechanistic explanation relating it to $\mathbf{y}$. Employing these criteria, there will be one predictor, call it $\mathbf{x}_1$, that is the first to be granted entry into $\mathbf{A}$, so that at this point $\mathbf{A} = \mathbf{x}_1$. Note that the choice of j with which $\mathbf{x}_1 = \mathbf{x}_j$ maximizes $r_{\mathbf{x}_j\mathbf{y}}$ is geometrically tantamount to identifying the predictor whose orientation in $\mathbb{R}^N$ is as nearly parallel to $\mathbf{y}$ as available.

9.6.4.2 Step 2

Next, it will be reasonable to check scaled projections,

$$\zeta_{1i} := \frac{\mathbf{x}_i'^T \mathbf{x}_1'}{\sqrt{\mathbf{x}_i'^T \mathbf{x}_i \mathbf{x}_1'^T \mathbf{x}_1'}}, \tag{9.190}$$

of remaining potential predictors on the chosen predictor. Predictors with low ζ_{1i} yet reasonably high $r_{\mathbf{x}_j\mathbf{y}}$ are particularly desirable, because they enrich geometrically $\mathbf{A}$'s column space, thus expanding the scope of $\mathbf{y}$'s $\mathbf{A}$ can be expected to skillfully predict. This is a subtle point, because theoretically, any $\mathbf{x}_{i\neq 1}$ linearly independent of $\mathbf{x}_1$ should expand $\mathcal{R}(\mathbf{A})$ to the same extent. In reality, however, that is not so, because if some $\mathbf{x}_{i\neq 1}$ is nearly parallel to $\mathbf{x}_1$, the angle between them can be as small as the noise levels characteristic of the problem. In this case, $\mathbf{A}$'s rank increase will be largely artificial (noise rather than signal related), because $\mathbf{x}_1 \approx \mathbf{x}_i$ and the difference in $\mathbf{y}$ projections on $\mathbf{x}_1$ and $\mathbf{x}_i$ will be largely spurious. The most important shortcoming of this situation will be that the added predictor will not yield better skill.

To demonstrate this issue, consider the simple problem of explaining $\mathbf{y} = (1, 2, 3) - 2$ in terms of either $\mathbf{x}_1 = (1.01, 1.99, 3.01) - 2.003$ in case 1, or $\mathbf{x}_1$ and $\mathbf{x}_2 = (0.98, 2.02, 2.99) - 1.997$ in case 2 (to simplify matters, all vectors have zero mean). The two problems are

$$\begin{aligned} \text{case 1:} \quad & \mathbf{A}_1\mathbf{p}_1 = \mathbf{x}_1 p_{11} = \mathbf{y} \\ \text{case 2:} \quad & \mathbf{A}_2\mathbf{p}_2 = (\mathbf{x}_1 \quad \mathbf{x}_2)\mathbf{p}_2 = (\mathbf{x}_1 \quad \mathbf{x}_2)\begin{pmatrix} \mathbf{p}_{21} \\ \mathbf{p}_{22} \end{pmatrix} = \mathbf{y}, \end{aligned} \tag{9.191}$$

and we wish to compare the performance of $\mathbf{p}_1$ and $\mathbf{p}_2$, evaluated by the percent scaled scalar error

$$e_i := 100\frac{\|\mathbf{A}_i\hat{\mathbf{p}}_i - \mathbf{y}\|}{\|\mathbf{y}\|}. \tag{9.192}$$

Clearly, $\mathbf{x}_1$ and $\mathbf{x}_2$ are both essentially $\mathbf{y}$ plus noise with characteristic magnitudes roughly 100 times smaller than the signal. Therefore, case 1's correct answer is $\hat{p}_{11} \approx 1$, with which the hindcasting error is $\|\mathbf{x}_1\hat{p}_{11} - \mathbf{y}\|/$

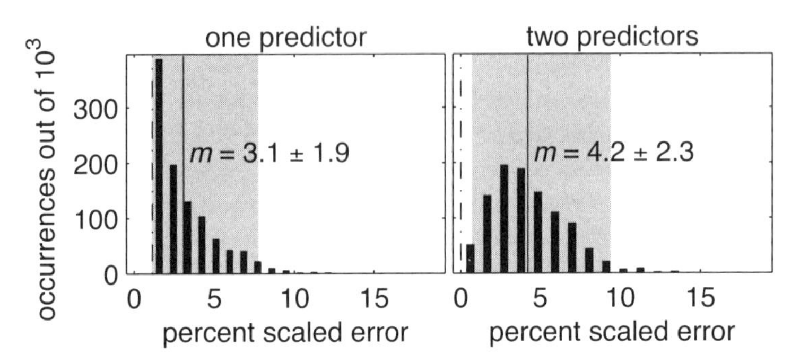

Figure 9.16. Histograms of scaled prediction error $100\|\hat{\mathbf{y}} - \mathbf{y}\|/\|\mathbf{y}\|$ using models with one (left) and two (right) predictors. The shaded regions show the respective scaled error range spanned by percentiles 2.5–97.5 of 10^3 noise realizations. The means and standard deviations of the entire populations are shown near the vertical full height bars, which show the respective means. The noise-free prediction errors are shown by the vertical dash-dotted lines.

$\|\mathbf{y}\| = \|\mathbf{x}_1 - \mathbf{y}\|/\|\mathbf{y}\| \sim \mathcal{O}(10^{-2})$. Given this result and $\mathbf{x}_1 \approx \mathbf{x}_2$, the solution of the second problem should be very nearly $\mathbf{p}_2 \approx (1, 0)$ or $(0, 1)$. Nonetheless, because case 2 has twice as many predictors (df) as case 1, it would be reasonable to expect $\mathbf{p}_2$ to better fit noise and thus yield a better hindcasting skill. This is indeed so, as the dash-dotted vertical lines in fig. 9.16 show; $e_1 \approx 1.2$, while $e_2 \sim \mathcal{O}(10^{-14})$.

But is this a good performance criterion? In the analysis of real data—in the process of trying to identify the correct model for as yet unexplained observations—our predictors and predictand(s) are specific realizations of random processes, yet it is the processes, not the specific realizations, we wish to explain. Let's therefore test the performance of cases 1 and 2 using a protocol expressly designed to test the process rather than the realizations. It is

loop on j between 1 and 10^3
  assign $\mathbf{y}_j = \mathbf{y} + N(0, 0.05^2)$
  calculate $\hat{\mathbf{p}}_{1j} = (\mathbf{A}_1^T\mathbf{A}_1)^{-1}\mathbf{A}_1^T\mathbf{y}_j$
  calculate $e_{1j} = 100\|\mathbf{A}_1\hat{\mathbf{p}}_{1j} - \mathbf{y}_j\|/\|\mathbf{y}\|$
  calculate $\hat{\mathbf{p}}_{2j} = (\mathbf{A}_2^T\mathbf{A}_2)^{-1}\mathbf{A}_2^T\mathbf{y}_j$
  calculate $e_{2j} = 100\|\mathbf{A}_2\hat{\mathbf{p}}_{2j} - \mathbf{y}_j\|/\|\mathbf{y}\|$
  save e_{1j} and e_{2j}
end of loop on j

where $N(0, 0.05^2)$ denotes zero mean normal distribution with 0.05^2 variance, i.e., noise with characteristic amplitude ~30 times smaller than **y**'s norm. The test design thus recognizes that real data vectors are never as pristine as **y**, yet a good model should be able to provide good predictions of the correct right-hand side **y**, not a specific, uninteresting, noise-contaminated realization thereof.

Figure 9.16 shows the histograms of $\{e_{1j}\}_{j=1}^{10^3}$ (left) and $\{e_{2j}\}_{j=1}^{10^3}$ (right). Unlike the pristine prediction error (dash-dotted vertical lines), which is larger in case 1 than in case 2, the statistics of the population that reflect more faithfully real-life applications proves model 1 superior to model 2. The mean errors satisfy $\bar{e}_2/\bar{e}_1 \approx 1.4$ (model 2's mean error is 40% larger than model 1's), and model 2's error population extends well further toward larger values. Finally, the majority of model 1's errors are very close to the error with the pristine **y**. Model 1's superior performance is actually even more convincing when considering the resultant F ratios, because model 2's $df = 2$ to model 1's $df = 1$.

This example's key elements are (1) a predictor very nearly parallel to **y**, the predictand (i.e., an excellent predictor); and (2) an additional predictor very nearly, but not quite parallel to the first. Conversely, the fact that **y** can be hindcasted virtually error free using the available predictors is secondary; we can add to **y** some non-$\mathcal{R}(\mathbf{A})$ piece, and all will remain the same, except that the full **y** will have to be replaced in such a case by $\mathbf{y}_r := \mathbf{A}^T\mathbf{y}$, the part of **y** from $\mathcal{R}(\mathbf{A})$ or the model reproducible part of **y**. Naively, we may expect that employing both predictors at once will improve the skill, because it adds a dimension to $\mathcal{R}(\mathbf{A}_2)$ relative to $\mathcal{R}(\mathbf{A}_1)$, thus expanding the subspace of $\mathbb{R}^3$ occupied by $\mathcal{R}(\mathbf{A})$. But since the signal (the meaningful part of **y**) is already nearly fully accounted for by $\mathbf{x}_1$, all that $\mathbf{x}_2$ does is overfit noise added to **y**. This has two outcomes: an improved hindcasting skill (the shift to the right of fig. 9.16's dash-dotted vertical lines in the left vs. right panel), accompanied by a decrease in what actually matters, the ability to explain (reproduce) the meaningful part of **y** while disregarding the meaningless noise. That is, not only is the addition of $\mathbf{x}_2$ to $\mathbf{A}_2$ not improving case 2's ability to explain the phenomenon of interest (**y**'s elements), it actually undermines it.

9.6.4.3 Steps > 2

Beyond the initial geometrical considerations offered in the above two sections, the issue of whether to add more predictors to the model, and, if so, which, is a basic statistical question discussed extensively in most regression texts, including the two (Ryan 1996; Draper and Smith 1998) mentioned earlier. The three most common selection protocols are forward and backward selection, and stepwise regression. There is no clearly superior strategy, and most employ some variant of the partial F test, which is best illustrated with a simple example.

Consider two competing models

$$\text{model 1:}\quad \mathbf{y} = p_{11}\mathbf{x}_1 \qquad \text{and} \qquad \text{model 2:}\quad \mathbf{y} = p_{21}\mathbf{x}_1 + p_{22}\mathbf{x}_2,$$

assuming, as before, that predictors and predictand are all centered (have zero mean) and that we have arranged the predictors so that $\mathbf{x}_1$ is already included in the model, and the question is whether $\mathbf{x}_2$ should enter the model as well (i.e., whether the latter model is superior to the former). Then

$$F = \frac{\Delta SS_R}{\Delta df_R \; MS_E} \equiv \frac{SS_R(\mathbf{x}_1,\mathbf{x}_2) - SS_R(\mathbf{x}_1)}{\Delta df_R \; MS_E}$$

$$= \frac{\sum_{i=1}^{N}(p_{21}x_{1i} + p_{22}x_{2i})^2 - \sum_{i=1}^{N}(p_{11}x_{1i})^2}{(2-1)\left[\sum_{i=1}^{N}(p_{21}x_{1i} + p_{22}x_{2i} - y_i)^2/(N-2)\right]} \tag{9.193}$$

$$= (N-2)\frac{\sum_{i=1}^{N}(p_{21}x_{1i} + p_{22}x_{2i})^2 - \sum_{i=1}^{N}(p_{11}x_{1i})^2}{\sum_{i=1}^{N}(p_{21}x_{1i} + p_{22}x_{2i} - y_i)^2},$$

where $\Delta df_R \equiv df_2 - df_1$ is the models' df difference, here 2 − 1, and MS_E is, as before, the mean squared residual characterizing the larger (more complex) of the two models (here model 2 with $df = N-2$). This calculated F is then compared with tabulated threshold values; if $F > F_p^{(2-1,N-2)}$, the hypothesis $H_0: p_{22}^{\text{true}} = 0$ is rejected at the desired significance level p, and $\mathbf{x}_2$ is declared a worthy predictor of $\mathbf{y}$.

Note that the number of possible models is not the number of predictors, but rather the number of predictor combinations (e.g., for three predictors the possibilities are $\mathbf{x}_1$, $\mathbf{x}_2$, $\mathbf{x}_3$, $(\mathbf{x}_1, \mathbf{x}_2)$, $(\mathbf{x}_2, \mathbf{x}_3)$, $(\mathbf{x}_1, \mathbf{x}_3)$, and $(\mathbf{x}_1, \mathbf{x}_2, \mathbf{x}_3)$). Given any order of testing potential models, at any decision juncture before the predictor pool has been exhausted the situation includes the most elaborate model thus far,

$$\text{model 1:} \quad \mathbf{y} = \sum_{i=1}^{n_1} p_{1i}x_i, \tag{9.194}$$

where n_1 is the number of predictors in model 1, and a suite of tested more elaborate alternative models comprising one additional predictor among the K remaining (not yet included) predictors,

$$\text{model } k\text{:} \quad \mathbf{y} = \sum_{i=1}^{n_1} p_{ki}\mathbf{x}_i + p_{k,n_1+1}\mathbf{x}_k, \qquad k = [1, K]. \tag{9.195}$$

At each such juncture, the challenge is to identify the most promising of the K candidates. A simple identifying criterion is based on model 1's residual,

$$\mathbf{e}_1 = \mathbf{y} - \sum_{i=1}^{n_1} p_{1i}\mathbf{x}_i, \tag{9.196}$$

which identifies the most reasonable next predictor as

$$\mathbf{x}_k: \quad \max_k |r_{\mathbf{x}_k\mathbf{e}_1}|, \tag{9.197}$$

where $\max_k$ denotes maximum over $k = [1, K]$, i.e., as the predictor most strongly correlated with model 1's residual. The reason this criterion is useful is that $\mathbf{e}_1$ represents the cumulative predictive or explanatory inadequacies of model 1's predictors. Those predictors span a subspace of $\mathbb{R}^N$, within which **y** fails to be fully contained. If we are to add another predictor to the model, what we want of it is to contribute to reducing the error, and the one predictor most capable of doing so is the one most nearly parallel to—most strongly correlated with—$\mathbf{e}_1$. In principle, the process must terminate when $\mathbf{e}_k$ is proven pure noise (as revealed from, e.g., a δ function like acf ($r_0 = 1$, $r_i \approx 0 \; \forall |i| > 0$) or more elaborate tests for whiteness) because beyond that point, a predictor can prove somewhat correlated with the residuals, yet it is purely by chance as the residuals have already been shown to have no structure.

9.7 Multidimensional Regression and Linear Model Identification

Identifying the optimal AR coefficient(s) of a scalar time series is a special case of model identification.[3] We envision the set of scalar data, $\{x_i\}$, to have arisen from a noisy AR process, whose coefficients we wish to optimally identify, thus unearthing the linear model governing the data behavior. Considering the AR(1) example, the process (model) we believe governs the data is

$$x_{i+1} = \alpha x_i + \xi_{i+1}, \tag{9.198}$$

and our job is to identify numerically the optimal α and possibly make some statements about the noise as well. In this section we will generalize model identification to vector processes.

Let's start by rederiving the scalar AR(1) coefficient α by following the procedure employed in section 8.1.1. Assuming a long enough, zero-mean, linearly evolving time series, we begin by multiplying (9.198) by x_i and applying the expectation operator. Here, invoking expectation simply means summing each term over all available pairs and dividing by the number of pairs (ignoring the fact that $\langle x_i x_i \rangle$ is one pair longer than $\langle x_{i+1} x_i \rangle$), and assuming the forcing term disappears whether or not it actually does, because the noise and the state are expected to be statistically unrelated (see section 8.1.1 for further details). This yields

$$\langle x_{i+1} x_i \rangle = \alpha \langle x_i x_i \rangle + \langle \xi_{i+1} x_i \rangle \quad \Longrightarrow \quad c_1 = \alpha c_0 \tag{9.199}$$

or $\alpha = c_1 c_0^{-1}$, where c_i denotes the ith autocovariance coefficient of the studied process, and α is outside the expectation operator because it is not a random variable. Note that, again assuming all terms share a common scaling factor, the expectation can also be written as

$$\mathbf{x}_+^T \mathbf{x}_0 = \alpha \mathbf{x}_0^T \mathbf{x}_0 + \xi_+^T \mathbf{x}_0 = \alpha \mathbf{x}_0^T \mathbf{x}_0, \tag{9.200}$$

where $\mathbf{x}_0$ holds $\{x_i\}_{i=1}^{N-1}$, $\mathbf{x}_+$ holds $\{x_i\}_{i=2}^{N}$, and ξ_+ holds $\{\xi_i\}_{i=2}^{N}$, so that

[3] Many thanks to Petros J. Ioannou, Physics Dept., National and Kapodistrian University of Athens, who generously and meticulously reviewed this section and offered many significant improvements.

$$\alpha = \frac{\mathbf{x}_+^T \mathbf{x}_0}{\mathbf{x}_0^T \mathbf{x}_0}. \tag{9.201}$$

While in section 8.1.1 we employed essentially the same procedure for estimating the acf of the process, here our motivation, and thus our interpretation of the above result, is different.

Here, we view α as the (admittedly very simple) finite-time, linear model governing x's evolution, with several key considerations. First, in nearly all cases x is a measurement of a physical property that evolves continuously in time, and the time-discrete nature of the data merely reflects such nonfundamental limitations as data collection protocols, transmission, or storage. The discrete x_is are thus best viewed as successive snapshots of the time-continuous $x(t)$, $x_i = x(t_i)$, with $\{x_i\}$ the discrete output of a forced nonlinear dynamical system of the general form $dx/dt = n[x(t)] + f[x(t), t]$, with a physically fundamental (and thus time-invariant), deterministic state-dependent internal model variability (given by $n[x(t)]$) and time- and state-dependent interactions with the external environment $f[x(t), t]$.

Let's consider the behavior of x over time intervals $[t_{\text{init}}, t_{\text{init}} + T]$ (where t_{init} is some initial time) well shorter than timescales over which x varies significantly under its governing nonlinearities. Over such relatively short timescales, $x = \langle x \rangle + x'$, i.e., x has a time-invariant or deterministically evolving mean $\langle x \rangle$ (where angled brackets denote time averaging), about which its fluctuations are $x'(t) \ll \langle x \rangle$. With this split, the full model becomes

$$\frac{d\langle x \rangle}{dt} + \frac{dx'}{dt} = n\big[\langle x \rangle + x'(t)\big] + f\big[\langle x \rangle + x'(t), t\big]. \tag{9.202}$$

To proceed, we expand the nonlinear terms in a Taylor series about $\langle x \rangle$ and—exploiting the fluctuations' smallness $x'(t) \ll \langle x \rangle$—truncate after the first order,

$$\frac{d\langle x \rangle}{dt} + \frac{dx'}{dt} \approx n\big[\langle x \rangle\big] + \frac{dn}{dx}\bigg|_{\langle x \rangle} x' + f\big[\langle x \rangle, t\big] + \frac{df}{dx}\bigg|_{\langle x \rangle} x'. \tag{9.203}$$

We proceed by noting that, by $\langle x \rangle$'s definition,

$$\frac{d\langle x \rangle}{dt} = n\big[\langle x \rangle\big] + f\big[\langle x \rangle, t\big]. \tag{9.204}$$

This equation represents the slow, stately evolution of the mean state under deterministic dynamics along a trajectory that is everywhere a solution of the nonlinear dynamics. An example may be highly predictable, well-understood warming by solar radiation from 5AM to 2PM. Thus, while f still varies (note its t dependence), it does so very slowly and in a well-understood manner. Subtracting the slowly varying evolution from the full evolution (subtracting eq. 9.204 from eq. 9.203) yields

$$\frac{dx'}{dt} = \frac{d(n+f)}{dx}\bigg|_{\langle x \rangle} x' + g(t), \tag{9.205}$$

where $g(t)$ combines state-independent external forcing and the imperfections of linearly approximating the nonlinear dynamics. Here t represents fast timescales, well distinct from the slow evolution of eq. 9.204. For example, if t in eq. 9.204 spans dawn to afternoon, as in the above example, then in eq. 9.205 t may mean seconds to minutes, governing (continuing with the temperature example) temperature variability due to physical processes other than increasing solar radiation (notably air turbulence).

Since locally (at $\langle x \rangle$), $d(n+f)/dx$ is a fixed number, we can write the fluctuation model more succinctly as

$$\frac{dx'}{dt} = m(t)x' + g(t), \tag{9.206}$$

where $m(t) := d(n+f)/dx$ evaluated at $\langle x \rangle$.

That is, the background to seeking an AR representation of the process is our view of x as locally *weakly* nonlinear, so that the nonlinear contributions can be lumped together with external forcing and viewed as at most time dependent but nearly state independent. We also employ a similar view of the model's nonautonomous behavior, considering $m(t) \approx m_0$ locally over relatively short time intervals.

With these simplifications, the approximate governing model for $0 \le t \le T$ is

$$\frac{dx'}{dt} = m_0 x(t) + g(t), \tag{9.207}$$

with the general solution over $t = [0, \tau < T]$

$$x_\tau = x_0 e^{m_0 \tau} + \int_0^\tau g(\xi)\, e^{m_0(\tau - \xi)} d\xi, \tag{9.208}$$

where $x_0 = x(t_0)$ is the initial state.[4] In eq. 9.208, the rightmost (integral) term represents the cumulative forcing history from time zero to time τ, in which forcing pulses $f(\xi)$ are propagated by the dynamics from the time they occurred, ξ, to the time of interest, τ. Over a specific time interval $[t_i, t_i + \tau]$, and viewing f as stochastic, with known statistical characteristics but unknown specific realizations, the linear model governing the evolution of $x(t)$ is

$$x_{t_i + \tau} = x_{t_i} e^{m_0 \tau} + \int_{t_i}^{t_i + \tau} g(\xi)\, e^{m_0(t_i + \tau - \xi)} d\xi. \tag{9.209}$$

With $x_i = x(t_i)$ and $x_{i+1} = x(t_i + \tau)$, this can be written as

$$x_{i+1} = x_i p_\tau + \psi_{i+1}, \tag{9.210}$$

where $p_\tau := e^{n_0 \tau}$ is the model *propagator*, the time-independent (inside $t = [0, T]$) linear operator progressing the system state τ time units forward, and

[4] E.g., Farrell, B. F. and P. J. Ioannou (1993) Stochastic dynamics of baroclinic waves, *J. Atmos. Sci.* **50**(24), 4044–4057.

$\psi_{i+1} := \int_{t_i}^{t_i+\tau} g(\xi)\, e^{m_0(\tau-\xi)} d\xi$, which we view as realized at time $t_i + \tau$ because we only observe the system at $t_i + \tau$ over the left-open interval $(t_i, t_i + \tau]$ (from which t_i itself is excluded on grounds that forcing up to and including that point has already been used in the most recent earlier model application, at t_i).

With $\alpha = p_\tau$ and $\xi_{i+1} = \psi_{i+1}$, eqs. 9.210 and 9.198 are the same, representing the same AR(1) process. We can thus take our interpretation of eq. 9.198's α as x's governing linear model a step further, into x's continuous time behavior, by converting the empirically obtained α—the process' τ-dependent propagator—to the τ- independent (and thus more fundamental) m_0 using $m_0 = \ln(\alpha)/\tau$.

Now let's generalize this to vector processes by considering a time sequence of M-vector observations $\{\mathbf{y}_i\}$, $\mathbf{y}_i \in \mathbb{R}^M$, with the above views about linearity and autonomy still in place. Generalizing the AR(1) process to M vectors, by analogy to eq. 9.200, we view the process as

$$\mathbf{y}_{i+1} = \mathbf{P}\mathbf{y}_i + \vec{\xi}_{i+1}, \tag{9.211}$$

where the rightmost vector is adorned with an arrow for clarity (as in boldface type Greek variables are fainter than Roman variables), $\mathbf{P} \in \mathbb{R}^{M\times M}$ is the process' single-step propagator, and all three terms are $M \times 1$ vectors. Postmultiplying the equation by $\mathbf{y}_i^T$, taking expectation, and envisioning the noise as statistically unrelated to (and thus on average orthogonal to) the state yields

$$\langle \mathbf{y}_{i+1}\mathbf{y}_i^T \rangle = \mathbf{P}\langle \mathbf{y}_i\mathbf{y}_i^T \rangle. \tag{9.212}$$

Assuming the time series of individual elements are centered (zero mean), introducing

$$\mathbf{D}_0 = \begin{pmatrix} \mathbf{y}_1 & \mathbf{y}_2 & \cdots & \mathbf{y}_{N-1} \end{pmatrix} \in \mathbb{R}^{M\times(N-1)} \tag{9.213}$$

and

$$\mathbf{D}_1 = \begin{pmatrix} \mathbf{y}_2 & \mathbf{y}_3 & \cdots & \mathbf{y}_N \end{pmatrix} \in \mathbb{R}^{M\times(N-1)}, \tag{9.214}$$

and ignoring a common scaling factor, eq. 9.212 becomes

$$\mathbf{D}_1\mathbf{D}_0^T = \mathbf{P}\mathbf{D}_0\mathbf{D}_0^T \tag{9.215}$$

or

$$\mathbf{C}_1 = \mathbf{P}\mathbf{C}_0, \tag{9.216}$$

where $\mathbf{C}_0$ and $\mathbf{C}_1$ are, to within a scaling factor, the process' lag 0 and 1 covariance matrices. Assuming the analysis has enough available realizations to render $\mathbf{C}_0$ full rank, we obtain the single step propagator by postmultiplying both sides by $\mathbf{C}_0^{-1}$,

$$\mathbf{P} = \mathbf{C}_1\mathbf{C}_0^{-1}. \tag{9.217}$$

In principle, we can go a step further, again calculating the τ independent dynamical operator governing $\mathbf{y}$'s evolution, $\mathbf{A}$ in

$$\frac{d\mathbf{y}}{dt} = \mathbf{A}\mathbf{y} + \mathbf{f}, \tag{9.218}$$

using

$$\mathbf{A} = \frac{1}{\tau}\ln\left(C_1 C_0^{-1}\right), \tag{9.219}$$

where the above ln means the matrix logarithm, the inverse of the matrix exponential and, as before, $\tau \stackrel{\text{def}}{=} t_{i+1} - t_i$ is the time step size.[5] In practice, $\mathbf{C}_1\mathbf{C}_0^{-1}$ may have nonpositive eigenvalues, which makes its logarithm formally undefined. This can arise if the studied process is poorly described as an AR(1), resulting in a rank-deficient $\mathbf{C}_1$, or if the process is dominated by waves, reflecting phase indeterminacy.

9.7.1 *Does This Method Work?*

Let's focus on identifying the propagator **P**, which—being the exponential of $\mathbf{A}\tau$—is a legitimate way of identifying a model.

9.7.1.1 Synthetic Example

Let's test the method using $t = 0, \tau, 2\tau, \ldots, N\tau$, $\tau = 4 \cdot 10^{-5}$, $N = 5 \cdot 10^4$, $M = 10$, and

$$\mathbf{A} = \begin{pmatrix} -2 & 1 & & & & & \\ 1 & -2 & 1 & & & & \\ & 1 & -2 & 1 & & & \\ & & & \ddots & & & \\ & & & 1 & -2 & 1 & \\ & & & & 1 & -2 & 1 \\ & & & & & 1 & -2 \end{pmatrix} + \left\{50 \sin\left(\frac{2\pi i}{M}\right)\cos\left(\frac{2\pi j}{M}\right)\right\}, \tag{9.220}$$

where $i, j = [1, M]$ are the row/column indices respectively. The first part of **A** is the diffusive operator, meant to bestow on **A** the adequate asymptotic decay with which $d\mathbf{y}/dt = \mathbf{A}\mathbf{y} + \mathbf{f} \approx 0$ represents an approximate energy balance between energy added by **f** and that removed by **A**'s inherent dissipation. The

[5] E.g., Penland, C. (1989) Random forcing and forecasting using principal oscillation pattern analysis, *Monthly Weather Review* **117**, 2165; Penland, C. and L. Matrosova (1994) A balance condition for stochastic numerical models with application to the El Niño–Southern Oscillation, *J. Climate* **7**, 1352; or DelSole, T. and B. F. Farrell (1996) The quasilinear equilibration of a thermally maintained, stochastically excited jet in a quasigeostrophic model. *J. Atmos. Sci.* **53**, 1782.

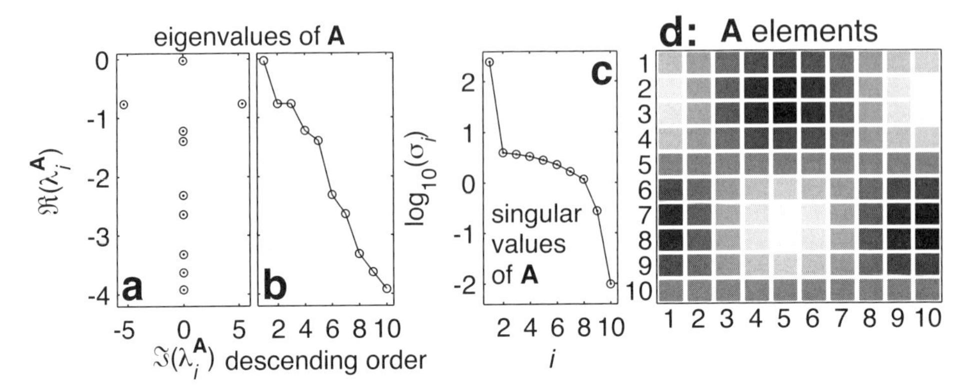

Figure 9.17. Some characteristics of the dynamical operator **A** discussed in section 9.7.1 and defined in eq. 9.220. The test **A**'s eigenvalues are shown on the complex plain (a) and arranged in descending real part (b), while its singular values are shown (c) as $\log_{10}$ because of their wide range. Panel d shows **A**'s elements.

second part of **A** (given element-wise in eq. 9.220 inside the curly brackets) is a checkerboard pattern that adds rotation to **A**'s action on vectors it premultiplies (or, put differently, guarantees that some of **A**'s eigenvalues will form complex conjugate pairs). Figure 9.17 shows **A**'s key attributes.

Using these t, **A** and $\mathbf{P} = e^{\mathbf{A}\tau}$, we then obtain synthetic data using

$$\mathbf{y}_{i+1} = \mathbf{P}\mathbf{y}_i + \vec{\xi}_{i+1}, \qquad i = 0, 1, 2, \ldots, N-1, \tag{9.221}$$

where the elements of both $\mathbf{y}_0$ and $\{\vec{\xi}_i\}$ are random draws from $N(0, 1)$, the standard normal distribution. We place all state vectors in

$$\mathbf{D} = \begin{pmatrix} \mathbf{y}_1 & \mathbf{y}_2 & \cdots & \mathbf{y}_N \end{pmatrix} \in \mathbb{R}^{M \times N} \tag{9.222}$$

and employ the above procedure to calculate $\hat{\mathbf{P}} = \mathbf{C}_1\mathbf{C}_0^{-1}$, calculating $\mathbf{C}_0$ and $\mathbf{C}_1$ using $\mathbf{D}_0$ and $\mathbf{D}_1$ defined in eqs. 9.213 and 9.214, where the hat, $\hat{\ }$, denotes "predicted" or "estimated," so that $\hat{\mathbf{P}}$ means the data-based (inverted) estimate of the true **P**.

To test the procedure's success (i.e., the degree to which **P** is retrieved in $\hat{\mathbf{P}}$), we employ several related metrics. First, in fig. 9.18a and b, we compare visually $\hat{\mathbf{P}}$ and **P** after subtracting **I** from each. There are two reasons for this subtraction. The first is merely mechanical, to facilitate better visual comparison. Given that the diffusive part of **A**, which dominates **A**'s three principal diagonals, is 50 times smaller than the checkerboard pattern (eq. 9.220), if the color range spans the diffusive part as well, all off-diagonal elements appear virtually identical and the ability to visualize the checkerboard pattern is lost. More importantly, the finite time propagator's key contribution is to leave things unchanged. For example, absent dynamical data or model information, the best

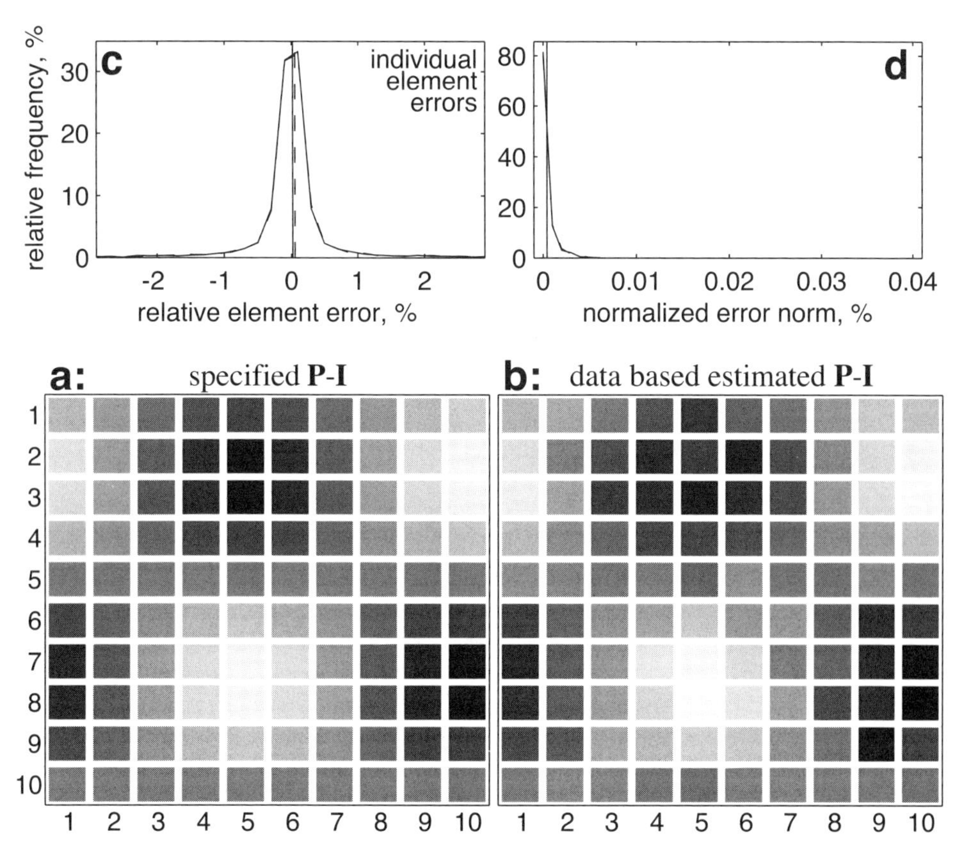

Figure 9.18. Model identification results for the problem discussed in section 9.7.1. Panels a and b show the dynamical operators as gray rectangles indicating element magnitudes, with the same gray scale shared by both. Panel a displays the specified $\mathbf{P} - \mathbf{I}$. Panel b shows the corresponding matrix with the estimated (inverted from data) matrix, $\hat{\mathbf{P}} - \mathbf{I}$. Panel c shows the distribution of normalized element errors, $e_{ij}^{(n)}$, $i = [1, 10^3]$, $j = [1, 10]$ for hindcasts (solid) and forecasts (dashed) with respective population means shown as vertical line segments. Panel d shows the distribution of normalized error norms, $n_i^{(e)}$, $i = [1, 10^3]$.

forecast of tomorrow's weather is today's. This null model skill, *persistence*, is the model's only skill when $\mathbf{P} = \mathbf{I}$, the zeroth-order term in the Maclaurin expansion of $e^{\mathbf{A}\tau}$. Since we are only interested in the model performance *beyond* the null skill of persistence, we eliminate the contribution of persistence by subtracting $\mathbf{I}$ from $\mathbf{P}$'s diagonal.

In the second performance metric, we use $\hat{\mathbf{P}}$ to predict various $\mathbf{y}_i$s; if the predicted vector was part of the inverted data set, we call $\hat{\mathbf{y}}_{i+1} := \hat{\mathbf{P}}\mathbf{y}_i$ "hindcast."

If, conversely, the predicted vector is *not* included in the inverted data set, we call $\hat{\mathbf{y}}_{i+1}$ "forecast." That is, e.g., if $\mathbf{D}_0$ comprises $\{\mathbf{y}_i\}_{i=1}^{499}$ and $\mathbf{D}_1$ holds $\{\mathbf{y}_i\}_{i=2}^{500}$, any $\hat{\mathbf{y}}_{[2,500]}$ (such as, e.g., $\hat{\mathbf{y}}_3 = \hat{\mathbf{P}}\mathbf{y}_2$) constitutes a hindcast, while $\hat{\mathbf{y}}_{503} = \hat{\mathbf{P}}\mathbf{y}_{502}$ is considered a forecast because we use $\hat{\mathbf{P}}$ based on $i = [1, 500]$ to predict observations at $i > 500$. Note that with these definitions, hindcast and forecast are multidimensional generalizations of interpolation and extrapolation, respectively.

Once a $\hat{\mathbf{y}}_i$ is obtained, its corresponding prediction error vector is

$$\mathbf{e}_i = \hat{\mathbf{y}}_i - \mathbf{y}_i, \tag{9.223}$$

its jth element's normalized prediction error is

$$e_{ij}^{(\mathrm{n})} = 100\,\frac{\hat{y}_{ij} - y_{ij}}{y_{ij}} \tag{9.224}$$

and its normalized squared prediction error norm is

$$n_i^{(\mathrm{e})} = 100\,\frac{\mathbf{e}_i^T\mathbf{e}_i}{\mathbf{y}_i^T\mathbf{y}_i}, \tag{9.225}$$

where the latter two measures are reported as a percentage of the actual (observed) value. Figure 9.18c,d shows those performance measures for the 10^3 hind- and forecasts $\{\hat{\mathbf{y}}_i\}$ from $i = N - 10^3 + 1 = 49{,}001$ to $i = N = 5 \cdot 10^4$.

Figure 9.18c shows the distributions and means of individual elements' normalized prediction errors (eq. 9.224, where solid and dashed lines correspond to hind- and forecasts, respectively, and the histograms are computed over the 10^4 element populations comprising all $M = 10$ elements of $\mathbf{e}_i$ times 10^3 such predicted vectors used for performance evaluation). Clearly, for both hind- and forecasts alike the errors are minuscule relative to the vector elements they predict. Figure 9.18d shows a different and complementary view (eq. 9.225) of essentially the same high skills. Importantly, the hind- and forecast curves and means in figure 9.18c and d are almost indistinguishable, which is an indication that the statistics are "saturated," meaning that the last $\sim 10^3$ data vectors, $\{\hat{\mathbf{y}}_i\}$, $i = [49{,}001, 5 \cdot 10^4]$, are essentially entirely redundant relative to earlier observations, so that the additional observed vectors add nothing to the model skill.

How many observations must we have to achieve such excellent predictive power? This is addressed by fig. 9.19. Clearly 30 realizations are not enough, and 50, while better, are still insufficient. But with even as few as 70 observed vectors, the model skill is quite high (the mean error is $<10\%$), and by the time the inversion comprises 160 realizations, the mean error is $<2\%$. These encouraging results are by no means general, partly reflecting the idealized experimental setup employed here. Real data-based identified linear models rarely behave this well.

9.7.1.2 A Real Data Example

Let's examine near surface air temperature over the Equatorial Pacific using 12-hour averages from 9AM–9PM, January 1, 1920 to 9AM–9PM, December 31,

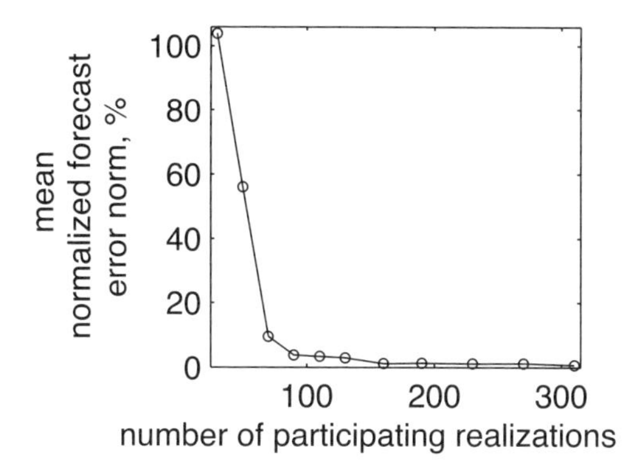

Figure 9.19. The dependence of model forecasting skill on the number of realizations participating in the inversion. Only forecasts are considered here, and the single performance measure used here is normalized squared prediction error norm (eq. 9.225) averaged over the 10^3 test vectors. The performance is evaluated 11 times, using 30, 50, 70, 90, 110, 130, 160, 190, 230, 270, and 310 realizations.

2007 ($N_t = 64{,}283$). To speed the analysis, we degrade the original data set[6] to a longitude resolution of 3° over 120°E–80°W so that there are $N_x = 54$ grid points per given latitude in the domain, and consider two latitudes, the equator and 6°N, to a total spatial grid of $M = 108$.

The first order of business is to devise a baseline, which we obtain by inverting for $\hat{\mathbf{P}}$ using $\mathbf{C}_0$ and $\mathbf{C}_1$ based on all but the last 10^3 observed vectors (i.e., using $N_t - 10^3 = 63{,}283$ observed vectors). Because of the much larger size of $\hat{\mathbf{P}}$ in this example, rather than show $\hat{\mathbf{P}}$'s elements, we characterize the inverted linear model by its two leading singular vectors (arising from the SVD representation $\hat{\mathbf{P}} = \mathbf{U}\Sigma\mathbf{V}^T$, see chapter 5), accounting for 17.2 and 2.5% of $\hat{\mathbf{P}}$'s total power and shown in fig. 9.20a–d. Given that $\mathbf{y}_i$ is first dotted with $\mathbf{V}$'s columns (in $\mathbf{V}^T\mathbf{y}_i$) and then transformed onto the direction of the corresponding column of $\mathbf{U}$, it is illuminating to see graphically $\mathbf{v}_{1,2}$ and $\mathbf{u}_{1,2}$ (fig. 9.20a,c and b,d, respectively). For both modes, to have a significant impact $\mathbf{y}_i$ must project strongly on fine scales (have high spatial frequencies) with particular importance to those waves' amplitudes near the domain's east and west boundaries and relative unimportance of the amplitudes near the domain's longitudinal center (the vicinity of the date line, 180°E/W). Both modes transform $\mathbf{y}_i$ energy in those fine scales into basin scale energy (fig. 9.20b,d) plus concentrated

[6] From the U.S. National Oceanic and Atmospheric Administration, Earth System Research Laboratory's 20th Century Reanalysis, e.g., Compo, G. P., J. S. Whitaker, and P. D. Sardeshmukh (2006) Feasibility of a 100 year reanalysis using only surface pressure data. *Bull. Am. Meteorol. Soc.* **87**, 175; or Whitaker, J. S., G. P. Compo, X. Wei, and T. M. Hamill (2004) Reanalysis without radiosondes using ensemble data assimilation. *Mon. Weather Rev.* **132**, 1190.

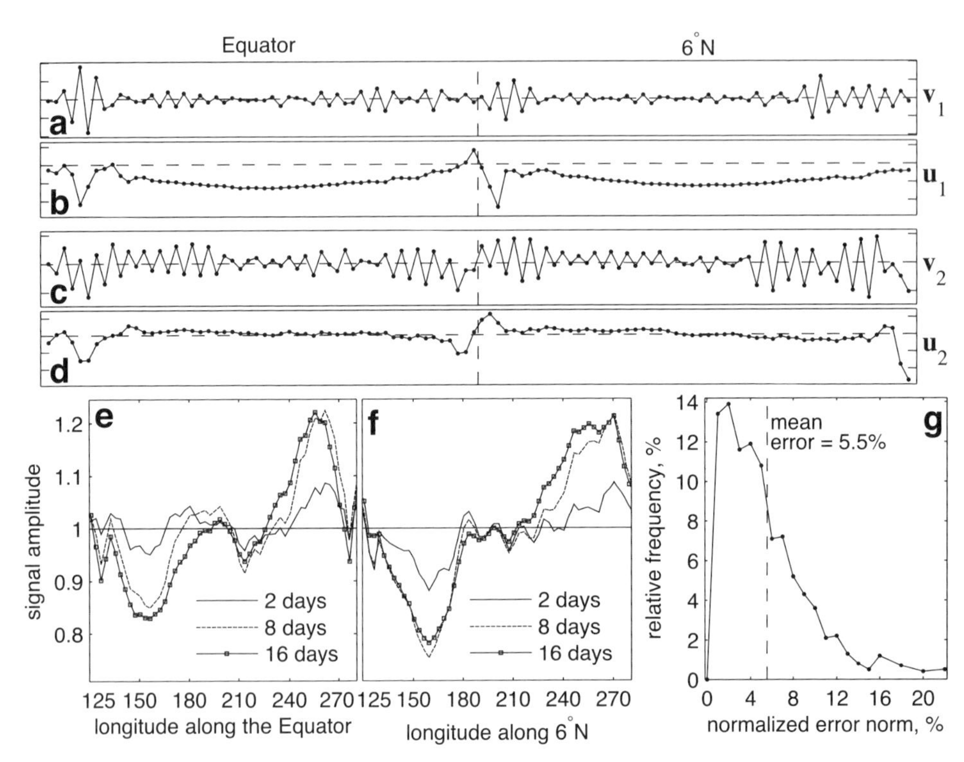

Figure 9.20. Some characteristics of the calculated $\hat{\mathbf{P}}$ discussed in section 9.7.1.2. Panels a–d show the leading singular vectors. The amplitude units (vertical axes) are arbitrary, and the spatial dependence (horizontal axes) reflects the structure of the observed vectors $\mathbf{y}_i \in \mathbb{R}^{M=2N_x=108}$, whose $[1, N_x]$ elements correspond to the equator (the left side of panels a–d, with the dashed vertical lines denoting the transition from the easternmost equatorial grid point to the westernmost grid point at 6°N), and $N_x + [1, N_x]$ elements correspond to 6°N (the right side of panels a–d). Panels e,f show the full action of repeat applications of $\hat{\mathbf{P}}$ to a uniform vector of 1s (solid horizontal lines) corresponding to passage of 2, 8, and 16 days (circles, diamonds, and squares, respectively). Panel g shows the distribution (solid) and mean (dashed) of forecast errors ($n_i^{(e)}$, eq. 9.225, with the grid point means removed to prevent unrealistically upward biased forecasting skill) derived from using the computed $\hat{\mathbf{P}}$ to forecast the last 3 years' worth of vectors ($365 \times 3 = 1095$).

additional energy near the east and west boundaries. Without diverting attention to El Niño dynamics, these results are intuitively appealing given the well-established[7] prominence of eastward (Kelvin) and westward (long Rossby) propagating linear shallow water waves in the Equatorial Pacific atmosphere.

[7] Sarachik, E. S. and M. A. Cane (2010) *The El Niño-Southern Oscillation Phenomenon*, Cambridge University Press, Cambridge, UK, ISBN 0521847869, 384 pp.; or Philander, S. G. (1989): *El Nino, La Nina, and the Southern Oscillation*, Academic Press, New York, ISBN 0125532350, 293 pp.

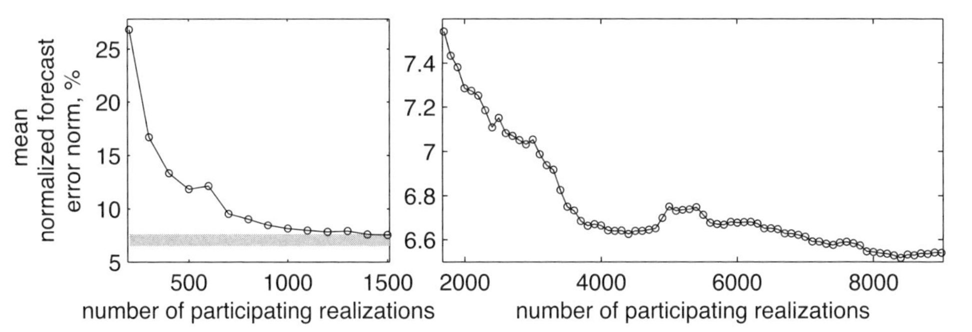

Figure 9.21. The dependence of the Equatorial Pacific temperature model forecasting skill on the number of realizations participating in the inversion. Like in fig. 9.19, the performance measure used here is normalized squared forecast error norm ($n_i^{(e)}$, eq. 9.225, with the grid point means removed) averaged over the 1095 test vectors. The left panel emphasizes the initial rapid error falloff, while the right panel emphasizes the subsequent slow error decay with increasing realization number. The right panel's vertical range is the left panel's shaded region.

With higher ($i > 2$) modes exhibiting qualitatively similar, but much lower amplitude, effects, fig. 9.20e,f shows what a spatially uniform $\mathbf{y}_i$ evolves into after the several shown time periods (because $\hat{\mathbf{P}}$ is asymptotically stable—all its eigenvalues' real parts are <1—the signal mean decays with number of $\hat{\mathbf{P}}$ applications, but in plotting fig. 9.20e,f this effect is suppressed by subtracting from each state its mean and adding the initial signal's mean, 1). Thus, $\hat{\mathbf{P}}$ transfers high-frequency energy into basin scale variability and transfers heat eastward.

How good is this model for forecasting 6 hours ahead? Figure 9.20g addresses this question, and shows errors 2–3 orders of magnitude higher than those reported in fig. 9.18d. Despite this, the errors are not devastatingly large; this certainly changes, for the worse, for long forecasting horizons (e.g., forecasting a week ahead), reflecting the model's great, only fleetingly warranted, simplicity.

And how rapidly does the model converge to the skill fig. 9.20g reports, with a ~5.5% mean forecasting error? Figure 9.21 addresses this question. The left panel shows rapid initial error decline, but the right panel shows that this effect quickly saturates, with a progressively smaller marginal added skill later realizations yield.

9.8 Summary

Regression is the algebraic technique most centrally located at the confluence of data analysis and modeling. When the analyst has a general idea what equation can be reasonably expected to govern the data behavior, and when that equation contains parameters not directly derivable from first principles, regression is the prime tool for choosing the best set of parameter values.

The assumed general model that gives rise to a regression problem is of the form

$$y = \alpha_1 f_1(x^{(1)}, \dots, x^{(n)}) + \cdots + \alpha_N f_N(x^{(1)}, \dots, x^{(n)}), \qquad (9.226)$$

where $\{x^{(i)}\}_{i=1}^n$ is the set of n observed explanatory variables on which y depends, and $\{f_i\}_{i=1}^N$ is the set of N (arbitrarily complex and not necessarily linear in $\{x^{(i)}\}_{i=1}^n$) functions that relate y to its predictors. When predictors $x^{(i)}$ and predictand y are sampled discretely M times and the model is applied to each discrete realization, the resultant system is

$$\begin{array}{ccccccc} y_1 & = & \alpha_1 f_1(x_1^{(1)}, \dots, x_1^{(n)}) & + & \cdots & + & \alpha_N f_N(x_1^{(1)}, \dots, x_1^{(n)}) \\ y_2 & = & \alpha_1 f_1(x_2^{(1)}, \dots, x_2^{(n)}) & + & \cdots & + & \alpha_N f_N(x_2^{(1)}, \dots, x_2^{(n)}) \\ & & \vdots & & \vdots & & \vdots \\ y_M & = & \alpha_1 f_1(x_M^{(1)}, \dots, x_M^{(n)}) & + & \cdots & + & \alpha_N f_N(x_M^{(1)}, \dots, x_M^{(n)}) \end{array}.$$

We then define $\mathbf{x} = (\alpha_1\ \alpha_2 \cdots \alpha_N)^T \in \mathbb{R}^N$, $\mathbf{b} = (y_1\ y_2 \cdots y_M)^T \in \mathbb{R}^M$, and

$$\mathbf{A} = \begin{pmatrix} f_1(x_1^{(1)}, \dots, x_1^{(n)}) & f_2(x_1^{(1)}, \dots, x_1^{(n)}) & \cdots & f_N(x_1^{(1)}, \dots, x_1^{(n)}) \\ f_1(x_2^{(1)}, \dots, x_2^{(n)}) & f_2(x_2^{(1)}, \dots, x_2^{(n)}) & \cdots & f_N(x_2^{(1)}, \dots, x_2^{(n)}) \\ \vdots & \vdots & & \vdots \\ f_1(x_M^{(1)}, \dots, x_M^{(n)}) & f_2(x_M^{(1)}, \dots, x_M^{(n)}) & \cdots & f_N(x_M^{(1)}, \dots, x_M^{(n)}) \end{pmatrix}$$
$$\in \mathbb{R}^{M \times N},$$

with which the system is succinctly

$$\mathbf{A}\mathbf{x} = \mathbf{b},$$

one of this chapter's key equations.

While it is possible in principle for $\mathbf{b}$ to be from $\mathbf{A}$'s columns space, so that an exact solution exists, this is highly unlikely and is extremely rarely realized. In nearly all practical application, no exact solution exists, and the best we can do is solve the system in a *least-squares* sense, which means obtaining a solution $\hat{\mathbf{x}}$ that renders the norm of the error vector $\mathbf{e} := \mathbf{A}\hat{\mathbf{x}} - \mathbf{b}$ a global minimum.

When the model is well chosen (it is physically plausible, and each f helps explaining y's variability beyond what other fs do, so that the columns of $\mathbf{A}$ are linearly independent), when observation number is ample ($M \gg N$), and when the observation are adequately mutually independent (so that $\mathbf{A}$ is full rank, $q = N \ll M$), then the least-squares solution is given by

$$\hat{\mathbf{x}} = \left(\mathbf{A}^T\mathbf{A}\right)^{-1}\mathbf{A}^T\mathbf{b},$$

another of this chapter's key equations. When some of these conditions are not met, other, imperfect yet practical, solution methods exist, as described in section 9.4.

In all cases, once a solution is obtained, statistical methods described in section 9.6 are used to quantify the model suitability and performance against alternative ones, and thus also the expected robustness of model predictions of future events not among the M events used to build the model.

TEN

The Fundamental Theorem of Linear Algebra

10.1 Introduction

THIS BRIEF CHAPTER, somewhat of a return to fundamentals, had to wait until we fully mastered regression. The chapter summarizes pictorially some of the linear algebraic foundations we have discussed thus far (most notably, section 3.3.1 and chapter 9) by revisiting the fundamental theorem of linear algebra, the unifying view of matrices, vectors, and their interactions. To make our discussion helpful and informal yet rigorous, and to complement the slightly more formal introduction of the basic ideas in section 3.3.1, here we emphasize the theorem's pictorial representation.

10.2 The Forward Problem

Figure 10.1 shows schematically what happens when $\mathbf{A} \in \mathbb{R}^{M \times N}$ maps an $\mathbf{x} \in \mathbb{R}^N$ from $\mathbf{A}$'s domain (upper left) onto $\mathbf{b} \in \mathbb{R}^M$ in $\mathbf{A}$'s range (lower right), how $\mathbf{A}$ transforms $\mathbf{x}$ into $\mathbf{b}$. This is the *forward* problem.

To enable visualization, both range and domain are depicted as three dimensional. This is, of course, just a metaphor, meant to make the schematic comprehensible by our limited sense of space; both domain and range can obviously be of arbitrary dimensions.

Both spaces are divided into two orthogonal subspaces: $\mathbb{R}^N$ (upper left) into the row and null spaces, $\mathcal{R}(\mathbf{A}^T)$ (red) and $\mathcal{N}(\mathbf{A})$ (blue), and $\mathbb{R}^M$ (lower right) into the column and left null spaces, $\mathcal{R}(\mathbf{A})$ (pink) and $\mathcal{N}(\mathbf{A}^T)$ (purple). The arbitrary $\mathbb{R}^N$ vector $\mathbf{x}$ comprises two orthogonal contributions: $\mathbf{x} = \mathbf{x}_n + \mathbf{x}_r$, where $\mathbf{x}_n \in \mathcal{N}(\mathbf{A})$ is the null space part, and $\mathbf{x}_r \in \mathcal{R}(\mathbf{A}^T)$ is the row space part.

When we carry out $\mathbf{Ax} = \mathbf{Ax}_n + \mathbf{Ax}_r$, a number of things happen. First, $\mathbf{x}_n$ is mapped to the zero vector in the target space, $\mathbf{x}_n \mapsto \mathbf{0} \in \mathbb{R}^M$; $\mathbf{Ax}_n = 0$ by construction. This is shown by the uppermost arrow, from $\mathbf{x}_n$'s tip in the upper left space to $\mathbf{0} \in \mathbb{R}^M$ in the lower right space. Conversely, $\mathbf{x}$'s row space part, $\mathbf{x}_r$, is mapped by $\mathbf{A}$ into the corresponding point $\mathbf{b}_r$ in the adjoint space: $\mathbf{Ax}_r = \mathbf{b}_r$, shown by the lower arrow from $\mathbf{x}_r$'s tip to that of $\mathbf{b}_r$. This mapping is exactly the same as $\mathbf{A}$ operating on the entire $\mathbf{x}$, because $\mathbf{x}$'s null space component $\mathbf{x}_n$ is mapped to zero, by construction, thus not changing the end result. This is shown by the middle arrow from $\mathbf{x}$'s tip to that of $\mathbf{b}_r$. Note also that there is no $\mathbf{A}$ and $\mathbf{x}$ combination that can give rise to $\mathbf{b}_n \in \mathcal{N}(\mathbf{A}^T) \in \mathbb{R}^M$.

You might ask yourself the following question: If we are solving here the *forward* problem, why choose an $\mathbf{x}$ with a nonzero null space component? Why

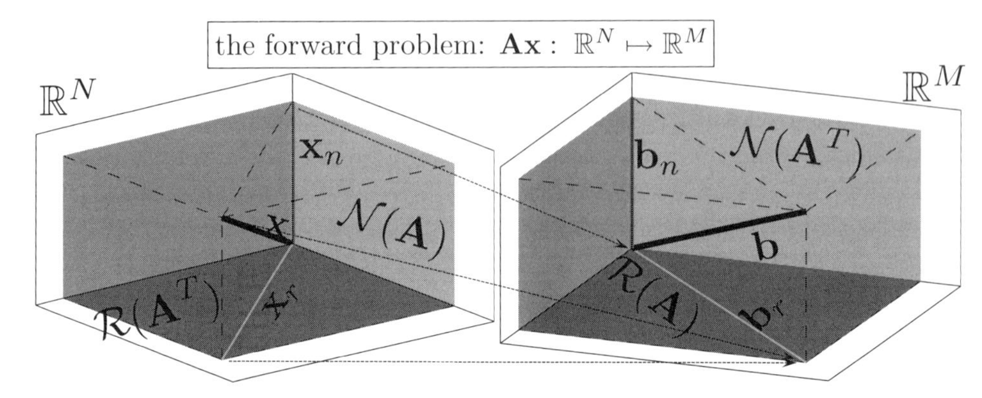

Figure 10.1. The forward problem, $\mathbf{A}$ operating on (premultiplying) $\mathbf{x}$, the first half of the fundamental theorem of linear algebra. The part of $\mathbf{x}$ from $\mathbf{A}$'s null space, $\mathbf{x}_n \in \mathcal{N}(\mathbf{A}) \subseteq \mathbb{R}^N$, is mapped by $\mathbf{A}$ to zero, as the top dashed arrow shows. The residual, $\mathbf{x}_r \in \mathcal{R}(\mathbf{A}^T) \subseteq \mathbb{R}^N$, as well as $\mathbf{x}$ in its entirety, is mapped by $\mathbf{A}$ to $\mathbf{b}_r \in \mathcal{R}(\mathbf{A}) \subseteq \mathbb{R}^M$, as the lowermost and middle dashed arrows, respectively, show.

choose $\mathbf{x}_n \neq \mathbf{0} \in \mathbb{R}^N$? This is a sensible yet easily answerable question. The reason is generality; arbitrary $\mathbb{R}^N$ vectors can contain $\mathcal{N}(\mathbf{A}^T)$ parts, and to be comprehensive, our schematic must therefore allow for this very realistic possibility.

10.3 The Inverse Problem

Figure 10.2 depicts the *inverse* problem, discussed in detail in chapter 9, section 9.4.1. While the spaces appear the same, the problem is very different. Here we start by obtaining a set of observations, the elements of $\mathbf{b}$ (dark yellow in the lower right space). Next, we envision certain physics that gave rise to $\mathbf{b}$, and, as before, construct $\mathbf{A}$ according to those physics. However, the parameters on which the solution depends (the components of $\mathbf{x}$) are now solved for optimally. The optimality of the solution means that $\mathbf{b}$'s left null space part, $\mathbf{b}_n$, which is the data noise or the part of the inverse problem we fail to reproduce, is as small as possible. However the solution is obtained (i.e., whether the $\mathbb{R}^M \mapsto \mathbb{R}^N$ mapping $\hat{\mathbf{x}} = \mathbf{Bb}$ is achieved using $\mathbf{B} = \mathbf{A}^+$ [derived in section 9.4.1 from $\mathbf{A}$'s SVD representation], or as $\mathbf{B} = (\mathbf{A}^T\mathbf{A})^{-1}\mathbf{A}^T$ [the least-squares solution derived in chapter 9]), there is nothing we can do about the (hopefully minor) inconsistencies $\mathbf{b}_n$ in the set of linear equations $\mathbf{Ax} = \mathbf{b}$; those inconsistencies are simply mapped onto the zero vector $\mathbf{Bb}_n = \mathbf{0} \in \mathbb{R}^N$, as the uppermost arrow shows. Given this, the best we can do is solve the closest problem to $\mathbf{Ax} = \mathbf{b}$ we know how to solve, $\mathbf{Ax} = \mathbf{b}_r$, where $\mathbf{b}_r = \mathbf{AA}^T\mathbf{b}$ is $\mathbf{b}$'s projection onto $\mathcal{R}(\mathbf{A})$.

Our troubles are not quite over yet. There still exists the all too realistic possibility that while we have reduced the insoluble problem $\mathbf{Ax} = \mathbf{b}$ to the tractable

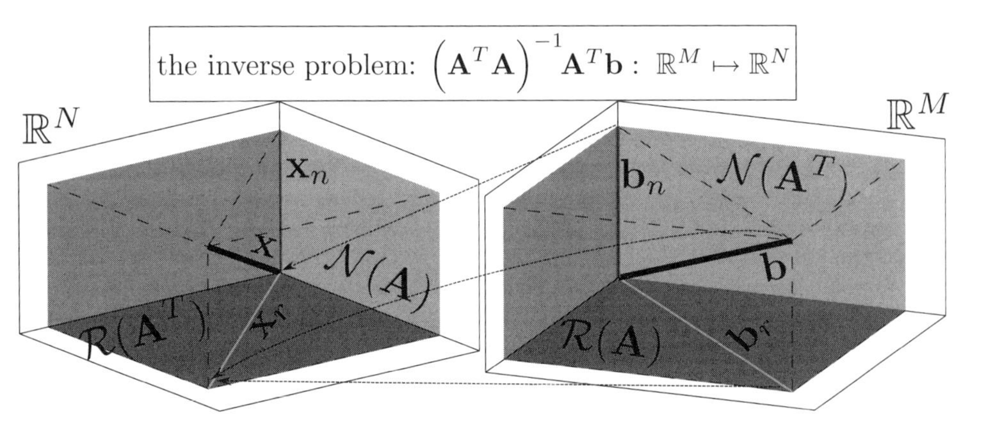

Figure 10.2. The inverse problem, **B** (see text in section 10.3) operating on (premultiplying) **b**, the second half of the fundamental theorem of linear algebra. The part of **b** from **A**'s left null space, $\mathbf{b}_n \in \mathcal{N}(\mathbf{A}^T) \subseteq \mathbb{R}^M$, is mapped by **B** to zero, as the top dashed arrow shows. The residual, $\mathbf{b}_r \in \mathcal{R}(\mathbf{A}) \subseteq \mathbb{R}^M$, as well as **b** in its entirety, are mapped by **B** to $\mathbf{x}_r \in \mathcal{R}(\mathbf{A}^T) \subseteq \mathbb{R}^N$, as the lowermost and middle dashed arrows, respectively, show.

$\mathbf{A}\hat{\mathbf{x}} = \mathbf{b}_r$, the solution may not be unique! This will happen when **A**'s rank q is smaller than its row dimension; $q < N$. In this case, **V**'s $q + 1 \to N$ columns, $\{\mathbf{v}_{q+1 \to N}\}$, span a nonempty $\mathcal{N}(\mathbf{A}) \subseteq \mathbb{R}^N$. In such a case, any null space component $\mathbf{x}_n$ can be added to the solution without any change in the goodness of the fit. Then, the solution $\mathbf{B}\mathbf{b}_r = \mathbf{B}\mathbf{b}$ performs a second, much less straightforward, optimization, minimizing $e = \|\mathbf{b}_n\|$: it simply picks the shortest possible **x**, the one with *no* null space component. That is, it sets $\mathbf{x}_n = \mathbf{0} \in \mathbb{R}^N$. This may be justified as the least conjectural, or most parsimonious, solution (given our imperfect information), but it is far from clear that these are the solution's most desirable attributes.

ELEVEN

Empirical Orthogonal Functions

11.1 Introduction

One of the most useful and common eigen-techniques in data analysis is the construction of empirical orthogonal functions, EOFs. EOFs are a transform of the data; the original set of numbers is transformed into a different set with some desirable properties. In this sense the EOF transform is similar to other transforms, such as the Fourier or Laplace transforms. In all these cases, we project the original data onto a set of functions, thus replacing the original data with the set of projection coefficients on the chosen new set of basis vectors. However, the choice of the specific basis set varies from case to case. In the Fourier case, for example, the choice is a set of sines and/or cosines of various frequencies, which is ideal for identifying the system's principal modes of oscillation. For example, if the signal projects strongly on sine waves with two specific frequencies, we will say that the signal is approximately the linear combination of these two frequencies. We will then attribute the remainder to other processes that are more weakly represented in the signal (the signal has small projection on them) and are thus assumed secondary. Another important and often desirable property for a basis is orthogonality (like Fourier sines of various frequencies), which permits accounting for any component of the signal only once (if a given part of the signal projects on one member of and orthogonal set, it has zero projection on the other set members). Alternative orthogonal bases to trigonometric ones include Legendre, Hermite, and Hough functions. (Orthogonality often holds only over a specific interval and sometimes requires "weighting functions." These are related to the metric choice, which I discuss briefly below.) The representation of the signal in terms of projection coefficients on a basis set is often very useful for separating cleanly various scales. For example, if our data comprises sea surface temperatures (SSTs) of a given ocean basin, we can think of the projection on the lowest-frequency full wave (with one crest and one trough within the domain) as representing the ocean's "basin scale," while that on wavelengths of order 10–100 km as representing "eddies."

In EOF analysis we also project the original data on a set of orthogonal basis vectors. However, unlike the Fourier or Legendre bases, the EOF basis is of empirical rather than analytic origin (hence the "E" in EOFs). The first EOF is chosen to be the pattern, without the constraint of a particular analytic form, on which the data project most strongly. In other words, with unusual exceptions, the leading EOF (sometime called the gravest or leading mode) is the pattern most frequently realized. The second mode is the one most commonly realized under the constraint of orthogonality to the first one, the third is the most frequently realized pattern that is orthogonal to both higher modes, and so on. The full set of normalized EOFs forms a complete orthonormal spanning set for the relevant space, with elements arranged in descending order of data projection on them.

11.2 Data Matrix Structure Convention

Data matrices can assume many, widely diverse forms and arrangements, and we can derive EOFs from just about any of them (how meaningful and revealing those will be is a different matter). To give our discussion the necessary specificity, however, we will envision our data matrices $\mathbf{A} \in \mathbb{R}^{M \times N}$ to conform with the following (easily transformed into when not originally true) structure.

Data matrices often have two distinct dimensions corresponding to different physical units. For EOF analysis to make sense, as will become clear soon, the order along one of those dimensions must be meaningful. Some examples will clarify this. Imagine administering IQ tests to M students at N locations. Whatever conclusions we may wish to draw from those tests, it is clear that the order of the visited locations is immaterial; $\mathbf{A}$'s columns can be interchanged at will with no loss of potential inferences. Conversely, imagine a data matrix whose (i, j) element is the county mean income for county i at time j. Now, clearly, the order of the columns matter: the leftmost is earliest, the subsequent one to its right is one time interval later, and so on, until the rightmost column, corresponding to the latest available income vector. Note that in both cases, however, the order along the columns (i.e., the order of the rows) is unimportant; whether we tested child m's IQ before or after child n's is entirely irrelevant, as is whether county p's income is reported before, or after, county q's. In our discussion, we will assume the analyzed data have at least one dimension along which order is important. We will call this dimension "time" (though space can also play the same role in cases where dependence of something on geography is examined) and assume it varies along the rows (i.e., we will assume that the rows of our data matrices $\mathbf{A}$ are time series at a particular location or for a particular item). If, e.g., $\mathbf{A}$ holds temperature data at M locations at N times, then $\mathbf{A}$'s jth column comprises the temperatures at all space points for the jth $(j = 1, 2, \dots, N)$ time point, and $\mathbf{A}$'s ith row holds the temperature time series at the ith spatial grid point $(i = 1, 2, \dots, M)$.

11.3 Reshaping Multidimensional Data Sets for EOF Analysis

While linear algebra knows only matrices, data sets obviously need not be two dimensional. For example, global atmospheric temperatures form the field $T(y,$

x, z, t), where x, y, z, and t denote the east–west, north–south, height, and time coordinates, and x follows y rather than the other way around because under Matlab or Octave this order simplifies the procedure. Consequently, $T(y, x, z, t)$ will be originally four dimensional. Since we do not know how to eigenanalyze multidimensional data sets, we need to be able to retrieve a matrix (or two-dimensional data set) from arbitrary-dimensional data sets. Unfortunately, this simple process often causes considerable and entirely needless confusion to some. Let's clarify the procedure using a hypothetical data set $F(y, x, t)$ depending on two space coordinates (x and y) and time, shown schematically in fig. 11.1.

We denote the indices of y, x, and t by $i = [1, m]$, $j = [1, n]$, and $k = [1, N]$ and introduce F's kth subset, the single time matrix representing the kth time point,

$$\mathbf{F}^k = \begin{pmatrix} \vdots & \vdots & & \vdots \\ \mathbf{f}_1^k & \mathbf{f}_2^k & \cdots & \mathbf{f}_n^k \\ \vdots & \vdots & & \vdots \end{pmatrix} \in \mathbb{R}^{m \times n}, \tag{11.1}$$

in which

$$\mathbf{f}_j^k = \begin{pmatrix} \mathbf{f}_{1j}^k \\ \mathbf{f}_{2j}^k \\ \vdots \\ \mathbf{f}_{mj}^k \end{pmatrix} \in \mathbb{R}^m \tag{11.2}$$

holds all m values of F at the jth x value and kth time (note that the superscripts k above denote the time index, not powers). The trick, if you wish to call it that, is to condense $\mathbf{F}^k$ into a vector by concatenating all N such vectors into a single matrix

$$\mathbf{A} = \begin{pmatrix} \vdots & \vdots & & \vdots \\ \mathbf{a}_1 & \mathbf{a}_2 & \cdots & \mathbf{a}_N \\ \vdots & \vdots & & \vdots \end{pmatrix} = \begin{pmatrix} \mathbf{f}_1^1 & \mathbf{f}_1^2 & \cdots & \mathbf{f}_1^N \\ \mathbf{f}_2^1 & \mathbf{f}_2^2 & \cdots & \mathbf{f}_2^N \\ & \vdots & & \vdots \\ \mathbf{f}_n^1 & \mathbf{f}_n^2 & \cdots & \mathbf{f}_n^N \end{pmatrix} \in \mathbb{R}^{M \times N}, \tag{11.3}$$

where $M = mn$ is the combined space dimension, and where $\mathbf{A}$'s kth column, $\mathbf{a}_k$, is implicitly defined above. If F has more space dimensions than two, the same procedure is repeated until the entire field at one time is a column vector whose dimension is the product of the dimensions of all space coordinates.

The order of the reshaping of an $\mathbf{F}^k$ matrix into a column vector is completely unimportant, but it is imperative to keep a record of the procedure and be able to undo it before plotting the EOFs. Given that Matlab and Octave reshape column-wise, the entire procedure requires a single instruction, `A= reshape(F,m*n,N)`, condensing F into $\mathbf{A}$.

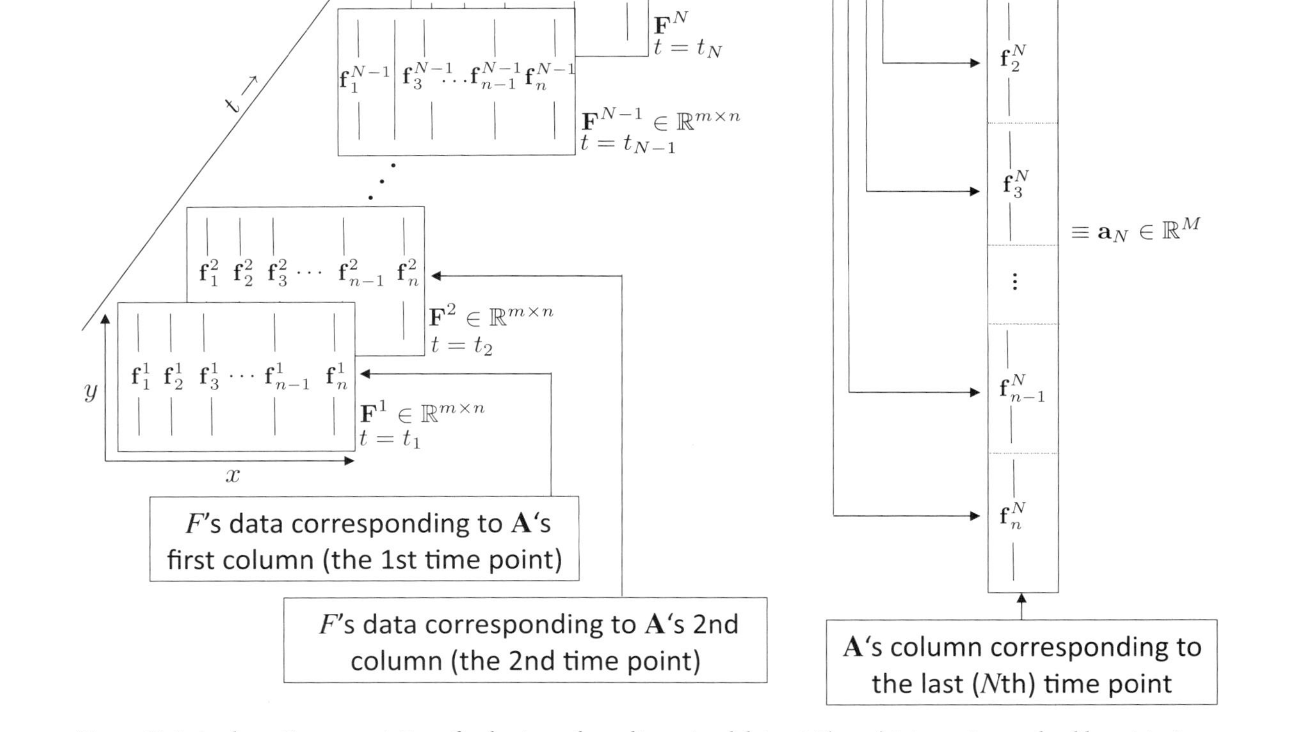

Figure 11.1. A schematic representation of reshaping a three-dimensional data set $F(y, x, t)$ into an eigenanalyzable matrix A. See text for details.

11.4 Forming Anomalies and Removing Time Mean

Once the data matrix **A** is created, the next issue is forming anomalies and removing the time mean of individual time series and possibly of the full data set. While the latter is trivial (requiring no more than subtraction of the grand mean, the average over all rows and columns), the former requires some care.

11.4.1 Anomalies

If individual time series (individual rows) are expected to be dominated by a seasonal cycle related to known physics, and if the focus of attention is not that cycle, but rather deviations from it, then the seasonal values of individual rows must be calculated and subtracted from each. As an example, and to remove ambiguity while allowing for an incomplete cycle (here at the end of the data temporal extent), let's assume the rows of our data matrix are monthly resolution time series, starting with January of year 1 and ending with July of year 92. The following Octave or Matlab code segment (which can be made much more efficient, as I wrote it with clarity, not efficiency, in mind) will compute and subtract the seasonal cycle, storing the monthly climatologies in the array `C` and the anomalies in `B`. Note that all lines of code end with a semicolon, which prevents the results of the row from being displayed in the screen's command window.

```
 1. C = zeros(M,12);
 2. B = A;
 3. for i = 1:M
 4.   a = A(i,1:12*91);
 5.   a = reshape( a , 12,91 );
 6.   m = mean(a,2)';
 7.   C(i,:) = m;
 8.   for j = 1:91
 9.     k = (j-1)*12;
10.     B(i,k+[1:12]) = A(i,k+[1:12]) - m;
11.   end
12.   k = k + 12;
13.   B(i,k+[1:7]) = A(i,k+[1:7]) - m(1:7);
14. end
```

Explanation of the above code's workings follow, aided by the identifying line numbers.

1. Allocate an array for the row climatologies.
2. Allocate an array for the anomalies.
3. Loop on **A**'s M rows.
4. Assign `a` = **A**'s ith row, years 1–91.
5. Turn `a` into a 12×91 matrix of calendar month by year.

6. Average over the 91 included years along the second dimension, and transpose (to make `m` a 1×12 row vector).
7. Record `m`, row *i*'s monthly climatologies.
8. Loop on the earlier 91 years.
9. Assign `k` to the index of the last previous December and use that as column index into `A` and `B`.
10. Subtract row *i*'s January–December climatologies from this row's *j*th year.
11. End loop on years 1–91.
12. Index of year 91's December.
13. Subtract row *i*'s January–July climatologies from this row's last (92nd) year.
14. End of loop on rows.

Remember that this step (calculating climatologies and removing them to get anomalies) is a matter of specific scientific considerations applied to a particular question asked of a particular data set, not a universal rule. If there is no a priori reason to expect a seasonal cycle, if plots of specific grid points' time series do not exhibit a clear, apparently deterministic, seasonality, or if that cycle is at the heart of the question being addressed, this step is skipped.

11.4.2 Removing in Time Means of Individual Time Series

Regardless of whether or not seasonal cycles have been removed from individual rows, the rows' time means must be removed at this point. While crucial, in Matlab or Octave this step requires a single line, `D = B - mean(B,2)*ones(1,N)`, which works as follows. The matrix `B` is the outcome of the seasonal cycle removal; if this step was deemed unnecessary, `D = A - mean(A,2)*ones(1,N)` replaces `D = B - mean(B,2)*ones(1,N)`. The `mean` operator is made to work on the second dimension, which is appropriate as we average time series, i.e., along the rows, the matrix's second dimension. The result of this is a column vector. We next postmultiply this column vector of row means by a row vector of ones, $(1, 1, \dots, 1) \in \mathbb{R}^{1 \times N}$. This step—which forms a matrix comprising the mean column vector duplicated N times—is necessary for rendering the subtraction dimensionality correct; we cannot subtract a vector from a matrix, but the dimensions of `mean(B,2)*ones(1,N)` or `mean(A,2)*ones(1,N)` are $M \times N$, the same as `B` and `A` themselves, producing a dimensionally correct subtraction.

11.5 Missing Values, Take 1

In many real-life applications, not all values of the data matrix are available because of instrument, transmission line, or recording device failure, reconfiguration of the observational array midstream, and so on. There are many different ways to deal with missing values, and some of those can be employed at this point.

The simplest filling procedure is useful when spatially neighboring values are available. For example, imagine a data field $v_{i,j}$ depending on two spatial dimensions indexed by i and j, and a missing value embedded within available values. Then the estimated filled in value (denoted by a tilde) can be, e.g.,

$$\tilde{v}_{i,j} = \frac{v_{i-1,j} + v_{i+1,j} + v_{i,j-1} + v_{i,j+1}}{4}, \tag{11.4}$$

$$\tilde{v}_{i,j} = \tfrac{1}{8}\big(v_{i-1,j-1} + v_{i-1,j+1} + v_{i+1,j-1} + v_{i+1,j+1} + v_{i-1,j} + v_{i+1,j} + v_{i,j-1} + v_{i,j+1}\big) \tag{11.5}$$

or

$$\tilde{v}_{i,j} = \frac{1}{4(1+w)}\big[v_{i-1,j-1} + v_{i-1,j+1} + v_{i+1,j-1} + v_{i+1,j+1} + w\big(v_{i-1,j} + v_{i+1,j} + v_{i,j-1} + v_{i,j+1}\big)\big]. \tag{11.6}$$

Above, w is some weight reflecting the larger distance to the focus point (i, j) of the peripheral points $(i \pm 1, j \pm 1)$ relative to points $(i, j \pm 1)$ and $(i \pm 1, j)$ (so (11.5) is (11.6) with $w = 1$, i.e., assuming all points are equally relevant to point (i, j)). More elaborate schemes, including more spatial points and more complex weighting conventions, are also possible. Protocols (11.4) and (11.6) are shown schematically in fig. 11.2's left and right, where, in the schematic for (11.6), the different weights of the arrows from immediate neighbors vs. those from diagonally distant ones reflect schematically the value of w.

A similar idea can be employed in time. For example, if the superscript k denotes the kth time point, and the temporally neighboring points are not missing, then

$$\tilde{v}^{k}_{i,j} = \frac{v^{k-1}_{i,j} + v^{k+1}_{i,j}}{2}, \tag{11.7}$$

$$\tilde{v}^{k}_{i,j} = \frac{v^{k-2}_{i,j} + v^{k-1}_{i,j} + v^{k+1}_{i,j} + v^{k+2}_{i,j}}{4} \tag{11.8}$$

and

$$\tilde{v}^{k}_{i,j} = \frac{v^{k-2}_{i,j} + v^{k+2}_{i,j} + w_t\big(v^{k-1}_{i,j} + v^{k+1}_{i,j}\big)}{2(1+w_t)}, \tag{11.9}$$

where w_t denotes the emphasis factor of points one time increment away over points two time increments away, are all reasonable choices.

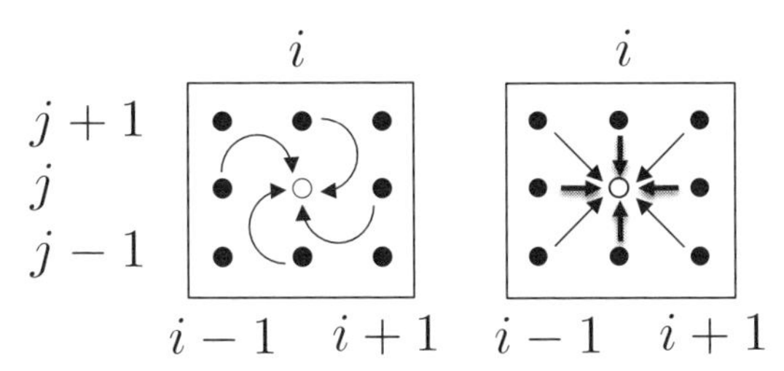

Figure 11.2. Schematic representation of the two missing value protocols (11.4) and (11.6). The grid point with missing data is denoted by the open circle. Full circles denote neighboring data points, assumed to possess full information.

Space–time combinations are likewise possible, e.g.,

$$\begin{aligned}\tilde{v}^k_{i,j} = & \frac{v^{k-2}_{i,j} + v^{k+2}_{i,j} + w_t\left(v^{k-1}_{i,j} + v^{k+1}_{i,j}\right)}{2(1+w_t)} \\ & + \frac{v^{k-2}_{i-1,j} + v^{k+2}_{i-1,j} + w_t\left(v^{k-1}_{i-1,j} + v^{k+1}_{i-1,j}\right)}{2(1+w_t)} \\ & + \frac{v^{k-2}_{i+1,j} + v^{k+2}_{i+1,j} + w_t\left(v^{k-1}_{i+1,j} + v^{k+1}_{i+1,j}\right)}{2(1+w_t)} \\ & + \frac{v^{k-2}_{i,j-1} + v^{k+2}_{i,j-1} + w_t\left(v^{k-1}_{i,j-1} + v^{k+1}_{i,j-1}\right)}{2(1+w_t)} \\ & + \frac{v^{k-2}_{i,j+1} + v^{k+2}_{i,j+1} + w_t\left(v^{k-1}_{i,j+1} + v^{k+1}_{i,j+1}\right)}{2(1+w_t)}.\end{aligned} \tag{11.10}$$

It is also possible to identify, for each spatial point with missing value(s), other point(s) strongly related to (e.g., temporally correlated with) the missing data point. The calculation is carried out over all time slices during which both the missing value point and the considered related point(s) data are present, and those correlations are then used as weights in the filling in procedure. For example, imagine a spatial point denoted by v_m with a missing value. Also imagine three other points, indexed (i, j, k), with values—at the same time as v_m is missing—of $v_{i,j,k}$, and temporal correlations with v_m of $r_{i,j,k}$, computed over all present values. Note that since we disregard latitude, longitude, or similar spatial coordinates, replacing them with a simple continuous enumeration of all the domain's grid points, there is only one index here. With this, we can set

$$\tilde{v}_m = \frac{r_i v_i + r_j v_j + r_k v_k}{r_i + r_j + r_k} \quad (11.11)$$

or some similar but more elaborate rendition of this containing time.

When the data time series exhibit clear seasonal dominance and the seasonal cycle is removed, it can be reasonably argued that the least conjectural filled in value for a missing datum is the corresponding climatological value; the most probable anomaly is zero. Then in **B** above, missing values are replaced by zeros. In addition, since the variance of the anomaly distribution whose mean is zero is known, we can perform repeated Monte Carlo simulations, in each of which anomaly values inserted in place of missing values are randomly drawn from the respective zero mean distributions, the calculations are repeated, and the collection of individual Monte Carlo results yield a range of EOF values.

11.6 Choosing and Interpreting the Covariability Matrix

The EOFs are the eigenvectors of the covariability matrix, arranged in descending order of the corresponding eigenvalues. Consequently, the choice of the covariability matrix is key. The section's title is deliberately ambiguous: It can mean literally the covariance matrix, but it can also mean matrices of other covariability measures, most notably correlation.

11.6.1 The Covariance Matrix

The simplest, and most commonly employed covariability matrix is the covariance matrix. By the time we obtained **D** we have already removed row means, as well as seasonal cycles if they were deemed important to eliminate. Consequently, the covariance matrix is simply

$$\mathbf{C} = \frac{1}{N-1}\mathbf{D}\mathbf{D}^T \equiv \langle \mathbf{d}\mathbf{d}^T \rangle, \quad (11.12)$$

where angled brackets denote both ensemble averaging, averaging over all available realizations, or, equivalently, expected values. Therefore,

$$\mathbf{C} = \langle \mathbf{d}\mathbf{d}^T \rangle = \frac{1}{N-1}\left(\mathbf{d}_1\mathbf{d}_1^T + \mathbf{d}_2\mathbf{d}_2^T + \cdots + \mathbf{d}_N\mathbf{d}_N^T\right) = \frac{1}{N-1}\sum_{i=1}^{N}\left(\mathbf{d}_i\mathbf{d}_i^T\right)$$

$$= \frac{1}{N-1}\sum_{i=1}^{N}\begin{pmatrix} d_{1i}d_{1i} & d_{1i}d_{2i} & \cdots & d_{1i}d_{Mi} \\ d_{2i}d_{1i} & d_{2i}d_{2i} & \cdots & d_{2i}d_{Mi} \\ \vdots & & & \vdots \\ d_{Mi}d_{1i} & d_{Mi}d_{2i} & \cdots & d_{Mi}d_{Mi} \end{pmatrix}, \quad (11.13)$$

where vectors' only index $i = [1, N]$ denotes time, and the scalar d_{ji} denotes $\mathbf{d}_i$'s jth element. Thus, **C** is an $M \times M$ square symmetric matrix whose elements kl

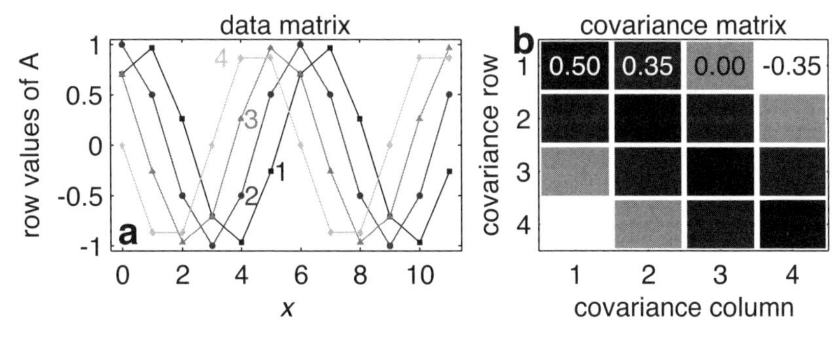

Figure 11.3. The simple 4×12 "data" matrix **S** discussed in the text. Panel a shows the matrix's four rows (annotated, 1 is in black and 4 is in the lightest gray shade). Panel b shows the 4×4 covariance matrix. The four distinct covariance values are indicated. Note that the vertical axis is reversed, to retrieve the familiar direction of the main diagonal (black entries corresponding to $C_{ii} = 0.5$, $i = 1, 4]$).

or lk hold the expected values of the product of the kth and lth elements of $\mathbf{d}_i$, $C_{kl} = C_{lk}$.

Because the covariability matrix, whose eigenvectors are the EOFs, is so central to EOF analysis, and because intuitive insights into its workings are not immediately obvious, let's spend some time trying to really understand it, focusing on the covariance matrix.

Let's start with a really simple example displayed in fig. 11.3. Imagine a 4×12 matrix **S** with rows

$$\mathbf{S} = \begin{pmatrix} \mathbf{s}_1 \\ \mathbf{s}_2 \\ \mathbf{s}_3 \\ \mathbf{s}_4 \end{pmatrix} = \begin{pmatrix} \cos\left(\frac{\pi x}{3} - \frac{\pi}{4}\right) \\ \cos\left(\frac{\pi x}{3}\right) \\ \cos\left(\frac{\pi x}{3} + \frac{\pi}{4}\right) \\ \cos\left(\frac{\pi x}{3} + \frac{\pi}{2}\right) \end{pmatrix}, \quad \begin{matrix} x = 0, 1, \ldots, 11 \\ \mathbf{s}_i \in \mathbb{R}^{1\times 12} \end{matrix} . \tag{11.14}$$

All rows have a vanishing mean and the same norm. As eq. 11.14 and fig. 11.3 show, **S**'s rows are essentially the same plus, importantly, an increasing phase shift. Because $\mathbf{s}_1$ and $\mathbf{s}_3$ (and, in general, $\mathbf{s}_i$ and $\mathbf{s}_{i+2}$) are $\pi/2$ out of phase, they are expected to be orthogonal. This is indeed borne out by $C_{13} = C_{31} = C_{24} = C_{42} = 0$ to within roundoff error. The two minor diagonals just off the main diagonal correspond to $\pi/4$ phase shift, which separates $\mathbf{s}_i$ and $\mathbf{s}_{i+1}$. Finally, $\mathbf{s}_i$ and $\mathbf{s}_{i+3}$ (of which we only have the single pair $(\mathbf{s}_1, \mathbf{s}_4)$) are $3\pi/4$ out of phase, which, because of the antisymmetry of $\cos(x)$ over $\delta x = \pi$, is the negative of $\text{cov}(\mathbf{s}_i, \mathbf{s}_{i+1})$.

This example shows the basic recipe governing the covariance matrix elements: for high covariability, rows must be as close to parallel as possible. Conversely, vanishing covariance characterizes nearly orthogonal rows. In more complex situation, the covariance matrix contains clusters of strongly covarying (or anti-covarying) elements, and other clusters within which individual grid points are largely mutually oblivious.

To show this, let's introduce the slightly more complex example

$$f(y,x,t) = a(t)\cos\left(\frac{\pi x}{30}\right)\cos\left(\frac{\pi y}{15}\right) + b(t)\cos\left(\frac{\pi x}{15}\right)\cos\left(\frac{\pi y}{7}\right) + \xi(y,x,t) \tag{11.15}$$

with $a(t) \sim N(0, 0.8)$, $b(t) \sim N(0, 0.4)$ and $\xi(y,x,t) \sim N(0, 0.1)$ over $x = 0, 1, \ldots, 30$ and $y = 0, 1, \ldots, 15$ (so $N_x = 31$ and $N_y = 16$). The signal is a linear combination of (primarily) the first rhs term (because a's variance is the largest), some of the second rhs term, and unstructured noise $\xi(y,x,t)$. The two deterministic patterns are shown in fig. 11.4. Note that they are mutually orthogonal (the cosines in both x and y are Fourier frequencies).

Let's observe this system for 50 time points, where each "time point" just means different random realizations of a, b, and ξ from their respective stated distributions. This is meant to simulate a physical situation in which a certain time-dependent field is generated by some dominant, large-scale physics with significant but variable amplitudes ($a(t)$ and $b(t)$), plus other, low-amplitude, small-scale processes, collectively represented here as $\xi(y,x,t)$. While perhaps sounding overly restrictive, this situation is actually very common. For example, imagine 50 daily temperature maps over the U.S. northeastern seaboard in the spring. The dominant pattern in most maps will be the starkly different response of land and ocean to the onset of summer, most notably the fact that it takes roughly 6 weeks to warm the upper ocean, while land warms in response to the same forcing on a few-day timescale. At the same time, however, weather still varies in the background, and particular days can differ dramatically from one another while still obeying the basic dictates of land–ocean contrasts.

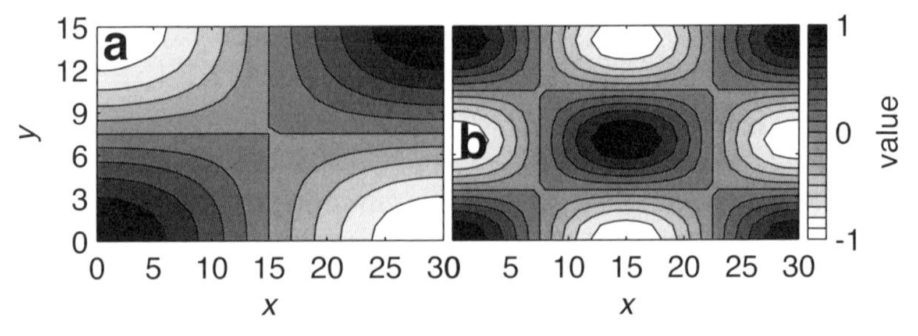

Figure 11.4. The spatial patterns used to generate the synthetic signals for the example discussed in the text.

With 50 realizations of this process, we can employ the steps outlined above (reshaping and removing means) and construct the process' full covariance matrix shown in fig. 11.5a, while four subregions, shown in panels b–e, help our interpretation. The spatial indexing convention of this problem, given in table 11.1, should also help.

Figure 11.5b displays the covariance of the rightmost (easternmost) column of the domain (grid points $(N_x - 1)N_y + [1, N_y]$, indicated on the full covariance, panel a, by the upper right thin white square), $\text{cov}(\mathbf{f}_{N_x}, \mathbf{f}_{N_x})$. Covariance is maximized near the diagonal ends and is negative near the upper left and lower right corners. To understand this pattern, recall the generating patterns, fig. 11.4a,b (hereafter p_1 and p_2). Along that edge, both p_1 and p_2 attain relatively high amplitudes (because of x extrema there dictated by the functional x dependence) and therefore are both important. At the same time, because $a \sim 2b$ in eq. 11.15, p_1 is roughly twice as important. The situation that gives rise to fig. 11.5b is summarized semi-schematically in fig. 11.6. Panels a–c and summarize the contributions of p_1 and p_2, respectively, while panel g, the appropriately weighted sum of panels c and f, is essentially fig. 11.5b, demystifying it.

Figure 11.5c shows very low covariances of grid points $15N_y + [1, N_y]$ and $22N_y + [1, N_y]$, which fig. 11.5a (rightmost of the panel's three white squares in the middle of the vertical range) also shows in their broader context. Referring back to fig. 11.4a shows that the $15N_y + [1, N_y]$ grid points are immediately adjacent to the vertical right-to-left sign reversal of p_1, so the p_1 values participating in these covariances are extremely small. Similarly, fig. 11.4b shows that the $22N_y + [1, N_y]$ grid points coincide with the right of the two vertical right-to-left sign reversals of p_2 separating the right half-nodes from the central full nodes. Consequently, the p_2 values participating in these covariances are also very small, resulting in fig. 11.5c's lowest amplitudes.

This is somewhat similar, though less extreme, in fig. 11.5d, which shows low, but not as low, covariances of grid points $15N_y + [1, N_y]$ and $15N_y + [1, N_y]$ (the domain's center). While in p_1 this region is near the mid-domain saddle point, and is thus mostly hovering around zero, it is a node center in p_2, attaining values near ± 1. Therefore, the products are low, but not as low as in fig. 11.5c.

Figure 11.5d's checkerboard pattern is also explained by the dominance of p_2 in this case (near the domain east–west center) and the particular structure of p_2 in the relevant region of $x \approx 15$, shown in fig. 11.7. Because fig. 11.5d addresses $x \approx 15$, p_2's central nodes, the vector whose covariances is addressed by fig. 11.5d is essentially the one fig. 11.7a,b displays, a top–bottom slice through fig. 11.4b's left–right center. When this pattern is multiplied by itself, the result is fig. 11.7c, essentially the checkerboard pattern of fig. 11.5d.

Finally, fig. 11.5e shows covariances of grid points $[1, N_y]$ and $15N_y + [1, N_y]$, up–down transects in the domain's western (left) edge and its center. Because in the domain center p_1's amplitudes are small, the center is dominated by p_2, but as both p_1 and p_2 have nodes along the western edge, both are important there, as shown by fig. 11.8.

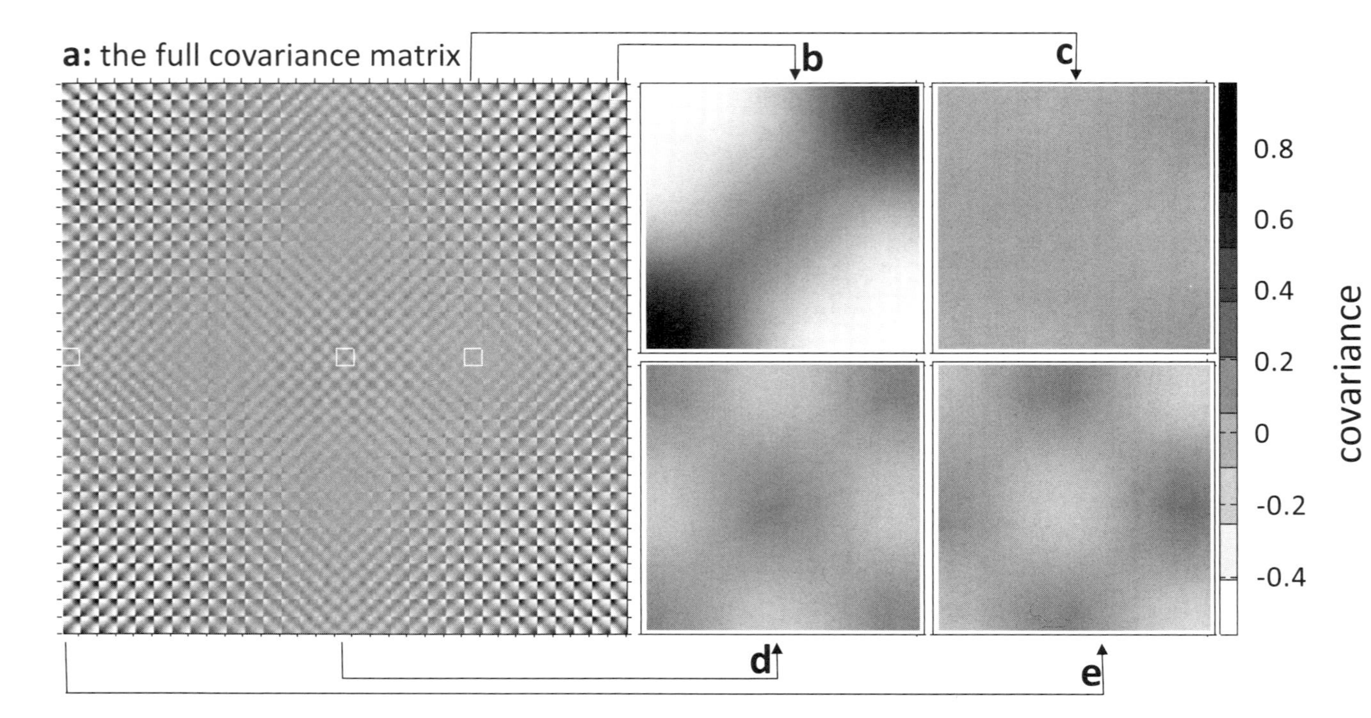

Figure 11.5. Several views of the covariance matrix of the example problem given by eq. 11.15. Panels b–e magnify the four indicated subregions. Tick marks are N_y apart; panel a comprises $N_xN_y \times N_xN_y$ covariances, while panels b–e comprise $N_y \times N_y$. All share the shown color bar.

Table 11.1

The spatial indexing convention of the elements of $f(y, x, t)$. The index of an element's value in the reshaped column vector is shown in the geographical $[(y, x)]$ location of that element. For example, the northwest element is element N_y, the northeast element's index is $N_x N_y$, and so on. The leftmost (westernmost) column is $\mathbf{f}_1$, and the rightmost (easternmost) column is $\mathbf{f}_{N_x}$.

N_y	$2N_y$	$\cdots$	$(N_x - 1)N_y$	$N_x N_y$
$N_y - 1$	$2N_y - 1$	$\cdots$	$(N_x - 1)N_y - 1$	$N_x N_y - 1$
$\vdots$	$\vdots$	$\vdots$	$\vdots$	$\vdots$
2	$N_y + 2$	$\cdots$	$(N_x - 2)N_y + 2$	$(N_x - 1)N_y + 2$
1	$N_y + 1$	$\cdots$	$(N_x - 2)N_y + 1$	$(N_x - 1)N_y + 1$
$\uparrow$	$\uparrow$		$\uparrow$	$\uparrow$
$\mathbf{f}_1(t)$	$\mathbf{f}_2(t)$		$\mathbf{f}_{N_x - 1}(t)$	$\mathbf{f}_{N_x}(t)$

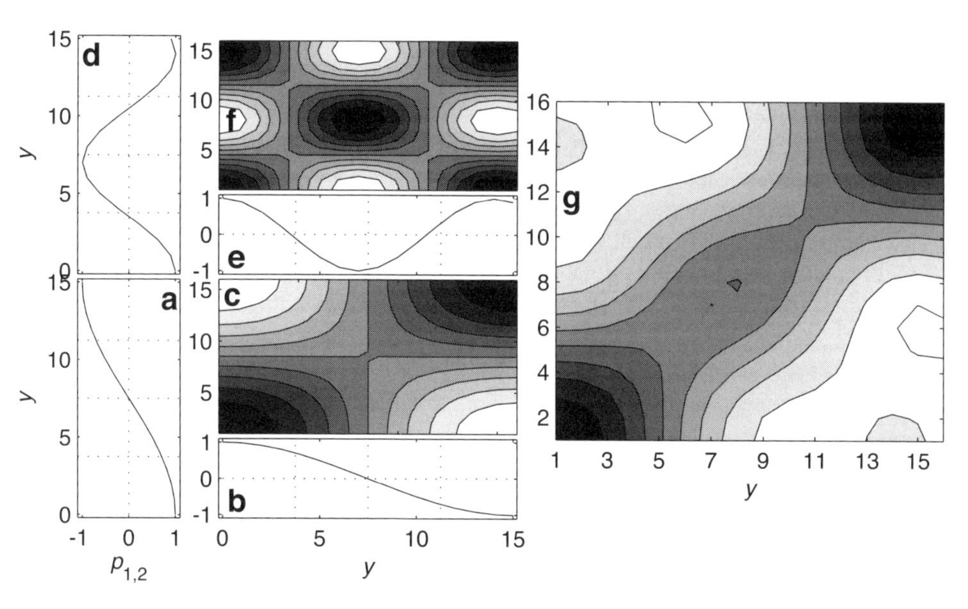

Figure 11.6. The shape of the generating patterns at the eastern edge, with black corresponding to most positive and white to most negative. Panel a shows p_1 there, and panel b shows the same pattern, transposed. This representation pictorially depicts the outer product of p_1 with itself at the eastern edge. Panel c shows the result of that outer product, $\text{cov}(p_1, p_1)$ at the eastern edge. Panels d–f do the same for p_2 in that location. Panel g shows the sum of twice panel c and panel f, the expected total covariance at the edge, given that the field comprises $\sim 2p_1 + p_2$.

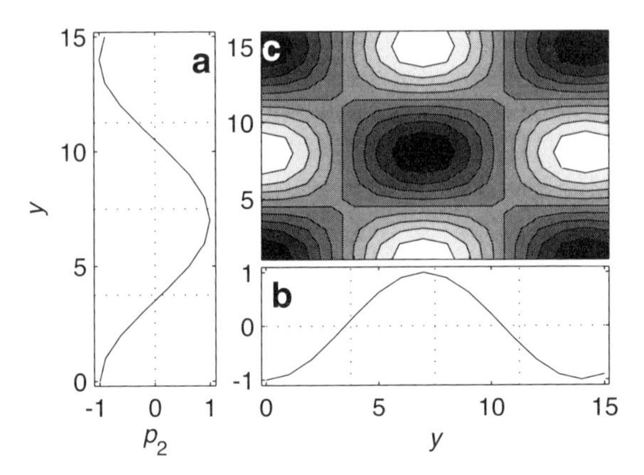

Figure 11.7. The outer product of the part of p_2 relevant to the domain center and thus to fig. 11.5d. Panels a and b give the relevant p_2 and its transpose, while panel c is their outer product, the covariance. Note the similarity of the checkerboard pattern to that of fig. 11.5d.

11.6.2 The Correlation Matrix

As mentioned earlier, most EOF analyses are based on decomposing the covariance matrix. In some situations, however, it is more illuminating to decompose the correlation matrix instead. This situation arises when the domain has a wide range of local variances. When these differences are considered well understood and not the center of attention, yet dominate the covariance matrix, they distract the method toward suboptimal patterns. This point is made clear by comparing the covariance (eq. 11.13, repeated below) and correlation matrix definitions,

$$\mathbf{C} = \frac{1}{N-1}\sum_{i=1}^{N}\begin{pmatrix} d_{1i}d_{1i} & d_{1i}d_{2i} & \cdots & d_{1i}d_{Mi} \\ d_{2i}d_{1i} & d_{2i}d_{2i} & \cdots & d_{2i}d_{Mi} \\ & \vdots & & \vdots \\ d_{Mi}d_{1i} & d_{Mi}d_{2i} & \cdots & d_{Mi}d_{Mi} \end{pmatrix} \tag{11.16}$$

and

$$\mathbf{R} = \begin{pmatrix} \frac{\sum d_{1i}d_{1i}}{\sqrt{\sum d_{1i}^2 d_{1i}^2}} & \frac{\sum d_{1i}d_{2i}}{\sqrt{\sum d_{1i}^2 d_{2i}^2}} & \cdots & \frac{\sum d_{1i}d_{Mi}}{\sqrt{\sum d_{1i}^2 d_{Mi}^2}} \\ \frac{\sum d_{2i}d_{1i}}{\sqrt{\sum d_{2i}^2 d_{1i}^2}} & \frac{\sum d_{1i}d_{2i}}{\sqrt{\sum d_{2i}^2 d_{2i}^2}} & \cdots & \frac{\sum d_{1i}d_{Mi}}{\sqrt{\sum d_{2i}^2 d_{Mi}^2}} \\ & \vdots & \vdots & \\ \frac{\sum d_{Mi}d_{1i}}{\sqrt{\sum d_{Mi}^2 d_{1i}^2}} & \frac{\sum d_{Mi}d_{2i}}{\sqrt{\sum d_{Mi}^2 d_{2i}^2}} & \cdots & \frac{\sum d_{Mi}d_{Mi}}{\sqrt{\sum d_{Mi}^2 d_{Mi}^2}} \end{pmatrix}, \tag{11.17}$$

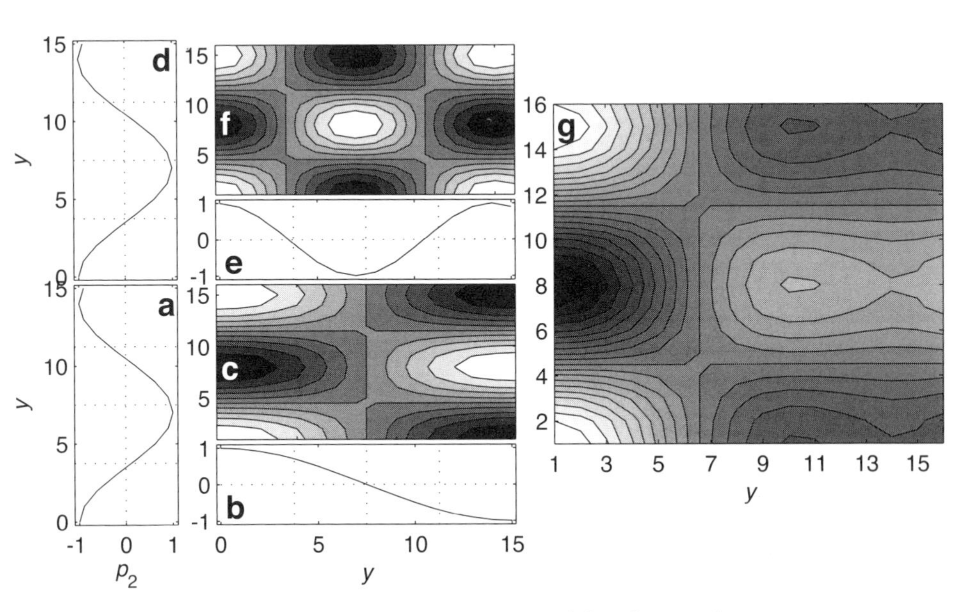

Figure 11.8. Panels a and d: p_2 pattern at $x = 15$. Panels b and e: p_1 and p_2 at $x = 0$, respectively. Panels c and f: the two respective outer products. Panel g is the appropriately weighted sum of panels c and f. Note the similarity of the pattern to that of fig. 11.5e.

respectively, where, recall, **D**'s rows are mean-removed and all sums in **R** run over $[1, N]$.

Thus, the difference between **C** and **R** is that in **R**, each covariance is scaled by the square root of the product of the two participating variances,

$$R_{ij} = \frac{\operatorname{cov}(d_i, d_j)}{\sqrt{\operatorname{var}(d_i)\operatorname{var}(d_j)}} = \frac{C_{ij}}{s_i s_j}, \tag{11.18}$$

where, conformal with eq. 11.13, d_i denotes the ith row in **A** and **D** or the time series of the ith grid point, and s_i is grid point i's sample standard deviation. Consequently, **R**-based EOF calculations can differ markedly from **C**-based ones only if the considered geographical domain contains significant spatial variability of local variance.

Appreciable spatial variability of local variance is hardly unusual, however. In climate work it is very common, the rule rather than the exception. In some cases, this spatial variability arises from land–ocean contrasts. For example, hourly to daily variance of most near surface climatic variables over land is far higher than over the ocean (e.g., fig. 11.9, where even the relatively small Great Lake system is clearly apparent as a local minimum). These differences are mostly due to the ocean's far higher heat capacity of the upper layer active on these timescales (10–100 m in the ocean, 3–5 m on land). In addition,

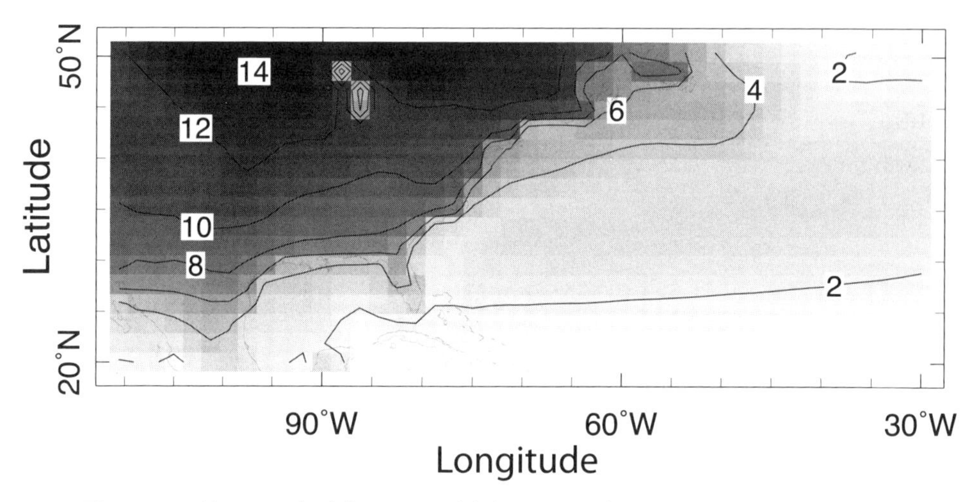

Figure 11.9. Time standard deviations of daily mean surface temperatures, in K, computed over January 1, 1948 through December 30, 2010 (23,010 daily values). Data from the NOAA NCEP-NCAR global reanalysis (www.esrl.noaa.gov/psd/data/reanalysis/reanalysis.shtml). All time points, irrespective of season, are considered together. Long-term time means, as well as straight-line best-fit trends, are removed prior to standard deviation calculations.

especially over land, the variance increases poleward (as evident from the ~10-fold standard deviation increase over fig. 11.9's shown land mass between 20°N and 50°N). These differences are also maintained, and compound with other geographical and climatological factors, when addressing lower temporal frequencies, as shown in fig. 11.10, which shows time standard deviations of annual mean anomalies. In addition to general land–sea contrasts, the figure emphasizes the dependence of local variability on climatic provinces. For example, interannual variability near the equator—mostly variability related to the El Niño system of the Equatorial Pacific between the International Date Line and the Peruvian coast—is more pronounced than in the rest of the tropics; arid regions (e.g., the Australian interior, Sahara, and Middle East) are more variable than heavily vegetated areas (equatorial Africa, Amazonia, and southeast Asia); and mountainous regions (the Himalayas and Tibetan plateau, the Andes) are more variable than lowlands.

Thus, in climate analysis, as well as in many other realistic situations, **R** and **C** are rather distinct. Deriving the EOFs from either matrix is acceptable, but will likely yield rather different results. The question of which of the two matrices is the preferred foundation of the EOF analysis is entirely a matter of the scientific question being asked, as discussed in the context of an example in the following section.

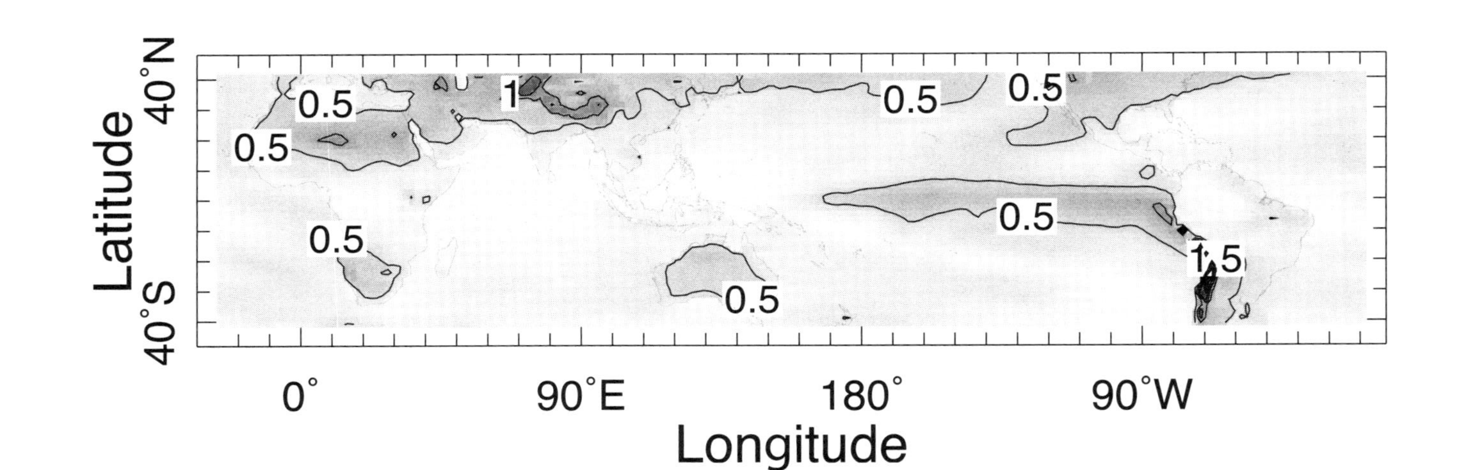

Figure 11.10. Time standard deviations of annual mean surface temperature anomalies (mean over January–December of each calendar year), in K, computed from monthly means. Data from the NOAA NCEP-NCAR global Reanalysis, spanning January 1949–December 2009 (732 monthly means, averaged into 61 annual means). Seasonality, long-term time means, and straight-line best-fit trends are removed prior to standard deviation calculations.

11.7 Calculating the EOFs

Calculating **A**'s EOFs means eigen-decomposing **D**'s **C** (or **R**, if correlation is deemed the more revealing covariability measure); **C**'s (or **R**'s) eigenvectors are **A**'s EOFs. Let's examine briefly some properties of the EOFs $\{\mathbf{e}_i\}$, assuming they are **C** based. Because $\mathbf{e}_i \in \mathbb{R}^M$ are **C**'s eigenvectors (where, recall, $M = N_x N_y$ is the state dimension), they satisfy the canonical eigenvalue–eigenvector equation

$$\mathbf{C}\mathbf{e}_i = \lambda_i \mathbf{e}_i, \quad i = [1, M]. \tag{11.19}$$

First, note that for many data sets, $N \ll M$, the number of participating time points is far smaller than the state dimension. Therefore, $q_c \ll M$, **C** is rank deficient. Yet, because **C** is symmetric (and, of course, square), it is guaranteed to have a complete orthonormal set of eigenvectors (i.e., even if **C** has repeated eigenvalues, their geometric multiplicity is full). Consequently, span $(\{\mathbf{e}_i\}_{i=1}^{q_c}) = \mathbb{R}^M$ (**A**'s full variance–covariance structure is spanned by **C**'s leading q_c eigenvectors), and span$(\{\mathbf{e}_i\}_{i=1}^{M}) = \mathbb{R}^M$ (**C**'s set of eigenvectors is complete, fully spanning $\mathbb{R}^M$). Rearranging eq. 11.19 and taking norms gives

$$\frac{\|\mathbf{C}\mathbf{e}_i\|}{\|\mathbf{e}_i\|} = \|\mathbf{C}\mathbf{e}_i\| = |\lambda_i| = \lambda_i, \tag{11.20}$$

where the two simplifications above recognize that $\|\mathbf{e}_i\| = 1$ by construction and that **C**'s eigenvalues are positive semi-definite because **C** is a quadratic. Because $\|\mathbf{C}^T\mathbf{e}_i\| = \lambda_i$ (as **C** is symmetric) and $\lambda_{i-1} > \lambda_i$ (the eigenvalues are arranged in descending order), $\lambda_1 = \|\mathbf{C}^T\mathbf{e}_1\|$ is the largest amplification factor of all the eigenvectors; $\mathbf{e}_1$ is the $\mathcal{R}(\mathbf{C})$ direction on which **A**'s variance–covariance structure has the largest projection. Most importantly, because of $\{\mathbf{e}_i\}$'s completeness, λ_1 is also the largest amplification factor of all $\mathbb{R}^M$ vectors; $\mathbf{e}_1$ is also the $\mathbb{R}^M$ direction on which **A**'s variance–covariance structure has the largest projection. In other words, $\mathbf{e}_1$ is the $\mathbb{R}^M$ direction spanning the most (the largest fraction) of **A**'s variance–covariance, $\mathbf{e}_2$ is the $\mathbb{R}^M$ direction spanning the largest fraction of **A**'s variance–covariance orthogonal to $\mathbf{e}_1$, and $\mathbf{e}_i$ is the $\mathbb{R}^M$ direction spanning the largest fraction of **A**'s variance–covariance orthogonal to $\mathbf{e}_{[1,i-1]}$. Because of the orthogonality of the eigenvectors, the percent of **A**'s total variance var(**A**) mode i accounts for is simply

$$100\frac{\lambda_i}{\sum_{1=1}^{M}\lambda_i} = 100\frac{\lambda_i}{\sum_{1=1}^{q_c}\lambda_i} = 100\frac{\lambda_i}{\mathrm{var}(\mathbf{A})}. \tag{11.21}$$

11.7.1 *Displaying the EOFs*

To display the EOFs in geographical space, we must exactly undo the reshaping of the high-dimensional data into **A** (see section 11.3, where we discuss the example of a data set $F(y, x, t)$ that depends on two spatial coordinates whose discrete representations are $y = y_i = (y_1, y_2, \ldots, y_m)$ and $x = x_j = (x_1, x_2, \ldots, x_n)$).

Using section 11.3's notation, **A**'s kth EOF is reshaped onto the (y, x) plane for plotting using

$$\mathbf{e}^k = \begin{pmatrix} e^k_{1,1} \\ \vdots \\ e^k_{m,1} \\ e^k_{1,2} \\ \vdots \\ e^k_{m,2} \\ \vdots \\ \vdots \\ e^k_{1,n} \\ \vdots \\ e^k_{m,n} \end{pmatrix} \implies e^k(y,x) = \begin{pmatrix} e^k_{1,1} & e^k_{1,2} & \cdots & e^k_{1,n} \\ e^k_{2,1} & e^k_{2,2} & \cdots & e^k_{2,n} \\ \vdots & \vdots & & \vdots \\ e^k_{m,1} & e^k_{m,2} & \cdots & e^k_{m,n} \end{pmatrix},$$

where $e^k_{\{i,j\}}$ is the value of the kth EOF at space point (y_i, x_j), so the columns correspond to fixed longitudes and the rows correspond to fixed latitudes, and where (m, n) are the y and x spatial dimensions, also often denoted (N_y, N_x). Note that while here the southernmost row of data points is row 1 in $e^k_{i,j}$, consistent with Matlab and Octave convention, this is a matter of the row indexing convention of the plotting program used, and may thus be reversed in some applications.

11.7.2 A Real Data Example

Let's introduce an example that will illuminate the EOF calculation process and its details. Consider monthly mean surface temperature anomalies (monthly mean deviations from monthly mean climatologies) between January 1949 and December 2009 (732 values) over the Equatorial Pacific domain (112°E–68°W, 30°S–30°N, comprising 2821 grid points), home court of the El Niño–Southern Oscillation (ENSO) system (data taken from the NCAR-NCEP Reanalysis again, with linear time trends removed). The monthly anomaly standard deviations, expressed as $\log_{10}$, are shown in fig. 11.11a. Clearly, the domain is widely variable in terms of characteristic temporal variability. While the familiar ENSO SSTA (sea surface temperature anomaly) pattern is apparent over the Equatorial Pacific ocean (compare this pattern of monthly anomalies to the corresponding one for annual anomalies shown in fig. 11.10), variability over the bounding land masses towers over the oceanic ones. This is maximized in the Andes, where the surface is several kilometers above sea level. Based on this, we expect the leading patterns of the covariance matrix to differ appreciably from their correlation matrix-based counterparts.

Perhaps surprisingly, the leading normalized eigenvalues of **C** and **R** (the covariance and correlation matrices) shown in fig. 11.11b appear similar. Yet

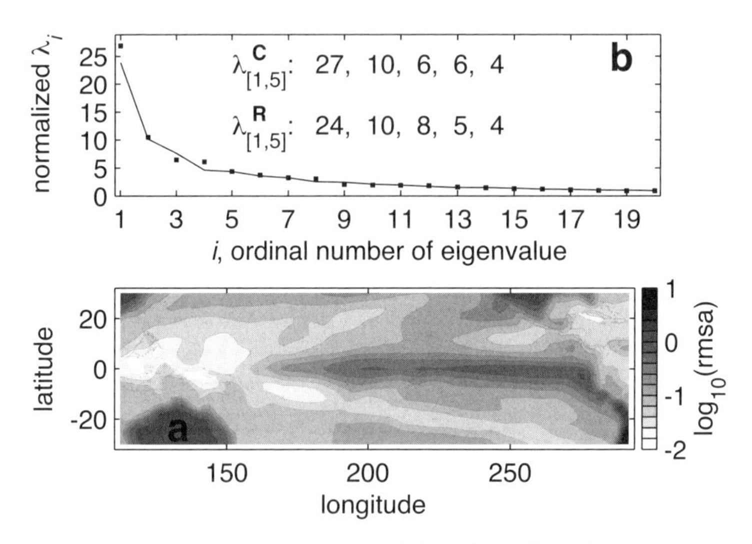

Figure 11.11. Panel a: temporal standard deviations of monthly mean surface temperature anomalies in the Equatorial Pacific. Because of the wide range of values, the standard deviations are expressed as $\log_{10}$. Panel b: leading 20 eigenvalues of the corresponding covariance and correlation matrices (full squares and solid curve, respectively). The approximate numerical values of the top five are given numerically. Eigenvalues are normalized by the total variance, i.e., the shown values are $100\lambda_i^{C,R}/\sum_i \lambda_i^{C,R}$.

examining the spatial patterns (the eigenvectors, fig. 11.12) reveals clearly distinct patterns. While the covariance-based leading mode (top left panel) is sharply defined (the pattern's high positive values are mostly centered on the equator, with a small coastal extension off South America), the **R**-based pattern (top right) is much more latitudinally expansive along the west coast of the Americas.

If we reexamine fig. 11.11a and eqs. 11.17 and 11.18, the differences between fig. 11.12's top panels become better understood. The oceanic variance along a narrow equatorial strip is high, dropping poleward, especially northward, rapidly. Therefore, when i, j or (i, j) are off equatorial, $C_{i,j}$ is small because of the low characteristic amplitude of at least one of the two grid points. Yet the corresponding $R_{ij} = C_{ij}/s_i s_j$ is not small, because s_i, s_j, or both are small, characterizing the low variance in off equatorial regions. Thus, those small $C_{i,j}$s yield larger $R_{i,j}$s due to the amplification that results from division by a small denominator. Having larger numerical values, these elements have more sway on **R**'s eigenvectors than on **C**'s as these eigenvectors reform to fit the extra power now present in those previously small, relatively powerless, elements.

Lower eigenvalue modes (fig. 11.12's bottom two panel rows) are even more mutually distinct. While both second patterns emphasize the equatorial strip

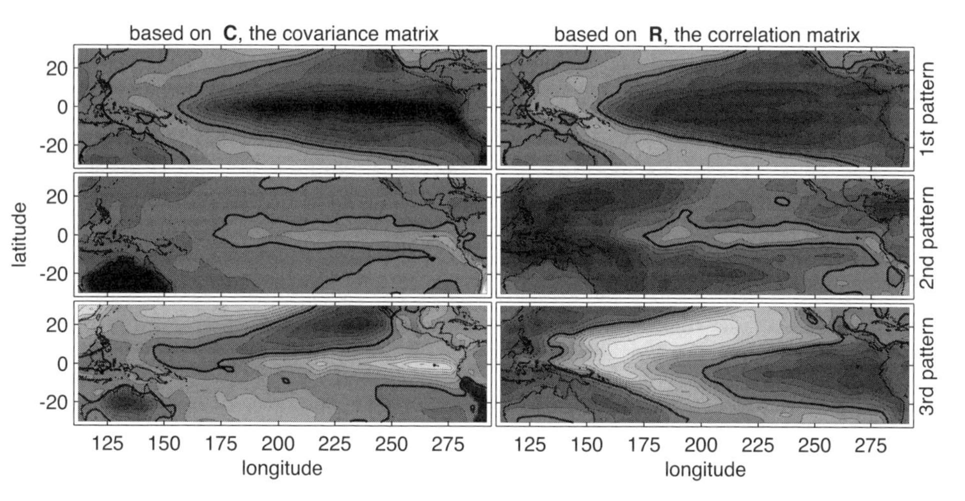

Figure 11.12. The leading three eigenvectors of the covariance (**C**, left) and the correlation (**R**, right) matrices of the anomalous Equatorial Pacific surface temperature example discussed in the text. Patterns are arranged, from the top down, in descending order of corresponding eigenvalue, with the top panels displaying the leading mode of either matrix. Contour values are arbitrary, but all panels share the same values, interval, and color range, with the zero contour emphasized by a thicker line.

and coastal South America, the **C**-based one is essentially a reversed polarity pattern between interior Australia and that equatorial strip, while in the **R**-based pattern the pole opposite to the equatorial node is all about ocean surface temperatures in the western Equatorial Pacific so-called Warm Pool and the Gulf of Mexico. You can see the high local variance of these areas in fig. 11.11a, but, because the variance cannot be negative, revealing the polar relationships between those variance centers requires the eigenanalysis summarized in fig. 11.12. The third patterns also differ, with the **C**-based one showing mostly a reversed polarity pattern between the eastern Equatorial Pacific and the northern outer tropics and continental Australia, in the **R** one both hemispheres' outer tropical oceans have the same polarity.

Thus, analyzing **C** and **R** can and regularly does give clearly different results. Which should you use? There are compelling arguments in favor of either. Because **C** is sensitive to actual amplitudes, while in **R** those are scaled away by the division by the individual variances, if you are interested in variability of such physically absolute quantities as, e.g., actual heat anomalies in joules, you would most likely favor the **C**-based analysis. For example, in the analysis of tropical temperature anomalies, the fact that off-equatorial latitudes covary with the equatorial strip may (depending on the question motivating the analysis) be of little interest when addressing absolute heat content anomalies; there may simply be too little variability, however structured, in off-equatorial latitudes to matter to the total heat content anomaly.

On the other hand, small amplitudes need not imply unimportance, and correlation patterns may prove more physically revealing. This is particularly likely when there are physical reasons to expect low amplitudes in parts of the domain. For example, in the above anomalous tropical surface temperature case, vanishing at the equator of latitude-dependent horizontal earth rotation rate governs the unique dynamics along the equator and predicts larger anomalies there. This may constitute a reason to treat equatorial anomalies as comparable to much smaller higher latitude anomalies. In that case, analyzing **R**—in some sense the statistical analog to the dynamically motivated scaling by rotation rate—may prove more desirable.

11.7.3 More Synthetic Analyses

Let's continue to analyze the synthetic example of the 50 $f(x, y, t)$ fields generated from the patterns given by eq. 11.15 and shown in fig. 11.4a,b, focusing on decomposing the covariance matrix. Figure 11.13 shows the three leading EOFs. Both generating patterns (fig. 11.4a,b) are well reproduced by the two leading EOFs, despite the noise and the random blending of the the signals by the amplitudes. The nodes of fig. 11.4's patterns and the EOF in fig. 11.13a,b have the same signs. That is, the nodes of the leading generating pattern (fig. 11.4a) and the leading EOF (fig. 11.13a) are both (clockwise from northwest) (−, +, −, +), as are the perimeter nodes of the second generating pattern (fig. 11.4b) and the second EOF (fig. 11.13b), both being (clockwise from north west) (+, −, +, −, +, −, +, −). This is not the rule, just happenstance, and is completely immaterial—at any given time the pattern can be multiplied by any real scalar amplitude, positive or negative, so the negative rendition of a positive pattern is just as valid as the pattern itself. However, sign changes within a pattern (e.g., the fact that fig. 11.3a's northwest area is negative while the northeast area is positive, or vice versa) is crucial, because it means that when the northwest features positive anomalies, the northeast and southwest anomalies

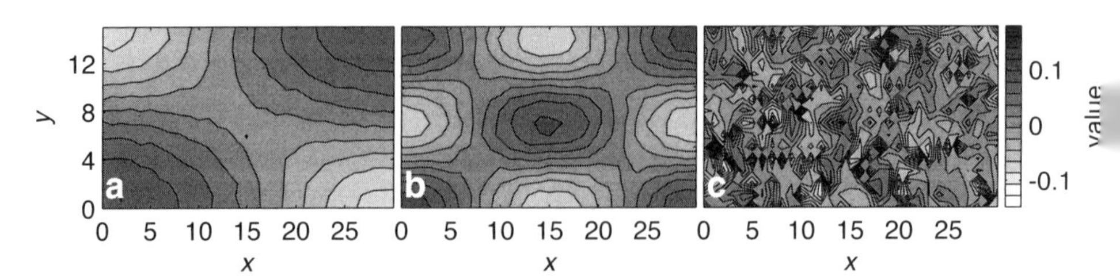

Figure 11.13. The leading spatial modes (EOFs) of the cosine signal given by eq. 11.15 and shown in fig. 11.4a,b.

are negative. Clearly, the third pattern (fig. 11.13c), a pale Jackson Pollock imitation, is structureless noise as expected, because there are only two structures generating the field, so all that is left for the third pattern to do is span the noise (ξ in eq. 11.15).

A possible limitation of the previous example is that by using mutually orthogonal generating patterns we made the method's job particularly (and artificially) easy. This is true—if the method is designed to turn arbitrary signals into an orthogonal decomposition of those signals, the real test of the method is with nonorthogonal signals.

To test this, let's use the signal

$$\begin{aligned} f(y,x,t) = {} & a(t)\cos\left(\frac{\pi x}{30}\right)\cos\left(\frac{\pi y}{15}\right) \\ & + b(t)\cos\left(\frac{\pi x}{15}\right)\cos\left(\frac{\pi y}{7}\right) \\ & + c(t)\exp\left(-\frac{(x-15)^2}{80}-\frac{(y-7.5)^2}{22}\right) \\ & + d(t)\left[\exp\left(-\frac{(x-7)^2}{8}-\frac{(y-4)^2}{5}\right)\right. \\ & + \quad \exp\left(-\frac{(x-16)^2}{8}-\frac{(y-8)^2}{5}\right) \\ & + \quad \left.\exp\left(-\frac{(x-25)^2}{8}-\frac{(y-12)^2}{5}\right)\right] \\ & + \xi(y,x,t) \end{aligned} \tag{11.22}$$

with a, $c \sim N(0, 0.8)$, $b, d \sim N(0, 0.4)$, and $\xi \sim N(0, 0.1)$. While this expression certainly looks complicated, it is, in fact, simply a noisy randomized blend of the patterns of figs. 11.4 (the terms whose amplitudes are $a(t)$ and $b(t)$) and 11.14 (the terms whose amplitudes are $c(t)$ and $d(t)$). Thus, the only difference from the previous synthetic EOF analysis is that, to address the possible

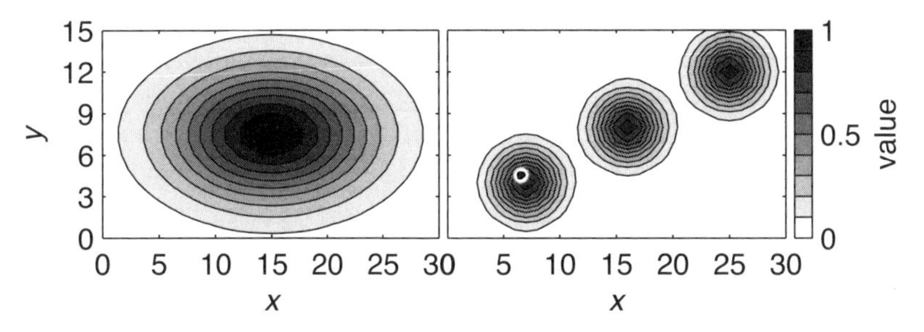

Figure 11.14. The spatial patterns used to generate the synthetic signals for the example discussed in the text.

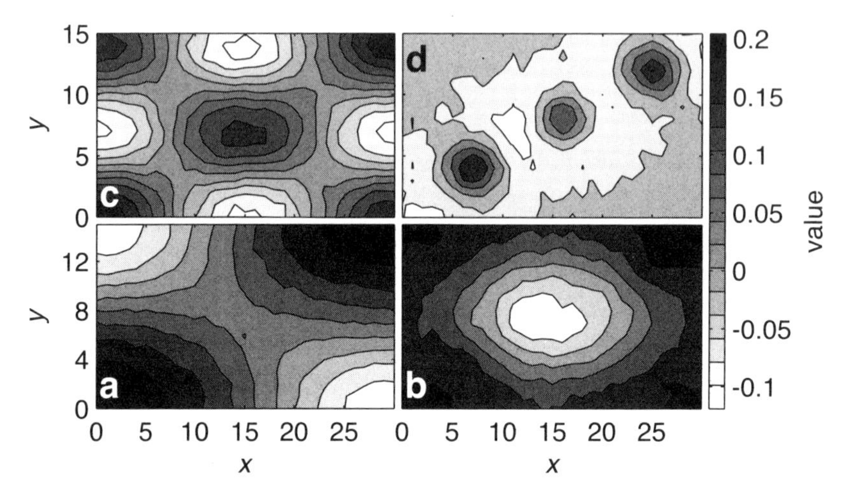

Figure 11.15. The leading spatial modes (EOFs) of the mixed cosine plus exponential signal shown in fig. 11.14a,b.

weakness of that analysis, we now add nonorthogonal components to the orthogonal ones we previously analyzed. Figure 11.15 shows that the method functions well even when the input signal is not artificially orthogonal, with all four generating patterns nicely reproduced with good fidelity (compared with the generating patterns). The patterns of fig. 11.15, especially b and d, are distorted relative to fig. 11.4 and fig. 11.14, as the EOF machinery cannot perform magic, replacing mutually projecting patterns with orthogonal ones without distortion. But, because the analysis focuses on maximally spanning the observed variance, the distortions are mostly concentrated in low-variance, and thus less crucial, areas.

The fraction of the total variance individual modes account for is important. For the two synthetic cases above, fig. 11.16 displays the leading 9 modes of the 50 total. As expected (because of the known 2–4 generating patterns plus noise) the rest are near zero in both cases. The eigenvalues corresponding to the raw cosine signals (modes 1 and 2) are very similar in both cases. Higher modes differ. In the cosine only, where the only remainder beyond mode 2 is noise, it is roughly equally distributed over the entire spectrum. Conversely, in the case of the added exponentials, the remainder has two structured modes (the two exponential terms), and indeed eigenvalues 3 and 4 are nonzero (while it is hard to see that for λ_4, its value is 2.2, roughly 14 times larger than the null eigenvalues—eigenvalues representing approximate equal distribution of the noise variance over all nonstructured modes—which are 0.15–0.16.

The order of the reproduced modes in fig. 11.15 is consistent with the relative magnitudes of a, c, b and d. The decrease in amplitude with mode

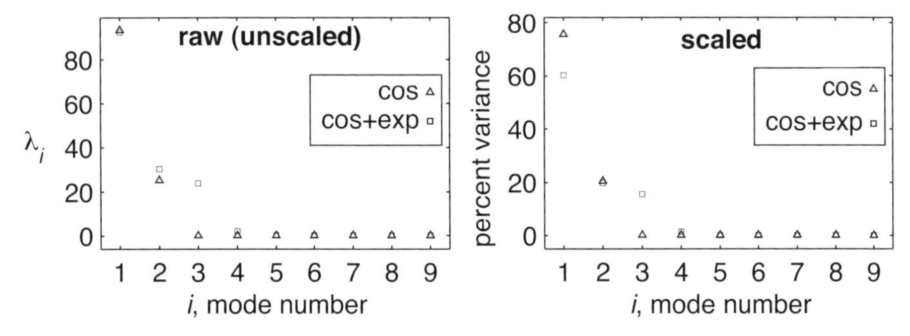

Figure 11.16. The eigenspectra of the two synthetic cases discussed in the text. Left: the leading raw eigenvalues of **C**. Right: the same eigenvalues, scaled by the total variance, $100\lambda_i / \sum_{j=1}^{M} \lambda_j$.

number (the falloff of the eigenspectrum in fig. 11.16) requires some thought. The eigenspectrum of the cosine only signal satisfies $\sqrt{\lambda_1/\lambda_2} \approx 2$, consistent with $a \sim 2b$ and $\lambda_{i>2} \approx 0$, as expected. The situation is less straightforward with the cosine plus exponential signal. Based on the (a, b, c, d) values, we expect this signal's eigenspectrum to satisfy $\lambda_a \approx \lambda_c$ and $\lambda_b \approx \lambda_d$ (where the subscript denotes the coefficient of the term in eq. 11.22 to which the pattern corresponds). This is clearly not borne out by fig. 11.16's squares (with values 92.4, 30.3, 23.9, and 2.2). This is an important, general, lesson: the eigenspectrum falloff is a property of the analysis, a corollary of the $\lambda_{i-1} \geq \lambda_i$ arrangement, so the order of neighboring modes need not be particularly physically meaningful. An important consideration is the scale of the mode; the larger the scale, the more likely it is to be pushed backward, toward the leading end of the eigenspectrum. The reason for this is that the smaller the scale, the larger the fraction of near zero grid points in the domain (due to either multiple zero crossings or the presence of numerous grid points, accounting for significant fraction of the domain, simply not being part of the pattern). For example, the number of grid points whose magnitudes are smaller than 2% of the domain maximum is 20 (~4%) for fig. 11.4a, and 36 (>7%) for fig. 11.13a. It is because of this abundance of small amplitude points in fig. 11.13a relative to fig. 11.4a that $\lambda_a \gg \lambda_c$ exhibited by fig. 11.16's squares instead of the expected $\lambda_a \approx \lambda_c$.

11.8. Missing Values, Take 2

Once we have obtained the EOFs it is possible to revisit the missing values problem. Suppose you overcame the initial missing values challenge (using, e.g., the methods suggested in section 11.5) and obtained a complete set of EOFs. Suppose the ith time slice of your data matrix $\mathbf{A}$, $\mathbf{a}_i$, contains m valid data points and $M - m$ missing values (we assume $M - m \ll m$ and $m/M \approx 1$), with zeros replacing missing elements. We can project this deficient vector on any of the EOFs, e.g.,

$$p_{ij} = \frac{M(\mathbf{a}_i^T \mathbf{e}_j)}{m} \tag{11.23}$$

on the jth EOF, where the M/m factor sets the projections on $\mathbf{e}_j$ of all vectors $\mathbf{a}_i$ on equal footing *if* we assume that the missing values, if they were available, would have behaved (in terms of projection on the EOFs) roughly like the present elements. With this assumption, the full jth contribution to the filled in $\mathbf{a}_i$ is

$$\tilde{\mathbf{a}}_i^j = p_{ij}\mathbf{e}_j = \frac{M(\mathbf{a}_i^T \mathbf{e}_j)}{m}\mathbf{e}_j, \tag{11.24}$$

and the quasi-complete filled in $\mathbf{a}_i$ is

$$\tilde{\mathbf{a}}_i = \sum_{j=1}^{c} p_{ij}\mathbf{e}_j = \frac{M}{m}\sum_{j=1}^{c}(\mathbf{a}_i^T \mathbf{e}_j)\mathbf{e}_j = \frac{M}{m}\mathbf{E}_c\mathbf{E}_c^T\mathbf{a}_i \tag{11.25}$$

where c is some cutoff and

$$\mathbf{E}_c = \begin{pmatrix} \mathbf{e}_1 & \mathbf{e}_2 & \cdots & \mathbf{e}_c \end{pmatrix} \tag{11.26}$$

comprises $\mathbf{E}$'s first c columns.

If $\mathbf{C}$'s rank q is well defined, so that $\mathbf{C}$'s eigenspectrum permits a clear distinction between $\mathcal{R}(\mathbf{C})$ and $\mathcal{N}(\mathbf{C})$ (between the column and null spaces of $\mathbf{C}$), then $c = q$ is the most sensible. In most realistic cases, however, $\mathbf{C}$'s rank is not so clearly defined. In such cases, it will make sense to employ a less unambiguous cutoff criterion, such as

$$c = \min_k \left(100 \frac{\sum_{j=1}^{k} \lambda_j}{\sum_{i=1}^{M} \lambda_i} \geq 90\% \right) \tag{11.27}$$

(the minimal upper summation bound needed to account for 90% or more of the total variance) or some other numerical value for the somewhat arbitrary required fraction of the total variance spanned, or

$$c = \max_i \left\{ \lambda_i \geq \lambda_1 \frac{p}{100} \right\}, \tag{11.28}$$

where c is set to the largest (most trailing) i satisfying the criterion that $\lambda_i \geq$ a specified percentage p of the leading (largest) eigenvalue.

With c and $\mathbf{E}_c$ evaluated, we can use eq. 11.25 to fill in each column of $\mathbf{A}$ or $\mathbf{D}$ containing missing values. If c is chosen properly, the truncated EOF set is still very nearly complete, and any $\mathbb{R}^M$ vector can be faithfully represented as a linear combination of the EOFs. The above filling in procedure is as complete as is $\{\mathbf{e}_i\}_{i=1}^{c}$.

11.8.1 Filling in Missing Values: An Example

Let's use the two 50 realization synthetic examples discussed earlier (based on the patterns of figs. 11.4 and 11.14, with the leading EOFs and eigenspectra of figs. 11.13, 11,15, and 11.16) to test and evaluate the performance of the filling in procedure. Let's refer to the cosine only data set as set a and the cosine plus exponentials data set as set b. For each of data sets a and b, we calculate the covariance matrix, eigenanalyze it, and truncate the EOF set using eq. 11.28 with $p = 1$. This results in $c_a = 2$ and $c_b = 4$, as expected based on eqs. 11.15 and 11.22. We then deploy a Monte Carlo randomization protocol, assuming the missing values are too few compared to the overall data set size to affect strongly the EOFs; otherwise, the test should have included repeated calculation of the EOF based on each deficient data subset. We test each column individually 100 times, in each withholding randomly chosen ~10% of the column elements as a proxy for missing values (because $M = 496$ is indivisible by 10, $m = 446$ and $M - m = 50$). We then calculate the two data sets' $M\mathbf{E}_c\mathbf{E}_c^T/m \approx 1.11\mathbf{E}_c\mathbf{E}_c^T$ and carry out the test using the following Matlab or Octave code segment.

```
1.  m = round(M*.9);
2.  Mmm = M-m;
3.  EEta = (M/m)*Ea*Ea';
4.  EEtb = (M/m)*Eb*Eb';
5.  z = zeros(Mmm,1);
6.  Nmc = 100;
7.  Ma = zeros(Nmc,Nt);
8.  Mb = zeros(Nmc,Nt);
9.  for i = 1:50
10.   for k = 1:Nmc
11.     j = sort(randi(M,1,Mmm));
12.     a = A(:,i); a(j) = z;
13.     b = B(:,i); b(j) = z;
14.     at = EEta*a;
15.     bt = EEtb*b;
16.     m1 = mean(abs(A(j,i) - at(j)));
17.     m2 = mean(mean(abs(A(j,:)),2));
18.     n1 = mean(abs(B(j,i) - bt(j)));
19.     n2 = mean(mean(abs(B(j,:)),2));
20.     Ma(k,i) = m1/m2;
21.     Mb(k,i) = n1/n2;
22.   end
23. end
24. Ma = 100*Ma(:);
25. Mb = 100*Mb(:);
```

The workings of the above code are as follows.

1. Number of valid data points in each Monte Carlo (MC) realization.
2. Number of missing values in each MC realization.
3. $M\mathbf{E}_c\mathbf{E}_c^T/m$ for data set a.
4. $M\mathbf{E}_c\mathbf{E}_c^T/m$ for data set b.
5. For simulating missing values.
6. Number of MC realization for testing each column.
7. For storing set a's MC results ($N_t = 50$).
8. For storing set b's MC results.
9. Loop on the time points (columns).
10. The MC loop.
11. The randomly chosen "missing values" (`randi` stands for "random integer").
12. Assign set a's "missing values."
13. Assign set b's "missing values."
14. Obtain set a's full, filled in, column.
15. Obtain set b's full, filled in, column.
16. Mean absolute difference between set a's withheld and filled in values.
17. Mean of set a's withheld values.
18. Mean absolute difference between set b's withheld and filled in values.
19. Mean of set b's withheld values.
20. Normalized error of set a's filled in values.
21. Normalized error of set b's filled in values.
22. End of MC loop.
23. End of loop on columns.
24. Combine all a's filled in values.
25. Combine all b's filled in values.

Thus, the test statistic, using set a as an example, is

$$e_p^a = 100 \frac{\overline{|\mathbf{a}_w - \tilde{\mathbf{a}}_w|}}{\overline{|\mathbf{a}_w|}}, \tag{11.29}$$

where overbars denote averaging over all withheld values of a particular MC realization addressing a specific column, and $\mathbf{a}_w$ is the withheld portion of any column $\mathbf{a}$, the filled values of which are $\tilde{\mathbf{a}}_w$.

The results of this test are the two histograms fig. 11.17 displays, derived from the populations of e_p^a and e_p^b (eq. 11.29). All results are quite acceptable, with characteristic absolute scaled errors in the 12–28% range. The filling in errors of set b, with four meaningful EOFs, are slightly smaller because the relative importance of the noise in that data set is smaller than in set a, with only two generating patterns.

11.9 Projection Time Series, the Principal Components

With the EOFs calculated, the amplitude of each mode at a specific point in time is given by the projection of the data vector at that time on the EOFs. It is

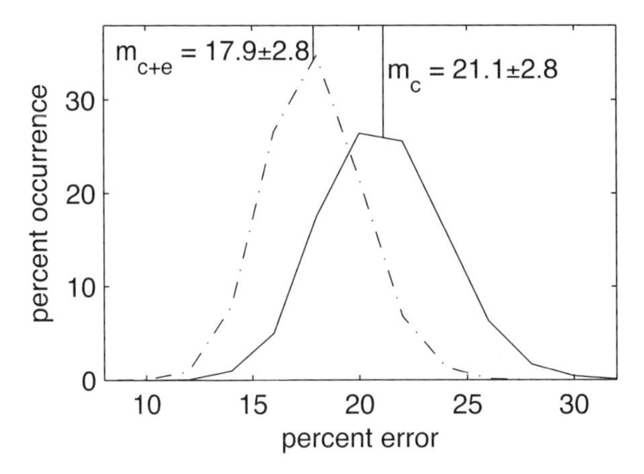

Figure 11.17. Histograms of scaled absolute error of filled in missing values in the two synthetic data set examples discussed in the text. The mean errors, given by eq. 11.29, take note of the withheld values only, not the full data vectors. The full error populations (combining all columns, all withheld values and all 10 MC realizations) are divided into 13 bins of width 2% error, and the percentage of all 5000 realizations that fall into each bin is the vertical value of the curve at the bin center percent error. The distributions' means are indicated as vertical solid lines, with numerical values (and standard deviations) annotated for set *a* (m_c) and *b* (m_{c+e}).

also customary, but not crucial, to normalize to unit norm by the eigenvalue's square root. For the *i*th EOF, $i = [1, c]$, this is

$$\mathbf{p}_i = \frac{1}{\sqrt{\lambda_i}} \mathbf{D}^T \mathbf{e}_i = \frac{1}{\sqrt{\lambda_i}} \begin{pmatrix} \mathbf{d}_1^T \mathbf{e}_i \\ \mathbf{d}_2^T \mathbf{e}_i \\ \vdots \\ \mathbf{d}_N^T \mathbf{e}_i \end{pmatrix} = \frac{1}{\sqrt{\lambda_i}} \begin{pmatrix} p_i(t_1) \\ p_i(t_2) \\ \vdots \\ p_i(t_N) \end{pmatrix} \in \mathbb{R}^N, \tag{11.30}$$

where, recall, N is the time dimension, the number of time points the data set comprises, so the notation above emphasizes the fact that each $\mathbf{p}_i$ is a time series, and $c \le \min(q, M, N)$. In some disciplines the projection time series are called principal components. Including all $[1, c]$ retained modes, this becomes

$$\mathbf{P}_c = \mathbf{D}^T \mathbf{E}_c \boldsymbol{\Lambda}_c^{-\frac{1}{2}}$$

$$= \left[\frac{1}{\sqrt{\lambda_1}} \begin{pmatrix} \mathbf{d}_1^T \mathbf{e}_1 \\ \mathbf{d}_2^T \mathbf{e}_1 \\ \vdots \\ \mathbf{d}_N^T \mathbf{e}_1 \end{pmatrix} \quad \frac{1}{\sqrt{\lambda_2}} \begin{pmatrix} \mathbf{d}_1^T \mathbf{e}_2 \\ \mathbf{d}_2^T \mathbf{e}_2 \\ \vdots \\ \mathbf{d}_N^T \mathbf{e}_2 \end{pmatrix} \quad \cdots \quad \frac{1}{\sqrt{\lambda_c}} \begin{pmatrix} \mathbf{d}_1^T \mathbf{e}_c \\ \mathbf{d}_2^T \mathbf{e}_c \\ \vdots \\ \mathbf{d}_N^T \mathbf{e}_c \end{pmatrix} \right] \tag{11.31}$$

$$= \begin{pmatrix} \mathbf{p}_1 & \mathbf{p}_2 & \cdots & \mathbf{p}_c \end{pmatrix} = \begin{pmatrix} p_1(t_1) & p_2(t_1) & \cdots & p_c(t_1) \\ p_1(t_2) & p_2(t_2) & \cdots & p_c(t_2) \\ \vdots & \vdots & & \vdots \\ p_1(t_N) & p_2(t_N) & \cdots & p_c(t_N) \end{pmatrix} \in \mathbb{R}^{N \times c},$$

where $\mathbf{E}_c \in \mathbb{R}^{M \times c}$ and $\mathbf{\Lambda}_c \in \mathbb{R}^{c \times c}$ are the truncated rectangular EOF and eigenvalue matrices implicitly defined above.

Note that if $N > M$ and $\mathbf{C}$ is full rank, $q = M$, then eq. 11.31 becomes

$$\mathbf{P} = \mathbf{D}^T \mathbf{E} \mathbf{\Lambda}^{-\frac{1}{2}}. \tag{11.32}$$

Two interesting observations can be made following result (11.32). First, transposing it gives

$$\mathbf{\Lambda}^{-\frac{1}{2}} \mathbf{E}^T \mathbf{D} = \mathbf{P}^T \tag{11.33}$$

$$\mathbf{E}^T \mathbf{D} = \mathbf{\Lambda}^{\frac{1}{2}} \mathbf{P}^T \tag{11.34}$$

$$\mathbf{D} = \mathbf{E} \mathbf{\Lambda}^{\frac{1}{2}} \mathbf{P}^T, \tag{11.35}$$

where we used the fact that, as a corollary of $q = M$, $\mathbf{E}_c = \mathbf{E} \in \mathbb{R}^{M \times M}$, so $\mathbf{E}^T \mathbf{E} = \mathbf{I}$. The above is exactly $\mathbf{D}$'s SVD! Along the same lines, the second, related, observation is about $\mathbf{P}^T \mathbf{P}$; given (11.32), and its transpose,

$$\mathbf{P}^T \mathbf{P} = \mathbf{\Lambda}^{-\frac{1}{2}} \mathbf{E}^T \mathbf{D} \mathbf{D}^T \mathbf{E} \mathbf{\Lambda}^{-\frac{1}{2}} = \mathbf{\Lambda}^{-\frac{1}{2}} \mathbf{E}^T \mathbf{E} \mathbf{\Lambda} \mathbf{E}^T \mathbf{E} \mathbf{\Lambda}^{-\frac{1}{2}}, \tag{11.36}$$

where the second stage employs $\mathbf{D}\mathbf{D}^T = \mathbf{E} \mathbf{\Lambda} \mathbf{E}^T$ (to within a scalar factor $N - 1$, which only elevates the overall power in $\mathbf{D}\mathbf{D}^T$ without changing the patterns). Taking advantage of $\mathbf{E}$'s orthonormality,

$$\mathbf{P}^T \mathbf{P} = \mathbf{\Lambda}^{-\frac{1}{2}} \mathbf{\Lambda} \mathbf{\Lambda}^{-\frac{1}{2}} = \mathbf{I}. \tag{11.37}$$

Thus, like $\mathbf{E}$, $\mathbf{P}$ is also orthonormal, i.e., the projection time series are mutually orthogonal.

The projection time series $\mathbf{p}_i$ quantifies the prominence, at a particular time, of the ith pattern (EOF) in the observed full field at that time. If at time j the ith pattern dominates the full field $\mathbf{d}_j$, then $p_i(j) = \mathbf{d}_j^T \mathbf{e}_i > \mathbf{d}_j^T \mathbf{e}_{k \neq i}$, the projection of $\mathbf{d}_j$ on $\mathbf{e}_i$ is larger than on any other retained EOFs. This may be very physically revealing if some of the EOFs have firm physical associations (as, e.g., in fig. 11.12, in which the leading patterns are clearly ENSO related), because in that case the projection time series identify physical mechanisms that are particularly active, or dormant, at specific times.

Let's clarify this using the example of winter (December–February) North Atlantic surface pressure anomalies. Figure 11.18 shows EOFs 1 and 3 of those anomalies, calculated over the 61 winters of 1949–2010. The modes are clearly distinct; while the leading one (fig. 11.18a) depicts mostly a meridional—midlatitudes vs. the subtropics—dipole, the third mode is mostly a zonal

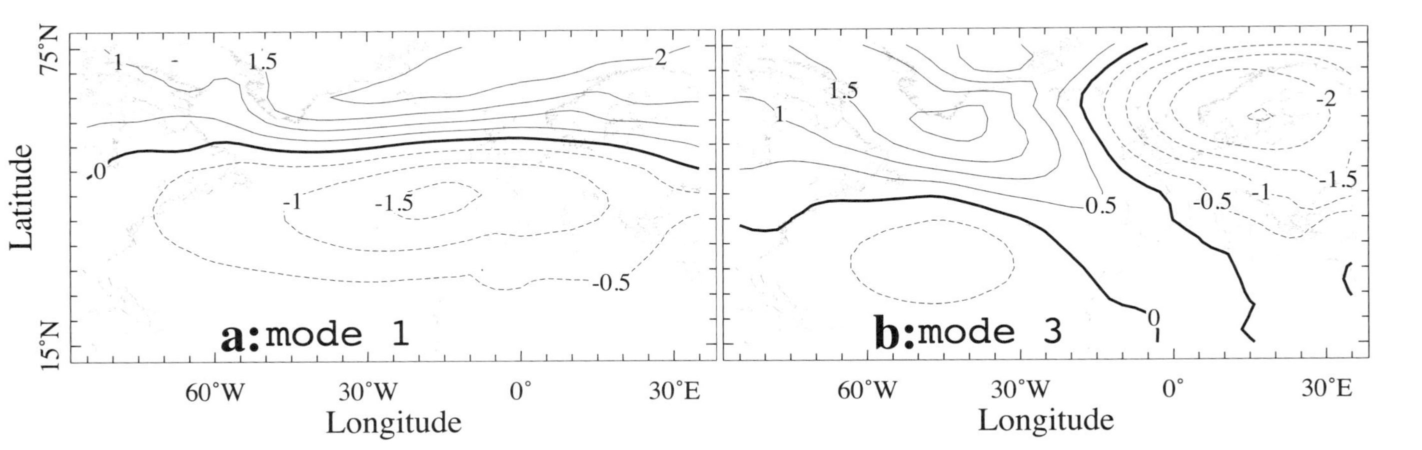

Figure 11.18. Modes 1 and 3 of North Atlantic winter (December–February mean) surface pressure anomalies. The data set comprises 61 seasons (1949/50–2009/10) and long-term linear time trend is removed. The shown modes account for approximately 50 and 13% of the total variance in the data set. Data from the NOAA NCEP-NCAR global Reanalysis (www.esrl.noaa.gov/psd/data/reanalysis/reanalysis.shtml). Units are arbitrary; contour interval is 0.5, with the zero contour highlighted.

midlatitude—Scandinavia vs. the Labrador Sea—seesaw. Not only do these differ in their orientation, meridional vs. zonal bipolarity, their spatial scales are radically different: very large, nearly planetary scale for the first, subbasin scale for the third. While outside the scope of this book, the province of dynamic meteorology and climate dynamics, variability on these starkly distinct spatial scales is most often forced by clearly distinct mechanisms. It is thus mechanistically revealing to identify times during which one mode or the other is disproportionally dominant.

To facilitate this identification, let's examine the difference in projection time series magnitudes, $|p_1(t)| - |p_3(t)|$, and examine the winters of this series' minimum (December 1971–February 1972) and maximum (December 2009–February 2010). Figure 11.19 shows the surface pressure anomaly maps of the two periods. The anomalies are clearly very different, which is expected given the way they were selected. What are less obvious, but very important, are the physical hints the maps reveal. While a full dynamical investigation of those anomalies is a climate dynamics question outside the scope of the book, some simple, relevant clues are rather apparent.

The early period (fig. 11.19a, 1971/2) is very similar to spatial pattern 3 (fig. 11.18b) and appears to be dominated by mid- to high-latitude land–sea contrasts, with below normal surface pressure over the northern North Atlantic and above normal pressure over Scandinavia. The latter period (fig. 11.19b, 2009/10), by contrast, is very similar to mode 1 (fig. 11.18a) and is a basin-wide weakening of the normal meridional pressure gradient with which the midlatitude jet is dynamically consistent. One can easily envision an almost planetary scale anomalous high over the entire North Atlantic and western Europe weakening and displacing meridionally the midlatitude jet. This may hint that planetary-scale atmospheric dynamics are key to the latter anomalies, while the early period's small scales are dominated by mostly one-dimensional (vertical) heat exchanges in which the widely disparate land–sea thermal properties play a key role.

Due to the methodological rather than dynamical focus, this discussion is rudimentary, preliminary, and incomplete. What is important is the demonstration of the way the projection time series and their relative magnitudes can focus the analyst's attention on particularly revealing periods, thus guiding subsequent analyses and facilitating mechanistic understanding.

11.9.1 The Complete EOF Analysis Using SVD and Economy Considerations

We can use the SVD representation $\mathbf{D} = \mathbf{E}\Sigma\mathbf{P}^T$ to get simultaneously the EOFs and principal components, **E**'s and **P**'s columns. Because the SVD modes are customarily arranged in descending order (the singular values, Σ's diagonal elements, satisfy $\sigma_i > \sigma_{i+1}$), $\mathbf{e}_1$, **E**'s first column, is the spatial pattern most frequently realized, the second is the spatial pattern most frequently realized while orthogonal to $\mathbf{e}_1$, and so on. The fraction of the total variance mode i accounts for is

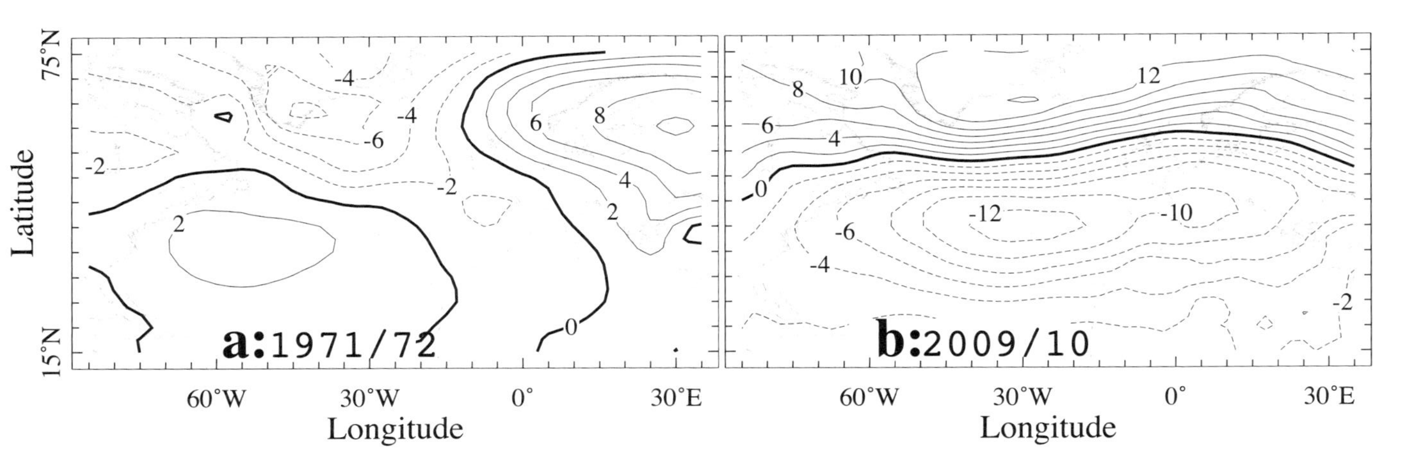

Figure 11.19. Mean observed surface pressure anomalies averaged over December–February 1971/2 (panel a) and 2009/10 (panel b) , in mb, with 2-mb contour interval. Data from the NOAA NCEP-NCAR global Reanalysis.

$$\frac{\sigma_i^2}{\sum_{j=1}^q \sigma_j^2} = \frac{\lambda_i}{\sum_{j=1}^q \lambda_j}. \tag{11.38}$$

The simple SVD representation of the EOF analysis hides important computational considerations. When $M \gg N$ and M is very large (as is often the case), it is needlessly, possibly prohibitively, expensive to calculate the full set of EOFs including EOFs $[q+1, M]$, which the SVD route entails. Instead, it will make sense to form and decompose the far smaller of the two covariance matrices, $\mathbf{D}^T\mathbf{D} \in \mathbb{R}^{N\times N}$, and eigenanalyze it,

$$\mathbf{D}^T\mathbf{D}\mathbf{p}_i = \lambda_i \mathbf{p}_i \qquad \Longrightarrow \qquad \mathbf{D}^T\mathbf{D}\mathbf{P} = \mathbf{P}\boldsymbol{\Lambda}. \tag{11.39}$$

Using $\{\mathbf{p}_i\}_{i=1}^c$ and the identity of the eigenvalues of $\mathbf{D}\mathbf{D}^T$ and $\mathbf{D}^T\mathbf{D}$ (section 5.1's eq. 5.4) we derive $\{\mathbf{e}_i\}_{i=1}^c$,

$$\mathbf{p}_i = \frac{1}{\sqrt{\lambda_i}}\mathbf{D}^T\mathbf{e}_i \tag{11.40}$$

$$\mathbf{D}\mathbf{p}_i = \frac{1}{\sqrt{\lambda_i}}\mathbf{D}\mathbf{D}^T\mathbf{e}_i \tag{11.41}$$

$$\sqrt{\lambda_i}\,\mathbf{D}\mathbf{p}_i = \mathbf{C}\mathbf{e}_i = \mathbf{E}\boldsymbol{\Lambda}\mathbf{E}^T\mathbf{e}_i, \tag{11.42}$$

where we use, slightly sloppily, $\mathbf{D}\mathbf{D}^T$ and $\mathbf{C}$ interchangeably. Because of the mutual orthonormality of the EOFs, $\mathbf{E}^T\mathbf{e}_i \in \mathbb{R}^M$ has all zero elements except element i, which is 1. When postmultiplying the diagonal $\boldsymbol{\Lambda}$, the result is an $\mathbb{R}^M$ vector with all zero elements except the ith, which is λ_i. Finally, when this vector postmultiplies $\mathbf{E}$, the result is $\lambda_i\mathbf{e}_i$. Equation 11.42 therefore becomes

$$\sqrt{\lambda_i}\,\mathbf{D}\mathbf{p}_i = \lambda_i\mathbf{e}_i \tag{11.43}$$

or

$$\mathbf{e}_i = \frac{1}{\sqrt{\lambda_i}}\mathbf{D}\mathbf{p}_i \qquad \Longrightarrow \qquad \mathbf{E}_c = \mathbf{D}\mathbf{P}\boldsymbol{\Lambda}_c^{-\frac{1}{2}}, \tag{11.44}$$

where, because $i = [1, c < M]$, this provides enough EOFs to fully span the variance, but not $\mathbb{R}^M$ in its entirety.

The computational savings this indirect EOF calculation approach affords can be vast, as fig. 11.20 shows. For $M \approx (2,4,6,8)\times 10^3$, the SVD calculation requires (32, 92, 159, 233) *times* the calculation time of the reduced EOFs. These ratios depend, of course, on N, the time length, because the smaller N/M, the smaller the fraction of the total number of EOFs obtained, and thus the larger the savings.

11.10 A Final Realistic and Slightly Elaborate Example: Southern New York State Land Surface Temperature

In this section, I present an EOF analysis example of a more realistic data set that presents several of the challenges this chapter addresses: it is larger

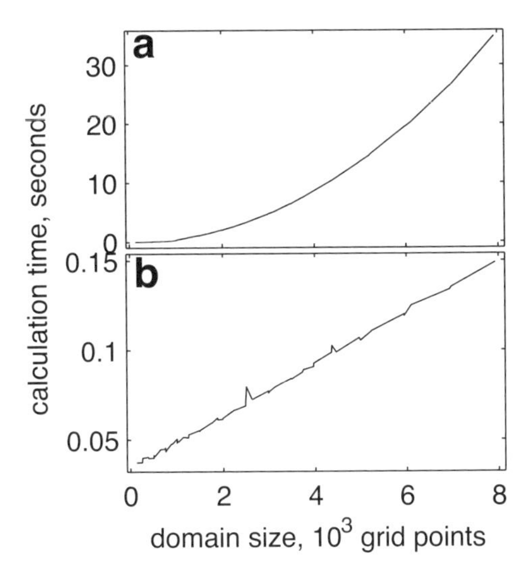

Figure 11.20. EOF calculation time on the same single processor using two distinct approaches. All calculations analyze the same problem, based on the patterns of fig. 11.4 (eq. 11.15) and similar to the one whose EOFs are shown in fig. 11.13, but with 200 instead of 50 time points. The same data set, containing the same patterns, is analyzed 64 times at increasing spatial resolutions that result in a larger number of grid points M the domain comprises, shown along the horizontal axis of both panels. In panel a, the full EOF and principal component sets are derived using the SVD of the data matrix. In panel b, the indirect approach is used for obtaining the variance spanning principal components and their respective EOFs (using eq. 11.39 to get the principal components, and then eq. 11.44 to get the EOFs). Since $M > N = 200$ throughout, in panel b's problems only the 200 variance spanning EOFs are obtained, i.e., $\mathbf{E} \in \mathbb{R}^{M \times 200}$ throughout. Note the different extents of the panels' vertical axes.

than previously analyzed data sets (comprising 3358 spatial grid points in the combined (x, y) spatial grid, but only 337 time points, so $M \gg N$); it contains missing values of two kinds (randomly and permanently missing values); it addresses a comparatively noisy physical field (land surface temperatures); and its covariance structure is complex yet can be readily explained physically. To describe completely unambiguously how each of the challenges is addressed, I favor here Octave or Matlab code and its detailed analysis rather than algebra, the language of previous sections.

The data set is a subset of NASA's MODIS[1] daytime 1-km spatial resolution land surface temperature. The full data set is delimited to southern New York (74.15°–73.15°W, 40.9°–42.2°N) and its resolution is degraded and unified (by "box averaging") to $0.022° \times 0.018°$ in longitude and latitude, respectively, resulting in $N_x = 46$, $N_y = 73$, and $= N_x N_y = 3358$. The original product's temporal resolution, 8 days, is retained. The domain includes parts of Long Island Sound, where obviously land surface temperatures are not available (yielding the data set's permanently missing values). Of the full domain's 3358 grid points, 179 points fall into this category. In addition, randomly missing values are scattered throughout the remainder of the data matrix. I call $\mathbf{T}_s \in \mathbb{R}^{M \times N}$ the matrix of raw temperatures, `Ts`, where missing values are marked by *NaN* ("not a number," on which all arithmetic operations are defined, producing *NaN*).

First, we eliminate fully or nearly fully empty time slices using

```
1. k = zeros(N,1);
2. for i = 1:N
3.   j = find(~isnan(Ts(:,i)));
4.   k(i) = length(j)>(M*.94);
5. end
6. Ts = Ts(:,find(k));
```

This code works as follows.

1. A vector to mark nearly empty times.
2. Loop on time points (`Ts`'s columns).
3. Find space indices of present values in `Ts` at time `i`.
4. `k(i)` becomes 1 if the number of missing values in column `i` exceeds 94% of the total (M), or 0 otherwise.
5. End of loop on time slices in `Ts`.
6. Retain in `Ts` only columns with more than 94% of the total (M) present.

The next task is to mark permanently empty grid points corresponding to Long Island Sound. This is done using the single Matlab line

```
I = find(~all(isnan(Ts')));
```

which works as follows. The inner `isnan(Ts')` takes $\mathbf{T}_s^T \in \mathbb{R}^{N \times M}$ and converts `NaN` (missing value) elements into 1's, and present values (real numbers) into 0's. Then, `all`—which, like most nonalgebraic Matlab operators, works down individual columns of a matrix—transforms the input $N \times M$ matrix of 1's and 0's into a $1 \times M$ row vector containing 1's for $\mathbf{T}_s$ rows (time series) containing nothing but missing values, and 0s for rows containing at least some

[1] modis.gsfc.nasa.gov/data/, conveniently available from iridl.ldeo.columbia.edu/expert/SOURCES/.USGS/.

real values. The logical "~", or "not", operator reverses this, producing 0s for $\mathbf{T}_s$ rows (time series) containing nothing but missing values, and 1's for rows containing at least some real values. Finally, find returns the indices among $[1, M]$ corresponding to $\mathbf{T}_s$ rows containing at least some real values. This is the variable I, the vector of indices into the combined space dimension marking data containing grid points outside of Long Island Sound. Using it will help us focus on available data, avoiding needless attempted calculations destined for failure.

Next, we must obtain climatologies in order to obtain anomalies. This requires a slight digression into time in Matlab. The relevant code is

```
t = t + datenum(2003,1,1);
T = datevec(t);
tn1 = datenum(T(:,1:3));
tn2 = [T(:,1) ones(N,2) zeros(N,2) ones(N,1) ];
tn2 = datenum(tn2);
T(:,4) = tn1-tn2;
```

The variable t, conforming to the MODIS time reporting convention, is given in days since January 1, 2003 and is loaded from a data file that is part of the data product we do not need to discuss any further. In the rest of the first line above, t is converted (by adding to it the day number of January 1, 2003) to days since the beginning of year zero. Then, T = datevec(t) converts each value in this time measure, days since the dawn of the AD era, into a time row vector containing (year, month, day, hour, minute, second). To handle the climatology calculations, we need a day-of-year column in T, accomplished by the last code line above, where we place in T's fourth column the difference, in full days and fractions, between the actual day, datenum(T(:,1:3)), and the first second of the same year, datenum([T(:,1) ones(N,2) zeros(N,2) ones(N,1)]), both expressed in days since the beginning of the AD scale. The input into datenum is the row vector containing (from left to right) [year (T(:,1)), 1 (for January), 1 (for January 1), 0 (hours), 0 (minutes), 1 (1 second after the year T(:,1) has begun)]. Thus, the difference, T(:,4), is the difference in days and day fractions between any time point and the year's beginning.

The raw climatologies are obtained using

```
1. Ni = length(I);
2. tc = 5:15:365;
3. Ct = nan*ones(Ni,length(tc));
4. for i = 1:Ni
5.   for k = 5:15:365
6.     j = find(~isnan(Ts(I(i),:)') & ...
          T(:,4)>(k-9) & T(:,4)<(k+9) );
```

```
7.     Ct(i,(k+10)/15) = mean(Ts(I(i),j));
8.   end
9. end
```

In line 1 above, `Ni` is the number of meaningful (not completely empty due to falling within Long Island Sound) grid points. Line 2 introduces `tc`, the time grid of the raw climatologies in day-of-year, and in line 3 `Ct` is the array that will contain the (meaningful) climatologies. In lines 4 and 5, we begin a double loop on grid point (`i`) and day-of-year (`k`). In line 6, for each space/time (`i`,`k`) combination, `j` holds the time indices of useful data in this row of `Ts` (grid point `I(i)`) (the first line, ending in "...", which tells the Octave parser that the instruction continues in the subsequent line) that falls within the `tc(k)` $\pm$ 8 days time interval (in the continuation line). Note that at 17 days, this interval is slightly wider than the spacing of `tc`. This enhances the robustness of each climatology estimate by increasing the number of data points it is based on, and also slightly smooths in time the sequence of estimates. Finally, in line 7, the climatology of grid point `i` at day-of-year index $(k+10)/15$, is the mean of all valid grid points within this interval, `Ct(i,(k+10)/15)`, and lines 8 and 9 close the loops.

Despite opening a somewhat redundantly wide interval, there are a few missing climatologies in `Ct`. We overcome that using a cosine fit. The rationale for this is simply that apart from noise, the annual cycle is by and large a cosine with a frequency of $\omega_a \equiv 2\pi/365.25$ day^{-1}, a minimum in the depth of winter, a maximum in the height of summer, and an amplitude that is the difference between the two. We thus use the following best-fit procedure to force cosine-like climatologies.

```
1. om = 2*pi/365.25;
2. Ct = [ Ct(:,end) Ct Ct(:,1) ];
3. tc = [ tc(end)-365 tc tc(1)+365 ];
4. tf = 0:366; of = ones(length(tf),1);
5. Cf = nan*ones(M,length(tf));
6. for i = 1:Ni
7.   c = Ct(I(i),:);
8.   j = find(~isnan(c));
9.   c = c(j)';
10.  th = tc(j)'; oh = ones(length(th),1);
11.  A = [ oh cos(om*th) sin(om*th) ];
12.  p = inv(A'*A)*A'*c;
13.  A = [ of cos(om*tf') sin(om*tf') ];
14.  Cf(I(i),:) = (A*p)';
15. end
```

This code works as follows. We take advantage of the periodicity of the annual cycle, whose frequency is defined in line 1. In line 2 we append the climatology with its last time point (the 365th day of the year) on the left, and the first point (the 5th day of the year) on the right, and update the time grid tc accordingly (line 3) to yield tc = [0 5 20 ⋯ 350 365 370]. In line 4 we define a finer climatology time grid, with a resolution of a single day, and in line 5 allocate the array Cf that will hold the temporally finer climatologies. Line 6 launches a loop on all relevant grid points (non-Long Island Sound ones, whose indices are held in the vector I defined earlier), and in line 7 we simply rename for succinctness the considered grid point's climatological time series c. In lines 8 and 9 we limit c to include only meaningful climatologies, excluding NaNs, and turn it into a column vector. In line 10 we take the corresponding measures on the climatology time grid tc, calling it th, a column vector. In line 11 we define the best-fit coefficient matrix, in line 12 we solve for the optimal parameter vector p, and in lines 13 and 14 we use this p to create a fine resolution, best-fit climatology, which we keep in Cf(I(i),:). After the loop is closed (line 15 above), only grid points excluded from I (i.e., ones that fall within Long Island Sound) are missing in the climatology Cf.

The next task is to obtain the anomalies, deviations of individual Ts values from their respective Cf climatologies, using

```
1. At = nan*ones(size(Ts));
2. for i = 1:N
3.   [~,j] = min(abs(tf-T(i,4)));
4.   a = Ts(I,i) - Cf(I,j);
5.   j = find(isnan(a));
6.   a(j) = zeros(length(j),1);
7.   At(I,i) = a;
8. end
```

In line 1 we declare the array to hold the temperature anomalies, and in line 2 we start a loop on the time points. Line 3 is slightly tricky. The "~" symbol is used here to indicate that while, in general, min has two output arguments, the actual minimum value (the first output argument) and the index in which this minimum is realized, we are only interested in the latter, which we store in j. Thus, line 3 identifies the time in tf, the time grid of the fine resolution climatologies obtained in the previous code segment, closest to the current day-of-year, T(i,4). In line 4, we create the raw anomaly vector a, which contains meaningful anomalies where Ts(I,i) data are available, and NaNs elsewhere. Lines 5 and 6 set those missing values to zeros. This is one option, not a universal rule, but it is not unreasonable for data sets with relatively few missing data, because it follows the basic logic that, absent any specific knowledge, the most probable value at any given time is that time's climatology. With this strategy for filling in the missing elements, implemented in lines 5 and 6, the resultant anomalies are exactly zero. In line 7 the anomalies are stored in At(I,i), and in line 8 the loop on time points is closed.

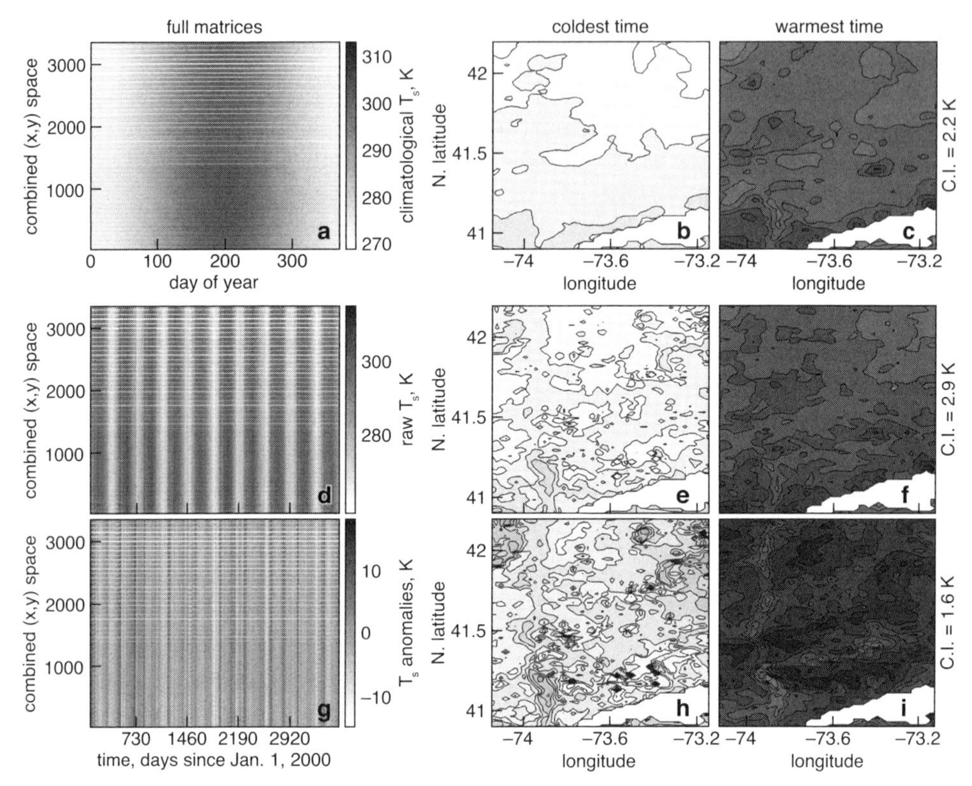

Figure 11.21. Some general characteristics of the $\mathbf{T}_s$ data example. In panels a, d, g, longitude and latitude are subsumed into a single combined space dimension, the panels' vertical axes. Panel a: climatological T_s, in K, as a function of day of year (horizontal axis, 1–366) and space (vertical axis). Panel d: raw T_s in K as a function of time (horizontal axis) and space (vertical axis). Panel g: T_s anomalies (deviations of panel d's full T_s from panel a's climatological ones), in K, as a function of time (horizontal axis) and space (vertical axis). The right panels (b, c, e, f, h, i) show, as examples, the two most extreme time fields in each of the panels to their left, transformed back to longitude–latitude space. Long Island Sound (where data are not available) is clearly visible in the right panels' lower right and as continuous empty rows in the left panels. Each row of panels share the same color scale, and each pair of left panels (e.g., e and f, or h and i) share the same contour values, with the contour interval (marked C.I.) shown on the very right.

Some of the results of these steps are shown in fig. 11.21. While all grid point climatologies (horizontal lines in fig. 11.21a) appear very similar, this is expected, as summer is warm and winter is cold throughout the considered region. Yet, panels b and c reveal the clearly structured spatial variability: higher T_s values to the south, and lower summer T_s along the Hudson, especially where it is unusually wide, just north of the Tappan Zee bridge. While panel d clearly shows the succession of summers and winters, which is entirely expected, well understood, and thus not often the focus, panel g reveals the less understood, and often the focus of attention, anomalies. While they too seem to be spatially uniform (note the linear vertical structures dominating panel g), panels h and i—where the Hudson is very clear in both summer and winter, as is high terrain east of the Hudson in summer—dispel this notion.

Despite the large M and small N, let's construct next the full covariance matrix using the numerically inefficient

```
C = At*At'/(N-1);
```

$$\Longleftrightarrow \qquad \mathbf{C} = \frac{1}{N-1}\mathbf{A}_t\mathbf{A}_t^T.$$

This full matrix is shown in fig. 11.22a, where white rows/columns correspond to spatial points falling within Long Island Sound. Two subsets of the full **C**,

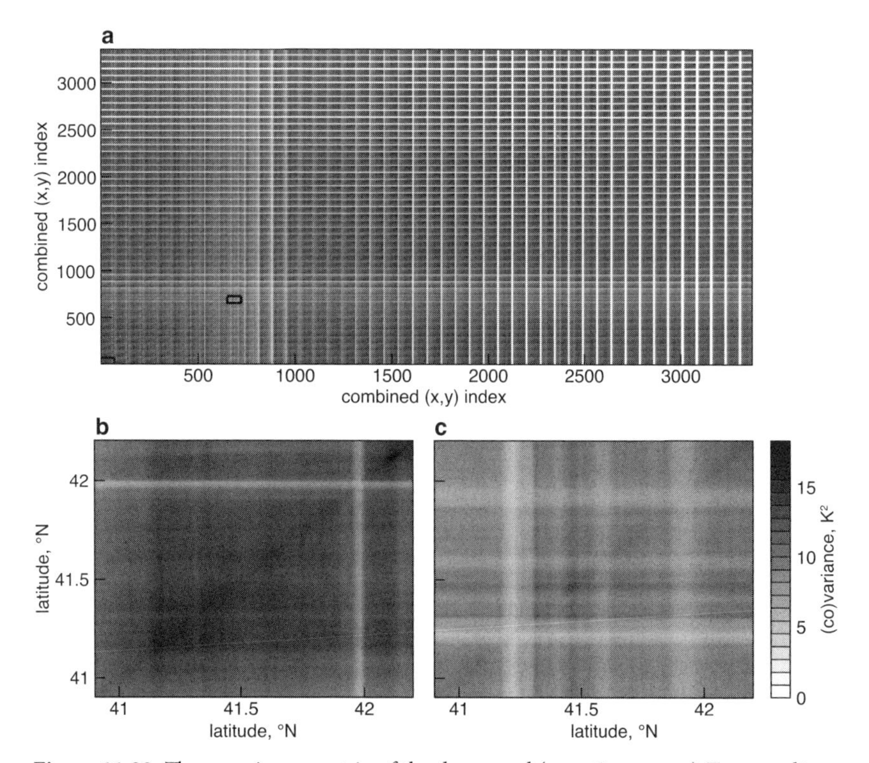

Figure 11.22. The covariance matrix of the demeaned (zero time mean) T_s anomalies held in `At`. In a, showing the full matrix, both axes are the combined (x, y) space coordinates, while two subsets framed in panel a are shown individually in the panels b and c. Panels b and c are $N_y \times N_y$. Panel b shows the covariance of the westernmost grid point column, at 74.15°W, with itself. Panel c shows the covariance of the tenth grid point column, at 73.95°W, with itself. All panels share the same shown color bar.

framed in thick black in fig. 11.22a, are magnified in panels b and c, and may seem initially puzzling. The two main features of panels b and c, and in numerous others not shown, are the local minima just south of 42°N that characterize all $N_y \times N_y$ tiles both on and off the diagonal, and the global minimum around the region covered by panel c (the upper right of the two framed regions in fig. 11.22a, showing the covariance of the column of grid points at 73.95°W with itself), with an extreme local minimum near 41.21°N.

To better understand these features of fig. 11.22, fig. 11.23 shows local (individual grid points') variance and topography (panels a–c and d, respectively). Let's first analyze the local covariance minima that occur just south of 42°N in fig. 11.22b, which corresponds to the westernmost longitude, 74.15°W. Figure 11.23a,b shows that at this location, the local variance attains a sharp minimum. This location occurs near the town of West Hurley, New York, in the foothills of the Catskill Mountains. The elevation there, ~600 ft (~200 m, upper left corner of fig. 11.23's leftmost panel), is not particularly high (compared with the >4000 ft, or >1200 m, to the immediate north), and the general rise of local variance with elevation (compare fig. 11.23's panels a and d) cannot be reasonably invoked. Instead, this location falls at the center of one of New York City's largest water reservoirs, the Ashokan Reservoir, with a surface area of ~33 km^2 and maximum capacity of ~0.5 km^3. Given the massive heat capacity of this reservoir, $\sim 2 \cdot 10^{12}$ J K^{-1}, and the fine spatial scale of the surface temperature data set used, it is little wonder that this large reservoir clearly affects both local variance

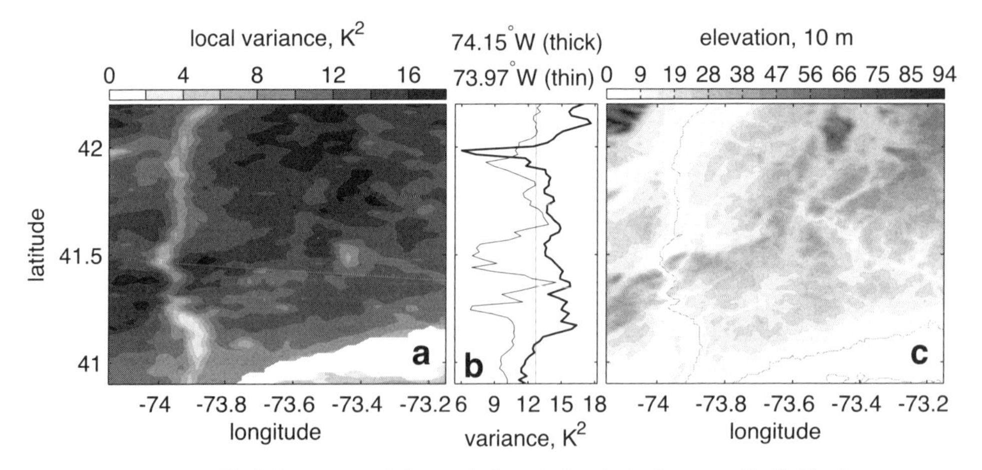

Figure 11.23. The leftmost panel shows the longitude—latitude map of individual spatial grid points' time variance in K^2. The local variance along the two fixed x columns whose covariance submatrices are shown in fig. 11.22b,c is shown in the two middle line graphs (with the longitude indicated in the title). The rightmost panel shows the region's topography.

in its immediate vicinity and the covariance of other points in the domain with points at or very near to the reservoir. This is the reason for the anomalously low (co)variance just south of 42°N so clearly visible in fig. 11.22b.

Next, let's address fig. 11.22c, showing covariance along 73.95°W. A glance at fig. 11.23d reveals that this longitude coincides either with the Hudson River itself or the flat, near sea level areas to its immediate east and west, where local variance throughout the north–south extent is much smaller than in the rest of the domain (fig. 11.23a and c). The generally suppressed variance along 73.95°W attains a global minimum at the intersection with 41.21°N, where the Hudson River is widest (~6 km, between Croton-on-Hudson and Haverstraw, both in New York, on the Hudson's east and west banks, respectively). It is again the anomalously high heat capacity of water surfaces relative to land surfaces that suppresses variance near this location, resulting in the general minima along 73.95°W, and the global minimum there near 41.21°N, fig. 11.22c displays.

The final step in the analysis is obtaining the EOFs. Because of the grid points with permanently missing data (corresponding to Long Island Sound), we cannot simply eigenanalyze the full covariance matrix (the previously introduced `C`), because eigenanalysis of all the matrix's values will return nothing but `NaN`s. Instead, temporarily throwing numerical efficiency considerations to the wind, we calculate

```
1. At(I,:) = At(I,:) - ...
              mean(At(I,:),2)*ones(1,N);
2. Ci = At(I,:)*At(I,:)'/(N-1);
3. [Ei,d] = eigs(Ci,3);
4. d = 100*diag(d)/trace(Ci);
5. E = nan*zeros(M,3);
6. E(I,:) = Ei;
7. E = reshape(E,Ny,Nx,3);
```

In line 1 we remove the temporal means from all non-Long Island Sound grid points, and in line 2, we compute the deficient covariance matrix comprising only these points. Line 3 is instead of the ostentatiously inefficient `[Ei,d] = eig(Ci);`, as the latter calculates a complete span for $\mathbb{R}^{\text{length}(I)}$, a very large, and mostly unnecessary, set. Instead, line 3 uses the so-called ARPACK package[2] to compute only the three eigenvectors of `Ci` corresponding to the leading three eigenvalues. In line 4, we compute the percentage of the variance accounted for by each one of the modes, and in lines 5–7 we prepare the leading EOFs for plotting by including them in a larger array, comprising all the domain's points (lines 5 and 6) and reshaping them back into the x–y plane. This results in fig. 11.24—which shows that the largest effect is elevation—and concludes the example.

[2] The Lanczoz algorithm implementation of the Arnoldi method applied to symmetric matrices.

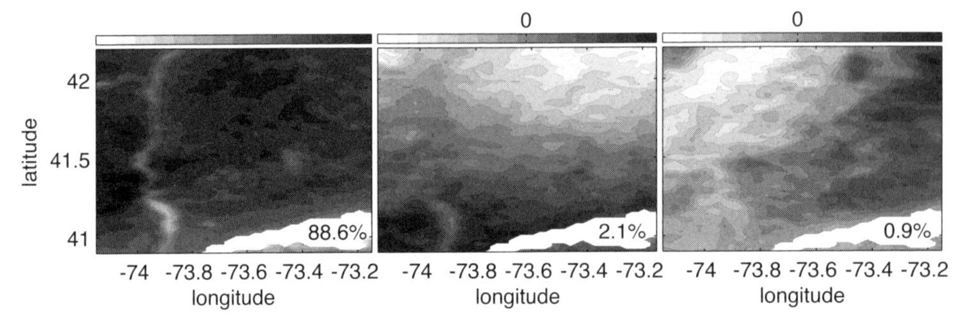

Figure 11.24. The analyzed data set's three leading EOFs in descending order of dominance from left to right. The percent variance a given mode accounts for is shown on the lower right (in the maps' Long Island Sound). Since the units are arbitrary, the color bars show explicitly only 0 if it is present inside the color range (i.e., if the EOF loadings are not of uniform sign throughout the domain).

11.11 Extended EOF Analysis, EEOF

The EOF procedure can be readily extended in a variety of ways, all amounting to analyzing simultaneously more than one data set. In many cases involving EEOF analysis, the distinct data sets are actually the same one, but at delayed times (more on that later). Other instances involve literally distinct data sets, say the two sets (though the analysis need not be limited to two or any other data set number) $A(x,y,t)=\mathbf{A}\in\mathbb{R}^{M_a\times N}$ and $B(p,t)=\mathbf{B}\in\mathbb{R}^{M_b\times N}$. Note that the space dimensions need not be the same and need not even correspond to the same physical coordinate; in the example above, field A depends on two spatial coordinates, x and y, and time, while field B depends on a single space coordinate, p, and time. As always, the time dimension—or an alternative coordinate replacing time with the important characteristic that order matters along it—does have to be shared by all participating data sets. To carry out the analysis, all we need is to form a single matrix from all participating data sets.

To be precise, let's assume the data sets have been modified in whatever ways they needed to be (e.g., removing seasonality, removing linear trends, removing grid point means, removing global mean, and so on). We will assume the first data set is structured as

$$\mathbf{A} = \begin{pmatrix} \mathbf{a}_1^1 & \mathbf{a}_1^2 & \cdots & \mathbf{a}_1^N \\ \mathbf{a}_2^1 & \mathbf{a}_2^2 & \cdots & \mathbf{a}_2^N \\ \vdots & \vdots & & \vdots \\ \mathbf{a}_{N_x}^1 & \mathbf{a}_{N_x}^2 & \cdots & \mathbf{a}_{N_x}^N \end{pmatrix} \in \mathbb{R}^{N_x N_y \times N}, \tag{11.45}$$

where $\mathbf{a}_i^j\in\mathbb{R}^{N_y}$ is the column of N_y grid points at $x=x_i$ and $t=t_j$, and $N_xN_y = M_a$. Similarly, we will assume the structure of the second data set is

$$\mathbf{B} = \begin{pmatrix} \mathbf{b}^1 & \mathbf{b}^2 & \cdots & \mathbf{b}^N \end{pmatrix} \in \mathbb{R}^{N_p \times N}, \tag{11.46}$$

where $\mathbf{b}^j \in \mathbb{R}^{N_p}$ is the column of N_p grid points at $t = t_j$, and $N_p = M_b$. We now combine the two sets into a single matrix

$$\mathbf{D} = \begin{pmatrix} \mathbf{A} \\ \mathbf{B} \end{pmatrix} = \begin{pmatrix} \mathbf{a}_1^1 & \mathbf{a}_1^2 & \cdots & \mathbf{a}_1^N \\ \mathbf{a}_2^1 & \mathbf{a}_2^2 & \cdots & \mathbf{a}_2^N \\ \vdots & \vdots & & \vdots \\ \mathbf{a}_{N_x}^1 & \mathbf{a}_{N_x}^2 & \cdots & \mathbf{a}_{N_x}^N \\ \mathbf{b}^1 & \mathbf{b}^2 & \cdots & \mathbf{b}^N \end{pmatrix}$$

$$= \begin{pmatrix} \mathbf{d}^1 & \mathbf{d}^2 & \cdots & \mathbf{d}^N \end{pmatrix} \in \mathbb{R}^{(N_x N_y + N_p) \times N}. \tag{11.47}$$

This "stacking" of the individual participating data sets into a single matrix is the reason all data sets must share the one ordered coordinate we call here "time."

In some situations, it may be necessary to transform all variables to have roughly equal sway over the eigenvectors. For example, if **A** holds temperature anomalies in K and **B** pressure anomalies in Pa, there may well be two orders of magnitude difference between the characteristic scales of the two variables. It will make sense then to replace **B** with $\mathbf{B}' := \mathbf{B}/100$. More sophisticated transformations are, of course, available and are neither universal nor standard. What is important is to keep in mind in all such situations that whatever transformation is employed likely affects the results, and that an alternative transformation will yield different results.

Some readers may find this stacking peculiar. There are several reasons why it makes sense. First, presumably the analyst combines mechanistically related variables, ones for which there are compelling physical reasons to expect cross-covariance. The stacking forces the analysis to devise patterns that optimally span the full variance–covariance–cross-covariance structure, taking note not only of the relationship between individual elements (most often grid points) of a single variable, but also of the cross-relationship between individual elements of distinct data sets. For example, think back to the days before Sir Gilbert Walker,[3] before Cane and Sarachik,[4] before we knew that the oceanic El Niño and the atmospheric southern oscillations phenomena are really two virtually inseparable sides of the same coin. In those days, one researcher could have reasonably analyzed tropical Pacific surface temperature anomalies (like we did in section 11.7.2), while another could have just as reasonably analyzed tropical Pacific surface atmospheric pressure anomalies. Each would have gotten nice,

[3] www.weatheronline.co.uk/reports/weatherbrains/Sir-Gilbert-Walker.htm.

[4] www.cambridge.org/catalogue/catalogue.asp?isbn=9780521847865\&ss=fro.

structured results, but absent a Walker or Wyrtki[5] class flash of deep insight, the two would most likely not talk to each other, and, more importantly, not consider their results in the context of the other's, thus unveiling the coupled picture we now have of the two fields. If, on the other hand, a single analysis would have combined both fields into a single EEOF calculation, the strong interlinked relationships between basin scale oceanic and atmospheric variability would have been immediately revealed.

Another reason the stacking together of distinct variables in the EEOF analysis is provided by the dynamical systems view point. Most dynamic variables we are likely to be interested in are part of a larger, interconnected system; the ocean, the land and the atmosphere are elements of the climate system; anthropogenic climate change and human behavior interact, and so on. If we think of the system as characterized by its state vector $\mathbf{x}$—the collection of values of all relevant variables at all nodes of an agreed upon numerical grid—then any model governing the evolution of the system can be described in the general form

$$\frac{d\mathbf{x}}{dt} = \mathbf{F}(\mathbf{x}) + \mathbf{f}, \tag{11.48}$$

where the left-hand term denotes the state's rate of change, $\mathbf{F}(\mathbf{x})$ is the state-dependent (in general, nonlinear) model, and $\mathbf{f}$ combines all external forcing mechanisms affecting $\mathbf{x}$ and its evolution. Because $\mathbf{x}$ comprises many variables (e.g, in a climate model parts of $\mathbf{x}$ will hold atmospheric temperature, humidity, and momentum values at specified nodes of the numerical mesh, while other parts can hold such land-related variables as vegetation cover, soil moisture, and so on, possibly on a different grid), the nonlinear model $\mathbf{F}$ takes note of numerous variables as it advances the state forward in time. In light of this, not only is stacking variables sensible, it is indeed imperative for the highly realistic simulations we grew to expect from our models of reality.

Just like in the individual data set EOF analysis, the crux of the matter is forming the covariance matrix (because there are several data sets, here it is the cross-covariance matrix that plays the same role the covariance matrix plays in ordinary—single data set—EOF analysis). Assuming the row means have been appropriately removed, this means

$$\begin{aligned}
\mathbf{C} &= \frac{1}{N-1}\mathbf{D}\mathbf{D}^T = \langle \mathbf{d}^i \mathbf{d}^{iT} \rangle \\
&= \begin{pmatrix}
\langle \mathbf{a}_1 \mathbf{a}_1^T \rangle & \langle \mathbf{a}_1 \mathbf{a}_2^T \rangle & \cdots & \langle \mathbf{a}_1 \mathbf{a}_{N_x}^T \rangle & \langle \mathbf{a}_1 \mathbf{b}^T \rangle \\
\langle \mathbf{a}_2 \mathbf{a}_1^T \rangle & \langle \mathbf{a}_2 \mathbf{a}_2^T \rangle & \cdots & \langle \mathbf{a}_2 \mathbf{a}_{N_x}^T \rangle & \langle \mathbf{a}_2 \mathbf{b}^T \rangle \\
\vdots & \vdots & & \vdots & \vdots \\
\langle \mathbf{a}_{N_x} \mathbf{a}_1^T \rangle & \langle \mathbf{a}_{N_x} \mathbf{a}_2^T \rangle & \cdots & \langle \mathbf{a}_{N_x} \mathbf{a}_{N_x}^T \rangle & \langle \mathbf{a}_{N_x} \mathbf{b}^T \rangle \\
\langle \mathbf{b} \mathbf{a}_1^T \rangle & \langle \mathbf{b} \mathbf{a}_2^T \rangle & \cdots & \langle \mathbf{b} \mathbf{a}_{N_x}^T \rangle & \langle \mathbf{b} \mathbf{b}^T \rangle
\end{pmatrix}
\end{aligned} \tag{11.49}$$

[5] www.soest.hawaii.edu/Wyrtki/.

$$
= \begin{pmatrix} \mathbf{C}_{11} & \mathbf{C}_{12} & \cdots & \mathbf{C}_{1N_x} & \mathbf{C}_{1\mathbf{b}} \\ \mathbf{C}_{21} & \mathbf{C}_{22} & \cdots & \mathbf{C}_{2N_x} & \mathbf{C}_{2\mathbf{b}} \\ \vdots & \vdots & & \vdots & \vdots \\ \mathbf{C}_{N_x1} & \mathbf{C}_{N_x2} & \cdots & \mathbf{C}_{N_xN_x} & \mathbf{C}_{N_x\mathbf{b}} \\ \mathbf{C}_{\mathbf{b}1} & \mathbf{C}_{\mathbf{b}2} & \cdots & \mathbf{C}_{\mathbf{b}N_x} & \mathbf{C}_{\mathbf{bb}} \end{pmatrix} \in \mathbb{R}^{(N_xN_y+N_p)\times(N_xN_y+N_p)},
$$

where, recall, angled brackets denote ensemble averaging, so that $\langle \mathbf{xy}^T \rangle$ is the matrix of expected values, whose ij element is the expected value of $x_i y_j$, the product of $\mathbf{x}$'s ith and $\mathbf{y}$'s jth grid points.

Just like the covariance matrix, this is a square symmetric matrix with a real, complete set of eigenvectors and real eigenvalues,

$$
\mathbf{Ce}_i = \lambda_i \mathbf{e}_i \in \mathbb{R}^{(N_xN_y+N_p)\times 1}, \qquad \mathbf{C} = \mathbf{E\Lambda E}^T. \tag{11.50}
$$

For plotting and interpretation purposes, the eigenvectors must be reshaped back to their physical domains. If we enumerate the elements of the kth eigenvector sequentially,

$$
\mathbf{e}^k = \begin{pmatrix} e_1^k \\ \vdots \\ e_{N_y}^k \\ e_{N_y+1}^k \\ \vdots \\ e_{2N_y}^k \\ \vdots \\ \vdots \\ e_{(N_x-1)N_y+1}^k \\ \vdots \\ e_{N_xN_y}^k \\ e_{N_xN_y+1}^k \\ \vdots \\ e_{N_xN_y+N_p}^k \end{pmatrix} \Longleftrightarrow \begin{pmatrix} \mathbf{a}_1 \\ \\ \mathbf{a}_2 \\ \vdots \\ \vdots \\ \mathbf{a}_{N_x} \\ \\ \mathbf{b} \end{pmatrix}, \tag{11.51}
$$

where the double arrow means "corresponding to," then

$$
\mathbf{e}_a^k = \begin{pmatrix} e_1^k \\ e_2^k \\ \vdots \\ e_{N_xN_y}^k \end{pmatrix} \Longrightarrow \begin{pmatrix} e_1^k & e_{N_y+1}^k & \cdots & e_{(N_x-1)N_y+1}^k \\ e_2^k & e_{N_y+2}^k & \cdots & e_{(N_x-1)N_y+2}^k \\ \vdots & \vdots & & \vdots \\ e_{N_y}^k & e_{2N_y}^k & \cdots & e_{N_xN_y}^k \end{pmatrix} \tag{11.52}
$$

(where the arrow denotes reshaping for displaying) and

$$\mathbf{e}_b^k = \begin{pmatrix} e^k_{N_xN_y+1} \\ e^k_{N_xN_y+2} \\ \vdots \\ e^k_{N_xN_y+N_p} \end{pmatrix}. \tag{11.53}$$

Using Matlab or Octave, this is straightforwardly accomplished for, e.g., EOFs $[1, K]$, using

```
Ea = reshape(E(1:Nx*Ny,1:K),Ny,Nx,K);, and
Eb = E(Nx*Ny+[1:Np],1:K);.
```

Note that unless the different data sets vary completely independently of each other (so that, using the two data set example, the only nonzero block in **C**'s last row/column of blocks is the diagonal one, $\langle \mathbf{bb}^T \rangle$, in which case stacking **A** and **B** makes no physical sense), these EOFs will differ from the EOFs obtained by eigenanalyzing the individual data sets separately. This can be understood narrowly—**C**'s eigenvectors are expected to differ depending on whether $\langle \mathbf{a}_i \mathbf{b}^T \rangle$ and $\langle \mathbf{b}\mathbf{a}_i^T \rangle$ (for $i = [1, N_x]$) in **C**'s last row and column blocks vanish or not—but it can be more insightfully viewed in terms of the eigenvectors' role. When **A** or **B** are analyzed individually, the sole role of the eigenvectors is to span most efficiently **A**'s or **B**'s variance–covariance structure. By contrast, when **C** combines **A** and **B**, the eigenvectors' job is to span most efficiently the full variance–covariance–cross-covariance structure of **A** and **B**. Since these jobs can be drastically different, so can the resultant eigenvectors.

Similarly, the eigenvalues have a somewhat different interpretation than before. They quantify how important a particular cross-covariance pattern $\mathbf{e}^k$ is to spanning *all* variables' cross-covariance, but says little about **A**'s spanning by $\mathbf{e}_a^k$ or **B**'s spanning by $\mathbf{e}_b^k$. These must be independently evaluated, as follows. After splitting $\mathbf{e}^k$ into $\mathbf{e}_a^k$ and $\mathbf{e}_b^k$ (eqs. 11.52 and 11.53) and normalizing them into $\hat{\mathbf{e}}_a^k$ and $\hat{\mathbf{e}}_b^k$ (by simply dividing by their norms, which will not be 1 because they are subsets of unit norm vectors), the fraction of the individual variance accounted for by cross-covariance mode k is

$$\zeta_a^k = \frac{(\hat{\mathbf{e}}_a^k)^T \mathbf{A}\mathbf{A}^T \hat{\mathbf{e}}_a^k}{\text{trace}(\mathbf{A}\mathbf{A}^T)} \tag{11.54}$$

and

$$\zeta_b^k = \frac{(\hat{\mathbf{e}}_b^k)^T \mathbf{B}\mathbf{B}^T \hat{\mathbf{e}}_b^k}{\text{trace}(\mathbf{B}\mathbf{B}^T)} \tag{11.55}$$

for data sets **A** and **B**, respectively. Note that $\{\zeta_a^k\}$ and $\{\zeta_b^k\}$ are *not* eigenvalue sets, and, relatedly, that, in general, the splits (11.52) and (11.53) render the sets $\{\hat{\mathbf{e}}_a^k\}$ and $\{\hat{\mathbf{e}}_b^k\}$ individually nonorthogonal (i.e., in general, $(\hat{\mathbf{e}}_a^m)^T \hat{\mathbf{e}}_a^n \neq 0$ and $(\hat{\mathbf{e}}_b^m)^T \hat{\mathbf{e}}_b^n \neq 0$), which means that some variance may well project on more than one mode, $\sum_k \zeta_a^k \neq \text{trace}(\mathbf{AA}^T)$ and $\sum_k \zeta_b^k \neq \text{trace}(\mathbf{BB}^T)$. A corollary of the above discussion is that it is entirely possible for an EEOF analysis to yield a robust $\mathbf{e}^1$ (in terms of its corresponding normalized eigenvalue), but on which individual data sets project rather poorly. This is the case when the data sets are strongly interrelated and thus exhibit significant mutual covariability, but the individual sets' variance related to the covariability with the other one(s) is a small fraction of their total individual variance.

Let's examine a synthetic example. As data set A, we use the vaguely "atmospheric"

$$\mathbf{A} = \left[\frac{1}{150} \exp\left(\frac{\mathbf{p}}{200} \right) \mathbf{f}^T + \cos\left(\frac{\pi \mathbf{p}}{500} \right) \mathbf{g}^T \right] \in \mathbb{R}^{11 \times 50}, \tag{11.56}$$

where $\mathbf{p} \in \mathbb{R}^{11}$ with elements $[0, 100, 200, \ldots, 1000]$ and $N_p = 11$, $\mathbf{f} \in \mathbb{R}^{50}$, and $\mathbf{g} \in \mathbb{R}^{50}$ are temporal coefficient vectors (i.e., $N_t = 50$) with elements $f \sim N(0, 0.8)$ and $g \sim N(0, 0.4)$ (fig. 11.25). As B, let's use the patterns generated by eq. 11.22 and shown in figs. 11.4 and 11.14, with individual leading patterns shown in fig. 11.15.

The analysis starts with

```
D = [ norm(B(:))*A/norm(A(:)) ; B ];
```

in which $\mathbf{B}$ is appended at the bottom of (the renormalized) $\mathbf{A}$, combining the two data sets while roughly equalizing their magnitudes by multiplying $\mathbf{A}$ by the ratio of B's and A's norms, considering all elements at once (achieved by "(:)"). Next, we remove the time mean of the three matrices

```
A = A - mean(A,2)*ones(1,Nt);
B = B - mean(B,2)*ones(1,Nt);
D = D - mean(D,2)*ones(1,Nt);
```

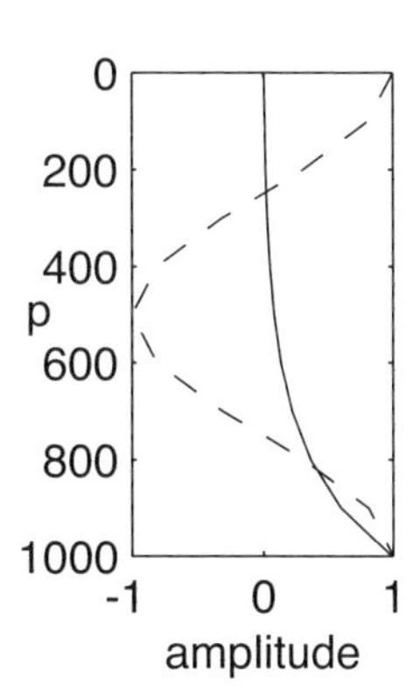

Figure 11.25. The p-dependent generating patterns used for data set A in the synthetic EEOF example. Referring to eq. 11.56, the solid curve is the pattern whose temporal coefficients are $\mathbf{f}$'s elements, while the dashed curve is the pattern whose temporal coefficients are $\mathbf{g}$'s elements.

and decompose the respective covariance matrices

```
[Ea,da] = eig(A*A'/(Nt-1));
[Eb,db] = eig(B*B'/(Nt-1));
[Ed,dd] = eig(D*D'/(Nt-1));
```

Because we need the modes in descending order of eigenvalues, we carry out, using the "a" data set as an example,

```
[da,i] = sort(diag(da),'descend');
Ea = Ea(:,i(1:5));
da = da(1:5)/sum(da);
```

where we retain the leading five eigenvectors for each data set, and where the last command normalizes the eigenvalues by the data set's total variance.

Next, a unique element of EEOF analysis, we split the eigenvectors of the combined matrix **D** into the parts corresponding to **A** (rows $[1, N_p]$) and **B** (rows $[N_p + 1, N_p + N_x N_y]$), using

```
Eda = Ed(1:Np ,:);
Edb = Ed(Np+[1:Nx*Ny],:);
```

and normalizing those split, or "partial," retained eigenvectors using

```
for i = 1:5
  Eda(:,i) = Eda(:,i)/norm(Eda(:,i));
  Edb(:,i) = Edb(:,i)/norm(Edb(:,i));
end
```

Finally, we obtain the projection coefficient sets $\{\zeta_a^k\}$ and $\{\zeta_b^k\}$ using

```
za = diag(Eda'*(A*A')*Eda)/trace(A*A');
zb = diag(Edb'*(B*B')*Edb)/trace(B*B');
```

The results of this analysis are summarized in fig. 11.26.

Figure 11.26a shows the eigenspectra. Because the combined (EEOF) analysis (filled triangles) is applied to a larger data set, and because the two participating data sets are not expressly designed to covary, the fraction of the total respective variance spanned by the combined analysis' first mode ($\lambda_1/\sum_j \lambda_j$) is smaller than either set A's or B's. Correspondingly, the combined spectrum's falloff is initially slower than that of A or B. Despite set B's four generating patterns, its $\lambda_4/\sum_j \lambda_j$ is almost zero, because—due to the heavy noise contamination—earlier modes have already spanned most structured variance, leaving little to λ_4. Because A is

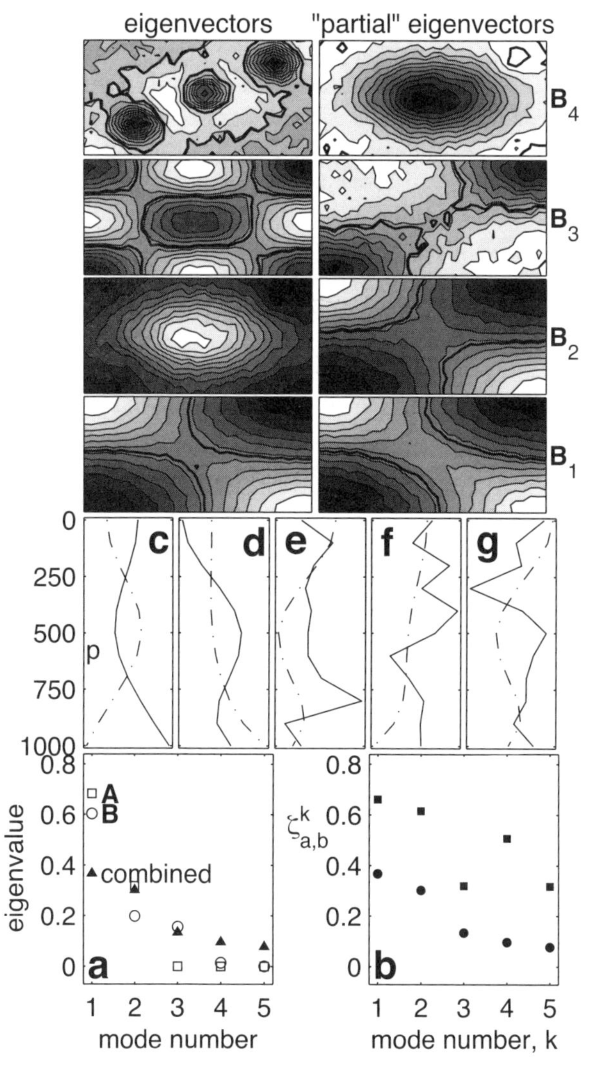

Figure 11.26. Results of the synthetic EEOF example. Panel a: leading eigenvalues of covariance matrices derived from the individual data sets (open squares and circles) and the combined one (black triangles). The projection coefficients $\zeta_{a,b}^{[1:5]}$ are shown in panel b using black squares (circles) for **A** and **B**. Panels c–g show the leading 5 eigenvectors (solid) and "partial" eigenvectors (dash-dotted) with panel c showing mode 1 and panel g mode 5. The top panels show **B**'s modes 1 ($\mathbf{B}_1$) through 4 ($\mathbf{B}_4$) as a function of x (horizontal axis) and y (vertical axis), eigenvectors on the left and corresponding "partial" eigenvectors on the right. The zero contour is emphasized.

smaller (only 11 spatial points, thus contaminated by relatively less noise) and simpler (depending on the single space coordinate p and only two generating patterns), its eigenspectrum faithfully represents the fact that it is based on only two generating patterns (empty squares at mode 3 and up are virtually zero).

Figure 11.26b shows the scaled projection coefficients (eqs. 11.54, 11.55). For the same reasons given above (A's simplicity, lower dimensions, a single independent coordinate, and only two generating patterns), A has fared better (had more sway over the analysis) than B in terms of choosing the combined analysis' leading patterns (e.g., $\zeta_a^1 \approx 0.8$ compared with $\zeta_b^1 \approx 0.4$). Also for the same reasons, the ζ_a set exhibits a peculiar property EEOF analysis permits, increasing relative importance of trailing modes ($\zeta_a^5 > \zeta_a^4$, both purely noise spanning for A, as the solid curves in fig. 11.26f,g show).

Figures 11.26c–g show A's spatial modes 1–5, plotted against A's only independent coordinate p. Clearly, both individually calculated relevant patterns (solid curves in fig. 11.26c,d) and the partial patterns (dash-dotted curves in fig. 11.26c–g) represent some combinations of A's two p-dependent generating patterns (fig. 11.25), essentially (using the "atmospheric" metaphor) ground-enhanced sinusoids. You may find it odd that while A is generated from only two patterns, all dash-dotted curves in panels a–g are structured. This is so for several reasons. First, like in the individual EOF analysis, the patterns often represent the noisy blend of the generating patterns, not those patterns' pristine rendition; the analysis—EOF or EEOF—has no way of telling apart noise from structure other than the frequency of being realized. Second, the random temporal coefficients f and g (eq. 11.56) combine a's generating patterns into many structured combinations, the most frequently realized while cross-covarying with B of which are the dash-dotted curves of fig. 11.26c–g. Third, and relatedly, the combined analysis' patterns represent a balancing act of spanning the most of A's structure while also spanning as much of B's structure as possible (or vice versa; the order of the data sets in the combined matrix makes no difference).

In fig. 11.26c, both $\mathbf{e}_a^1$ (A's first "partial" eigenvector; dash-dotted) and A's $\mathbf{e}^1$ (A's first individually calculated EOF; solid) are "ground"-enhanced sinusoids, combinations of A's two generating patterns. In the combined analysis, fig. 11.26c's dash-dotted curve ($\mathbf{e}_a^1$) is paired with fig. 11.26's $\mathbf{B}_1$'s right-hand pattern, also clearly a combination of B's generating patterns, mostly the two high-amplitude ones shown in figs. 11.4a and 11.14a. Thus, the leading combined mode is a blend of the two generating patterns of A and B, where in the former it is the *only* generating patterns, while in the latter it is the leading (high amplitude) ones.

Subsequent modes get even trickier to interpret in the absence of clear physical expectations, because, in this synthetic case, they reflect the randomness of the temporal coefficients and noise amplitudes progressively more prominently. Yet their spatial patterns—on both the A (fig. 11.26d–g) and B (fig. 11.26's $\mathbf{B}_{[2,4]}$) sides—maintain their structured nature, clearly representing combinations of the generating patterns that—due to random alignment of temporal coefficients—happened to dominate the individual sets while mutually cross-covarying.

11.11.1 The Dynamical Systems View of EEOFs

Many authors reserve the term EEOF analysis to situations in which the "stacked" variables in eq. 11.47 are the same physical variable at different times,

$$\mathbf{D} = \begin{pmatrix} \mathbf{a}_1^T & \mathbf{a}_1^{T+1} & \cdots & \mathbf{a}_1^{T+N-1} \\ \vdots & \vdots & & \vdots \\ \mathbf{a}_{N_x}^T & \mathbf{a}_{N_x}^{T+1} & \cdots & \mathbf{a}_{N_x}^{T+N-1} \\ \mathbf{a}_1^{T-1} & \mathbf{a}_1^{T} & \cdots & \mathbf{a}_1^{T+N-2} \\ \vdots & \vdots & & \vdots \\ \mathbf{a}_{N_x}^{T-1} & \mathbf{a}_{N_x}^{T} & \cdots & \mathbf{a}_{N_x}^{T+N-2} \\ \vdots & \vdots & & \vdots \\ \vdots & \vdots & & \vdots \\ \mathbf{a}_1^{2} & \mathbf{a}_1^{3} & \cdots & \mathbf{a}_1^{N+1} \\ \vdots & \vdots & & \vdots \\ \mathbf{a}_{N_x}^{2} & \mathbf{a}_{N_x}^{3} & \cdots & \mathbf{a}_{N_x}^{N+1} \\ \mathbf{a}_1^{1} & \mathbf{a}_1^{2} & \cdots & \mathbf{a}_1^{N} \\ \vdots & \vdots & & \vdots \\ \mathbf{a}_{N_x}^{1} & \mathbf{a}_{N_x}^{2} & \cdots & \mathbf{a}_{N_x}^{N} \end{pmatrix} \in \mathbb{R}^{N_x N_y T \times N}, \qquad (11.57)$$

where we assume for specificity that A depends on two spatial coordinates, x and y, N_x and N_y long, respectively, in addition to time. This or similar "stacking" procedures arise in various contexts (notably in singular spectrum analysis or in attractor reconstruction in delayed realization space; neither method is discussed in this book). In the EEOF context it may be understood in terms of the dynamical system that governs A's evolution.

Many physical nonlinear dynamical systems, left long enough to their own devices and allowed to evolve under statistically stable forcing, achieve a statistical steady state. In this state, elements of the system's state vector noisily fluctuate as the system evolves, but attain values randomly drawn from bounded distributions with time-independent statistical moments. Picture, for example, the weather in New York in August; some days may be marginally tolerable, but by and large most days' highs and lows fall within the 80–90 and 60–70 degrees Fahrenheit ranges with modest day-to-day departures. We do not expect an August day with a high of 134°F or an August night with a low of 14°F, because neither has been realized over the roughly two centuries for which we have records of New York weather. Long-term trends notwithstanding, this system can be reasonably said to be at or near a statistical steady state in which external and internal heat inputs and outputs more or less cancel out over suitably long timescales, so that individual fluctuations are kept in check.

Yet the system is constantly bombarded by what can be reasonably characterized as external forcing agents. For example, New York weather can be temporarily disrupted by an unusually active sun, an unusually powerful El Niño in the Equatorial Pacific, or a rare tropical storm making an unfortunate landfall. Under those circumstances, shortly after the perturbations have had a chance to exact their full toll, the system finds itself in the relatively unfamiliar territory of large, and possibly also unusually structured, deviations of its state vector elements from their long-term means. Provided the unusual event was not too large, shortly thereafter the system begins to relax back to its normal state. Depending on the state of the system and the forcing, the relaxation may unfold rapidly, or, in other circumstances, may take a long time. Thus, the relatively short period following the system displacement by external forcing away from its normal (statistical steady) state are of great import. Think, e.g., about an ordinary oppressive summer heat wave in Chicago, typically lasting a day or two followed by a healthy thunderstorm and return to normal, vs. the infamous July 1995 heat wave that lasted about a week and killed hundreds of people throughout the Chicago area and the Midwest. It is not only the unusual length of this event—humidity played a crucial role, and so-called blocking events routinely last this long—but, compounded with its deadliness, the anomalous stubbornness of the event emphasizes the great societal and fundamental scientific importance of stability of weather systems.

Figure 11.27 shows an example of this for a three interacting species Lotka-Volterra dynamical system. Following an imposed perturbation, the system oscillates in time as it gradually returns to normal, which is manifested in the shown space as spiraling inward toward the fixed point (the system's long-term normal state), the middle of the central empty circle. The length of time it takes to return to normal—here the number of cycles the system completes before its state returns to the center—depends on the system properties (parameters such as the primary producer's intrinsic growth rate or the grazer's grazing efficiency) and the particular state analyzed. A climate-related alternative example may constitute the return of the Equatorial Pacific system to roughly normal state following a major El Niño and the possible subsequent evolution toward the next La Niña.

The EEOF "temporal stacking" (eq. 11.57) allows each column—if it happened to contain a relaxation event, and if that relaxation occurred on a timescale shorter than or equal to T time units—to play the role of a single realization of the return process. Columns that do not contain such return events contain little, and presumably unstructured, time evolution and thus contribute relatively little, allowing relaxation containing columns to dominate **D**'s covariance structure and thus the eigenvectors. When the eigenvectors are obtained, they must be "unpacked," e.g., for the kth eigenvector,

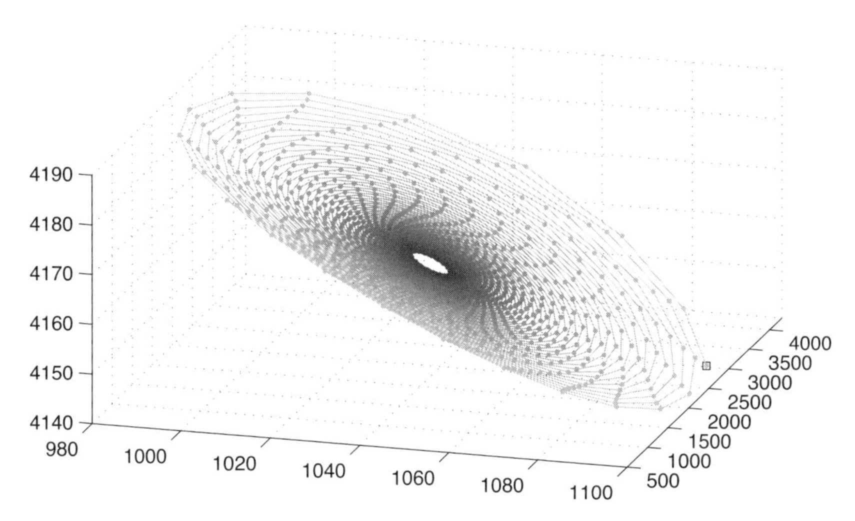

Figure 11.27. An example of dynamical system relaxation from a forced perturbation back to a steady state. The dynamical system is a three interacting species Lotka-Volterra system comprising a predator, a grazer, and a primary producer. The three species' abundances form the three dimensions of the shown space. The initial forced perturbation, from which the numerical integration begins, is marked by a black square (lower right). Each successive pair of points is connected by a straight-line segment. Time is not explicitly a dimension in the shown space, but its progression is shown implicitly by the gray shade, which darkens with time.

$$
\begin{aligned}
\mathbf{e}^k &= \begin{pmatrix} e_1^k & e_2^k & \cdots & \cdots & e_{N_x N_y T}^k \end{pmatrix}^T \\
&= \begin{pmatrix} \mathbf{e}_1^{k,T} \\ \vdots \\ \mathbf{e}_{N_x}^{k,T} \\ \mathbf{e}_1^{k,T-1} \\ \vdots \\ \mathbf{e}_{N_x}^{k,T-1} \\ \vdots \\ \vdots \\ \mathbf{e}_1^{k,2} \\ \vdots \\ \mathbf{e}_{N_x}^{k,2} \\ \mathbf{e}_1^{k,1} \\ \vdots \\ \mathbf{e}_{N_x}^{k,1} \end{pmatrix}
\Longrightarrow
\begin{matrix}
\begin{pmatrix} \mathbf{e}_1^{k,T} & \mathbf{e}_2^{k,T} & \cdots & \mathbf{e}_{N_x}^{k,T} \end{pmatrix} \\
\begin{pmatrix} \mathbf{e}_1^{k,T-1} & \mathbf{e}_2^{k,T-1} & \cdots & \mathbf{e}_{N_x}^{k,T-1} \end{pmatrix} \\
\begin{pmatrix} \mathbf{e}_1^{k,T-2} & \mathbf{e}_2^{k,T-2} & \cdots & \mathbf{e}_{N_x}^{k,T-2} \end{pmatrix} \\
\vdots \\
\begin{pmatrix} \mathbf{e}_1^{k,2} & \mathbf{e}_2^{k,3} & \cdots & \mathbf{e}_{N_x}^{k,3} \end{pmatrix} \\
\begin{pmatrix} \mathbf{e}_1^{k,2} & \mathbf{e}_2^{k,2} & \cdots & \mathbf{e}_{N_x}^{k,2} \end{pmatrix} \\
\begin{pmatrix} \mathbf{e}_1^{k,1} & \mathbf{e}_2^{k,1} & \cdots & \mathbf{e}_{N_x}^{k,1} \end{pmatrix},
\end{matrix}
\end{aligned}
\tag{11.58}
$$

where, assuming again that the original data set depends on x, y, and time, for each lag $l = [1, T]$,

$$\begin{pmatrix} \mathbf{e}_1^{k,l} & \mathbf{e}_2^{k,l} & \cdots & \mathbf{e}_{N_x}^{k,l} \end{pmatrix} = \begin{pmatrix} e_1^{k,l} & e_{N_y+1}^{k,l} & \cdots & e_{(N_x-1)N_y+1}^{k,l} \\ e_2^{k,l} & e_{N_y+2}^{k,l} & \cdots & e_{(N_x-1)N_y+2}^{k,l} \\ \vdots & \vdots & & \vdots \\ e_{N_y}^{k,l} & e_{2N_y}^{k,l} & \cdots & e_{N_xN_y}^{k,l} \end{pmatrix} \in \mathbb{R}^{N_y \times N_x} \quad (11.59)$$

is a physical domain map of the kth spatial pattern l time units into the T time units temporal evolution of interest. Using the El Niño system as an example again, if the studied variable is monthly mean Equatorial Pacific SSTAs and T corresponds to December (around the time most El Niño events attain maximum amplitude), the $N_y \times N_x$ map $\mathbf{e}_{[1,N_x]}^{1,T-2}$ is the leading pattern ($k = 1$) of October ($l = T - 2$) SSTAs participating in the coherent evolution toward maximally evolved December warm events. This pattern need not, and, in general, *will* not, optimally span October anomalies; this map, previously denoted, in general, $\mathbf{e}^1$ with no subscript, must be calculated from the covariance matrix of October anomalies only over all available years and, of course, cannot take note of any other month. The full kth optimal span of the Equatorial Pacific sea SSTA evolution leading to a fully evolved warm event, assuming as an example a 4-month evolution, is given by the map sequence

$$\mathbf{e}_{[1,N_x]}^{k,\text{Sep.}} \longrightarrow \mathbf{e}_{[1,N_x]}^{k,\text{Oct.}} \longrightarrow \mathbf{e}_{[1,N_x]}^{k,\text{Nov.}} \longrightarrow \mathbf{e}_{[1,N_x]}^{k,\text{Dec.}}. \quad (11.60)$$

Because, e.g., $\mathbf{e}_{[1,N_x]}^{1,\text{Nov.}}$ is the optimal November pattern taking note of neighboring months, all November data can be projected on this pattern, which will single out Novembers more or less likely to develop into a warm event in the following December, both retrospectively and into the future.

11.11.2 A Real Data EEOF Example

Let's consider monthly mean soil moisture anomalies in southern New England (data set A[6]) and western North Atlantic SSTAs (data set B[7]). The time grid shared by both data sets is monthly means from January 1982 to June 2010 (N_t = 342 months). The soil moisture data (A) spans 76°–72°W, 42°–47°N at $1° \times 1°$ resolution, with all 30 grid points containing data. The SSTA data set (B) covers

[6] Fan, Y. and H. van den Dool (2004) Climate Prediction Center global monthly soil moisture data set at 0.5° resolution for 1948 to present. *J. Geophys. Res.*, **109** D10102, doi:10.1029/2003JD004345; obtained from iridl.ldeo.columbia.edu\linebreak[3]/expert/SOURCES/.NOAA/.NCEP/.CPC/.GMSM/.w/.

[7] Reynolds, R. W., N. A. Rayner, T. M. Smith, D. C. Stokes, and W. Wang (2002) An improved in situ and satellite SST analysis for climate. *J. Clim.*, **15**, 1609–1625; obtained from iridl.ldeo.columbia.edu\linebreak[3]/expert/SOURCES/.NOAA/.NCEP/.EMC/.CMB/.GLOBAL/.Reyn_SmithOIv2/.

80°–64°W, 30°–44°N, at 2°×2° resolution, with 15 of the 72 grid points falling on land, thus containing no data, leaving 57 active Atlantic SSTA grid points (whose indices into the full 72×1 vector is kept in the index vector `io`). Let's denote the soil moisture and SSTA full data sets by $\mathbf{A} \in \mathbb{R}^{30\times342}$ and $\mathbf{B} \in \mathbb{R}^{57\times342}$, with respective month specific, annual resolution temporal subsets from t_i to t_j by $\mathbf{A}_{[i,12,j]} \in \mathbb{R}^{30\times[(j-i)/12])}$ and $\mathbf{B}_{[i,12,j]} \in \mathbb{R}^{57\times[(j-i)/12]}$.

Suppose we want to examine the connection, if any, between North Atlantic SSTAs and soil moisture anomalies in southern New England, and, in particular, the SSTA patterns that play a coherent role in the evolution of April New England soil moisture anomalies. To do so using EEOF, we let

$$\mathbf{D} = \begin{pmatrix} \mathbf{A}_{[4,12,342]} \\ \mathbf{B}_{[3,12,341]} \\ \mathbf{B}_{[2,12,340]} \\ \mathbf{B}_{[1,12,339]} \end{pmatrix} \equiv \begin{pmatrix} \mathbf{A}_4 \\ \mathbf{B}_3 \\ \mathbf{B}_2 \\ \mathbf{B}_1 \end{pmatrix} \in \mathbb{R}^{(30+3\cdot57)\times29}. \tag{11.61}$$

For brevity, let's denote the Octave variables containing the four data sets, from the top down, by `A4`, `B3`, `B2`, and `B1`. We then get the norm of the combined `B` data sets by carrying out

```
nB = norm([B1(:);B2(:);B3(:)]);
```

and assemble `D` with the rescaled `A4` that has roughly the same norm as the combined `B` data sets using

```
D = [ nB*A4/norm(A4(:)) ; B1 ; B2 ; B3 ];
```

As usual, we remove the long-term means of each time series,

```
D = D - mean(D,2)*ones(1,length(D(1,:)));
```

and then form and decompose the covariance,

```
C = D*D'/(length(D(1,:))-1);
[E,d] = eigs(C,3); d = 100*diag(d)/trace(C);
```

where we have calculated only the leading three eigenvectors and eigenvalues, normalizing the latter by the full variance to obtain the percent variance spanned by each of the modes.

It will be convenient to split the eigenvectors into their individual parts, using

```
Ea4 = E(          1:30 ,:);   ← Apr. soil moisture part
Eb3 = E(30+     [1:57],:);    ← ocean only Mar. SSTA part
```

```
Eb2 = E(30+   57+[1:57],:);   ← ocean only Feb.y SSTA part
Eb1 = E(30+2*57+[1:57],:);   ← ocean only Jan. SSTA part
```

Next, we can obtain the individual projection time series, e.g., using the example of `A4` and `B3`,

```
Pa4 = A4'*Ea4; Pb3 = B3'*Eb3;
```

and calculate the individual fractions of total variance spanned by the retained modes, e.g., for the same example data sets,

```
za4 = diag(Ea4'*(A4 *A4')*Ea4)/trace(A4*A4');
zb3 = diag(Eb3'*(B3 *B3')*Eb3)/trace(B3*B3');
```

To plot the spatial pattern, we must return them to geographical space. For the soil moisture, all of the grid points of which are active and present, this requires simply

```
Ea4 = reshape(Ea4,Nys,Nxs,3);
```

where `Nxs` and `Nys` are the soil moisture's x and y dimensions. These are the leading spatial patterns of April New England soil moisture anomaly that covary coherently with preceding January–March North Atlantic SSTAs. And what do the corresponding North Atlantic SSTA patterns look like? This is slightly tricky, as we have previously limited the SSTA domain to ocean points only, which resulted in reducing the full, 72-grid-point, domain into a 57-grid-point subset indexed by `io`. That is, the truncated "partial" eigenvectors (comprising ocean points only) cannot be reshaped back into geographical space (they are 57 elements long, while the product of the SSTA's space dimensions, `Nxt` = 9 and `Nyt` = 8, is 72). Instead, we carry out (using the `B3` example)

```
i = nan*ones(72,3);
i(io,:) = Eb3;
Eb3 = reshape(i,Nyt,Nxt,3);
```

using `i` as a temporary array and the ocean points' indices, `io`.

These patterns are shown in fig. 11.28, with panels a–g presenting the leading three SSTA patterns for the three indicated preceding months, and panels j–l presenting the corresponding soil moisture patterns. Table 11.2 presents the percent variance spanned by each mode. The SSTA patterns of the three preceding months clearly differ (e.g., panels a,d,g), describing the North Atlantic SSTA evolution that optimally cross-covaries with the April soil moisture patterns shown in panels j, k, l.

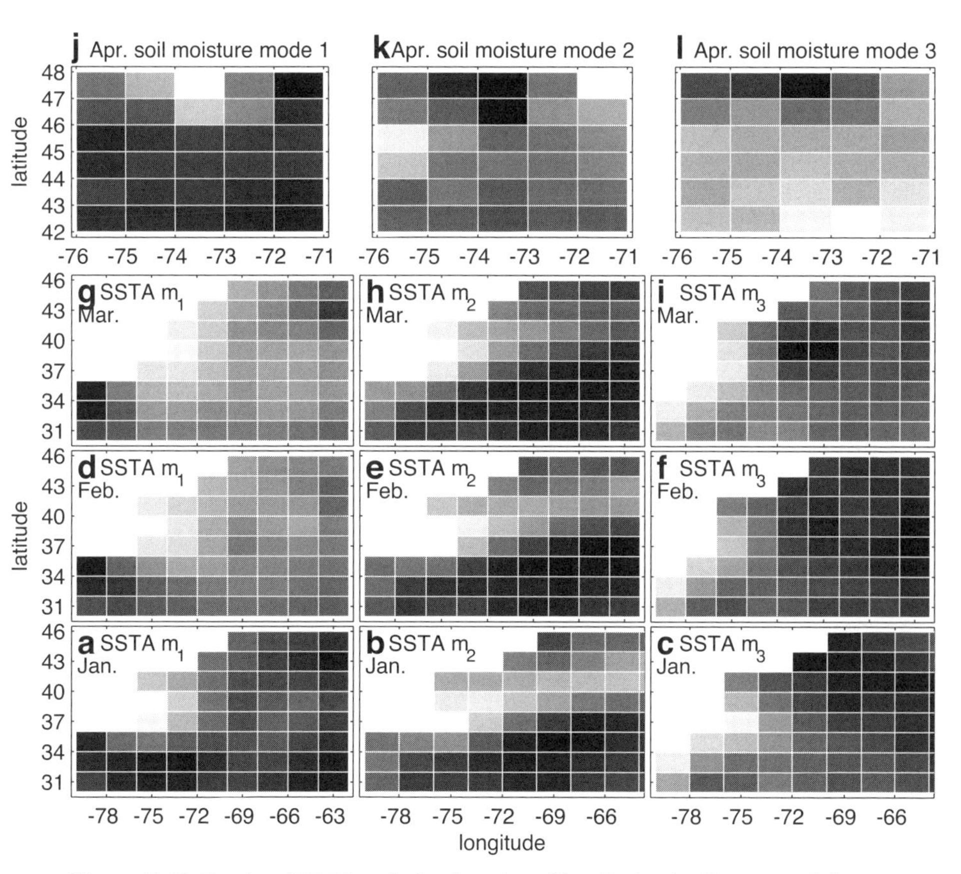

Figure 11.28. Results of EEOF analysis of southern New England soil moisture (whose three leading patterns are shown in panels j–l) and North Atlantic SSTAs (whose three leading patterns for January are shown in panels a–c, for February in panels d–f, and for March in panels g–i). Because there are so few grid points in both considered domains, the patterns are shown as checkerboards.

Table 11.2

Approximate percentage of spanned variance. The rows correspond to the three leading combined modes, with scaled eigenvalues given by the leftmost numerical column. The three rightmost columns show the percent variance of individual data sets spanned by the set's corresponding "partial" eigenvectors.

k	$100\lambda^k/\text{trace}(\mathbf{C})$	$100\zeta^k_{a_4}$	$100\zeta^k_{b_3}$	$100\zeta^k_{b_2}$	$100\zeta^k_{b_1}$
mode 1	25	36	2	2	2
mode 2	16	8	8	8	11
mode 3	14	19	1	3	2

11.12 Summary

Constructing the EOFs of a data set is arguably the most basic data reduction technique. It recasts an original data set comprising an N-element long time series of observed $\mathbf{d}_i \in \mathbb{R}^M$

$$\mathbf{D} = \begin{pmatrix} \mathbf{d}_1 & \mathbf{d}_2 & \cdots & \mathbf{d}_N \end{pmatrix} \in \mathbb{R}^{M \times N}$$

into a series

$$\mathbf{D} = \sum_{i=1}^{q} w_i \hat{\mathbf{e}}_i \hat{\mathbf{p}}_i^T,$$

where q is the rank of the covariability (either covariance or correlation) matrix employed; $\hat{\mathbf{e}}_i \in \mathbb{R}^M$, the ith EOF, is that matrix's ith normalized eigenvector; $\hat{\mathbf{p}}_i \propto \mathbf{D}^T \mathbf{e}_i \in \mathbb{R}^N$ is the ith normalized principal component; and $\{w_i\}_{i=1}^q$ is a set of scalar weights arranged in descending order of contribution to total information content, $w_{i+1} \leq w_i$.

The leading EOF, $\hat{\mathbf{e}}_1$, is thus the most dominant spatial pattern in the data set, in most realistic cases the spatial pattern most frequently realized. The second EOF, $\hat{\mathbf{e}}_2$, is the most dominant spatial pattern in the data set that is orthogonal to $\hat{\mathbf{e}}_1$; the third EOF, $\hat{\mathbf{e}}_3$, is the most dominant spatial pattern in the data set that is orthogonal to both $\hat{\mathbf{e}}_1$ and $\hat{\mathbf{e}}_2$, and so on,

$$\hat{\mathbf{e}}_i^T \hat{\mathbf{e}}_j = \hat{\mathbf{p}}_i^T \hat{\mathbf{p}}_j = \begin{cases} 1 & i = j \\ 0 & i \neq j \end{cases},$$

$$\|\mathbf{D}^T \hat{\mathbf{e}}_i\| \leq \|\mathbf{D}^T \hat{\mathbf{e}}_{i-1}\|, \quad i = 2, 3, \ldots, q.$$

The jth value of the ith principal component ($\hat{p}_{ji}$, the jth element of $\hat{\mathbf{p}}_i$) is a dimensionless measure of the relative dominance of pattern i ($\hat{\mathbf{e}}_i$) at time j, the degree to which $\mathbf{d}_j$ spatially resembles $\hat{\mathbf{e}}_i$.

TWELVE

The SVD Analysis of Two Fields

IN THE PRECEDING CHAPTER, we discussed one of the many methods available for simultaneously analyzing more than one data set. While powerful and useful (especially for unveiling favored state evolution pathways), the EEOF procedure has some important limitations. Notably, because the state dimensions rapidly expand as state vectors are appended end-to-end, EEOF analysis may not always be numerically tractable. For analyzing two data sets, taking note of their cross-covariance but not explicitly of individual sets' covariance, the SVD method is the most natural.

SVD analysis (which, unfortunately, shares its name with the related but separate mathematical operation introduced in chapter 5) can be thought of as a generalization of EOF analysis to two data sets that are believed to be related. Like EEOF analysis, SVD analysis requires the two data sets to share the dimension of one orderly independent variable they depend on, which we continue to call time t. We assume, as before, that spatial dependencies have been combined into a single space dimension, as described in section 11.3, so the multidimensional data sets A and B are of the form

$$\mathbf{A} = \begin{pmatrix} \vdots & \vdots & & \vdots \\ \mathbf{a}_1 & \mathbf{a}_2 & \cdots & \mathbf{a}_N \\ \vdots & \vdots & & \vdots \end{pmatrix} \in \mathbb{R}^{M_a \times N} \tag{12.1}$$

and

$$\mathbf{B} = \begin{pmatrix} \vdots & \vdots & & \vdots \\ \mathbf{b}_1 & \mathbf{b}_2 & \cdots & \mathbf{b}_N \\ \vdots & \vdots & & \vdots \end{pmatrix} \in \mathbb{R}^{M_b \times N}, \tag{12.2}$$

where the shorthand $\mathbf{a}_i \equiv \mathbf{a}(t_i)$ is used, and the respective combined space dimensions are M_a and M_b. We also assume that at this point $\mathbf{A}$ and $\mathbf{B}$ are centered in time (their row means vanish) and that taking anomalies—or any other transforms of the original data deemed necessary—has already taken place.

Next, we form the cross-covariance matrix of $\mathbf{A}$ and $\mathbf{B}$, proportional to either $\mathbf{AB}^T$ or $\mathbf{BA}^T$. These matrices—each other's transpose, with respective dimensions $M_a \times M_b$ and $M_b \times M_a$—are therefore entirely equivalent and their eigenanalyses exert the same computational burden. Let's choose the former representation,

$$\mathbf{C} \sim \mathbf{A}\mathbf{B}^T \in \mathbb{R}^{M_a \times M_b}. \tag{12.3}$$

Note that by constructing the covariance matrix **C**, the time dimension **A** and **B** share has been "consumed," and is no longer explicitly present in the problem; this is the reason **A** and **B** must share the time dimension N.

The cross-covariance matrix is the sum

$$\begin{aligned}
\mathbf{C} = \langle \mathbf{a}\mathbf{b}^T \rangle &= \frac{1}{N-1}\left(\mathbf{a}_1\mathbf{b}_1^T + \mathbf{a}_2\mathbf{b}_2^T + \cdots + \mathbf{a}_N\mathbf{b}_N^T\right) \\
&= \frac{1}{N-1}\sum_{i=1}^{N}\left(\mathbf{a}_i\mathbf{b}_i^T\right) \qquad (12.4) \\
&= \frac{1}{N-1}\sum_{i=1}^{N}\begin{pmatrix} a_{1i}b_{1i} & a_{1i}b_{2i} & \cdots & a_{1i}b_{M_b i} \\ a_{2i}b_{1i} & a_{2i}b_{2i} & \cdots & a_{2i}b_{M_b i} \\ \vdots & \vdots & \vdots & \\ a_{M_a i}b_{1i} & a_{M_a i}b_{2i} & \cdots & a_{M_a i}b_{M_b i} \end{pmatrix} \in \mathbb{R}^{M_a \times M_b},
\end{aligned}$$

where vectors' only index $i = [1, N]$ denotes time, and the scalar a_{ji} denotes $\mathbf{a}_i$'s jth element. Thus, **C** is no longer the square symmetric matrix it was in the EOF analysis. Recall that the angled brackets denote "ensemble" averaging. This is more than just a simple technicality; it represents our view of $\mathbf{C}_{ij}$ as the "most representative" product of the ith element of **a** and the jth element of **b** (after the means are removed), where here **a** and **b** are not indexed, as they are not specific realizations but rather hypothetical, "characteristic," realizations of the two sets. Since we don't know these "characteristic" products, the best we can do is estimate them by averaging available ones from realizations $\mathbf{a}_{i=1,2,\ldots,N}$ and $\mathbf{b}_{i=1,2,\ldots,N}$.

In the next (crucial) step, **C** is decomposed using SVD (chapter 5):

$$\mathbf{C} = \mathbf{U}\boldsymbol{\Sigma}\mathbf{V}^T. \tag{12.5}$$

Because of the way **C** is constructed, the sets $\{\mathbf{u}_i\}$ and $\{\mathbf{v}_i\}$ span **A**'s and **B**'s spatial spaces, respectively. (By "spatial spaces" I mean the algebraic spaces whose dimensions are M_a and M_b, $\mathbb{R}^{M_a}$ and $\mathbb{R}^{M_b}$, in which the spatial patterns of A and B reside.) Recall that the $\{\mathbf{u}_i\}$, $\{\mathbf{v}_i\}$ pairs are arranged in descending order of cross-covariance projection on them, given by $\boldsymbol{\Sigma}$'s decreasing diagonal elements $\sigma_{i-1} \geq \sigma_i$, $i = [2, \min(M_a, M_b)]$. Consequently, $\mathbf{u}_1$ is A's spatial structure whose temporal cross-covariance with *any* spatial pattern of B is the strongest. And the B spatial structure with which $\mathbf{u}_1$ cross-covaries the strongest is $\mathbf{v}_1$. The remaining pattern pairs—$(\mathbf{u}_{i>1}, \mathbf{v}_{i>1})$—are A's and B's most strongly cross-covarying patterns under the constraint $\mathbf{u}_i^T\mathbf{u}_j = \mathbf{v}_i^T\mathbf{v}_j = 0$, $i \leq j-1$, i.e., orthogonality with lower modes in the respective spatial spaces.

The last step of the analysis reintroduces time into the problem (recall that it was temporarily eliminated from the problem by taking $\mathbf{C} \sim \mathbf{A}\mathbf{B}^T$). For an individual mode $i \leq \min(M_a, M_b)$, the projection time series are

$$\mathbf{f}_i = \mathbf{A}^T\mathbf{u}_i = \begin{pmatrix} \mathbf{a}_1^T\mathbf{u}_i \\ \mathbf{a}_2^T\mathbf{u}_i \\ \vdots \\ \mathbf{a}_N^T\mathbf{u}_i \end{pmatrix} \in \mathbb{R}^N \tag{12.6}$$

and

$$\mathbf{g}_i = \mathbf{B}^T\mathbf{b}_i = \begin{pmatrix} \mathbf{b}_1^T\mathbf{v}_i \\ \mathbf{b}_2^T\mathbf{v}_i \\ \vdots \\ \mathbf{b}_N^T\mathbf{v}_i \end{pmatrix} \in \mathbb{R}^N. \tag{12.7}$$

For all [1, c] modes deemed interesting, the same is achieved using

$$\mathbf{F} = \mathbf{A}^T\mathbf{U}_c = \begin{pmatrix} \mathbf{a}_1^T\mathbf{u}_1 & \mathbf{a}_1^T\mathbf{u}_2 & \cdots & \mathbf{a}_1^T\mathbf{u}_c \\ \mathbf{a}_2^T\mathbf{u}_1 & \mathbf{a}_2^T\mathbf{u}_2 & \cdots & \mathbf{a}_2^T\mathbf{u}_c \\ & & \vdots & \\ \mathbf{a}_N^T\mathbf{u}_1 & \mathbf{a}_N^T\mathbf{u}_2 & \cdots & \mathbf{a}_N^T\mathbf{u}_c \end{pmatrix} \in \mathbb{R}^{N\times c} \tag{12.8}$$

and

$$\mathbf{G} = \mathbf{B}^T\mathbf{V}_c = \begin{pmatrix} \mathbf{b}_1^T\mathbf{v}_1 & \mathbf{b}_1^T\mathbf{v}_2 & \cdots & \mathbf{b}_1^T\mathbf{v}_c \\ \mathbf{b}_2^T\mathbf{v}_1 & \mathbf{b}_2^T\mathbf{v}_2 & \cdots & \mathbf{b}_2^T\mathbf{v}_c \\ & & \vdots & \\ \mathbf{b}_N^T\mathbf{v}_1 & \mathbf{b}_N^T\mathbf{v}_2 & \cdots & \mathbf{b}_N^T\mathbf{v}_c \end{pmatrix} \in \mathbb{R}^{N\times c}. \tag{12.9}$$

The cutoff c can be q, **C**'s effective rank, say the smallest q satisfying $\sum_{i=1}^{q} \sigma_i / \sum_{j=1}^{\min(M_a,M_b)} \sigma_j \geq 0.85$, the largest q satisfying $\sigma_q \geq 10^{-3}\sigma_1$, or a similar ad hoc criterion. The fraction of **A**'s and **B**'s variance mode i spans is given by

$$\frac{\mathbf{f}_i^T\mathbf{f}_i}{\text{trace}(\mathbf{A}\mathbf{A}^T)} = \frac{\mathbf{u}_i^T\mathbf{A}\mathbf{A}^T\mathbf{u}_i}{\text{trace}(\mathbf{A}\mathbf{A}^T)}, \qquad \frac{\mathbf{g}_i^T\mathbf{g}_i}{\text{trace}(\mathbf{B}\mathbf{B}^T)} = \frac{\mathbf{v}_i^T\mathbf{B}\mathbf{B}^T\mathbf{v}_i}{\text{trace}(\mathbf{B}\mathbf{B}^T)}. \tag{12.10}$$

The columns of **F** contain the projection time series of **A** on its spatial patterns (**U**'s columns), and those of **G** the projection time series of **B** on its spatial patterns (**V**'s columns), with, e.g., **F**'s ith column, $\mathbf{f}_i = \mathbf{A}^T\mathbf{u}_i$, being the projection of the full time-dependent data in **A** on $\mathbf{u}_i$. Unlike the EOF case, the SVD projection time series are in general not mutually orthogonal. Since $\mathbf{u}_1$ is **A**'s spatial pattern that cross-covaries with any **B** pattern the most, the first projection time series' jth element, $\mathbf{a}_j^T\mathbf{u}_1$, indicates how dominant the $\mathbf{u}_1$ pattern is over **A**'s state at time j, $\mathbf{a}_j$. Similarly, the **B** pattern with which $\mathbf{u}_1$ cross-covaries the strongest is $\mathbf{v}_1$, so the projection of the full time-dependent **B** on it, $\mathbf{g}_1$, indicates how dominant $\mathbf{v}_1$ is at any time point in spanning **B**'s variance.

The degree of temporal cross-covariance between the two time series corresponding to the $(\mathbf{u}_i, \mathbf{v}_i)$ pair is

$$\langle f_i(t), g_i(t)\rangle = \mathbf{f}_i^T \mathbf{g}_i = (\mathbf{A}^T\mathbf{u}_i)^T(\mathbf{B}^T\mathbf{v}_i) = \mathbf{u}_i^T\mathbf{A}\mathbf{B}^T\mathbf{v}_i = \mathbf{u}_i^T\mathbf{C}\mathbf{v}_i \qquad (12.11)$$

where we treat $\mathbf{AB}^T$ as $\mathbf{C}$, which is correct to within an unimportant multiplicative scalar (on the left, **f** and **g** are represented as scalar time series and are thus not boldfaced). This can be compared to

$$\begin{aligned} \mathbf{C} &= \mathbf{U}\boldsymbol{\Sigma}\mathbf{V}^T \\ \mathbf{U}^T\mathbf{C} &= \boldsymbol{\Sigma}\mathbf{V}^T \quad , \\ \mathbf{U}^T\mathbf{C}\mathbf{V} &= \boldsymbol{\Sigma} \end{aligned} \qquad (12.12)$$

which, mode-wise, is $\mathbf{u}_i^T\mathbf{C}\mathbf{v}_i = \sigma_i$. Equating the two results, we obtain the expression for the cross-covariance of the sister time series from the two spaces

$$\langle f_i(t), g_i(t)\rangle = \mathbf{u}_i^T\mathbf{C}\mathbf{v}_i = \sigma_i. \qquad (12.13)$$

The fraction of the total $(\mathbf{A}, \mathbf{B})$ cross-covariance a given $\{\mathbf{u}_i, \mathbf{v}_i\}$ pair accounts for is then given by

$$\frac{\sigma_i}{\sum_{i=1}^{c}\sigma_i} = \frac{\sigma_i}{\|\mathbf{C}\|}, \qquad (12.14)$$

where $c = \min(M_a, M_b)$ and we use here the Frobenius norm of $\mathbf{A} \in \mathbb{R}^{M\times N}$, a generalization of the L_2 norm of a vector,

$$\|\mathbf{A}\|^2 \equiv \sum_{j=1}^{M}\sum_{i=1}^{N} A_{ij}^2. \qquad (12.15)$$

A variant of the SVD analysis method outlined above is the decomposition of the cross-correlation matrix **R** instead of the cross-covariance **C** described above. This is a trivial difference computationally, but its implications can be very important if the spatial domains contain a wide range of local variance levels (recall that the cross-correlation matrix is simply the cross-covariance matrix of the normalized data). The latter is obtained by dividing the zero-mean rows of **A** and **B** by their respective temporal standard deviations (square root of temporal variance).

If all grid points in the domain have the same temporal variance, there will be no difference between the SVD of **C** or **R**, as the two matrices will be the same to within an unimportant multiplicative constant (the latter will obviously change **Σ**'s diagonal elements, but not the structures of **U** and **V**). However, it is clear that this will rarely be the case. Under more reasonable conditions, in which the temporal variance displays some nonnegligible spatial structure, the two analyses will yield different results. The choice of decomposing **C** or **R** depends on the questions asked and the involved data sets. Their use in conjunction can sometimes be revealing in its own right.

12.1 A Synthetic Example

Let's consider the example of $A(x_a, y_a, t)$ and $B(x_b, y_b, t)$ using the time grid $t = 1, 2, \ldots, 30$. With $x_a = 0, 1, \ldots, 20$, $y_a = 0, 1, \ldots, 10$, $x_b = 0, 1, \ldots, 10$, and $y_b = 0, 1, \ldots, 5$,

$$A = f_1(t) \sin\left(\frac{2\pi x_a}{20}\right) \sin\left(\frac{2\pi y_a}{10}\right) + \frac{1}{2} f_2(t) \sin\left(\frac{4\pi x_a}{20}\right) \sin\left(\frac{4\pi y_a}{10}\right) \quad (12.16)$$

and

$$B = g_1(t) \sin\left(\frac{2\pi x_b}{10}\right) \sin\left(\frac{2\pi y_b}{5}\right) + \frac{1}{2} g_2(t) \sin\left(\frac{4\pi x_b}{10}\right) \sin\left(\frac{4\pi y_b}{5}\right). \quad (12.17)$$

The fields' scalar coefficients $f_{1,2}$, $g_{1,2}$ evolve as AR(1) (order 1 autoregressive) processes, e.g., for f_1,

$$f_1(t+1) = 0.8 f_1(t) + N(0, 0.5). \quad (12.18)$$

Figure 12.1's upper four panels show the two basic pattern pairs, while that figure's lowermost panels show the time coefficients. Figure 12.2 shows the spatial patterns of temporal variance of the two data sets.

With the data sets reshaped into the matrices A and B, we first remove, as usual, the overall time mean,

```
A = A - mean(A,2)*ones(1,Nt);
B = B - mean(B,2)*ones(1,Nt);
```

and then form the covariance and correlation matrices

```
C = A*B'/(Nt-1);
R = zeros(Ma,Mb);
for i = 1:Ma
  for j = 1:Mb; R(i,j) = cor(A(i,:),B(j,:)); end
end
```

where `cor(a,b)` $= \mathbf{a}^T\mathbf{b}/\sqrt{\mathbf{a}^T\mathbf{a}\mathbf{b}^T\mathbf{b}}$ is a shorthand for the correlation of two vectors. Finally, we carry out the SVD analyses using

```
[Uc,Sc,Vc] = svd(C);
[Ur,Sr,Vr] = svd(R);
```

The leading three patterns of each analysis are shown in fig. 12.3. The decomposition of the cross-covariance and the cross-correlation yield, clearly, very different results, which is immediately understood given the highly spatially structured variance of the two fields, shown in fig. 12.2.

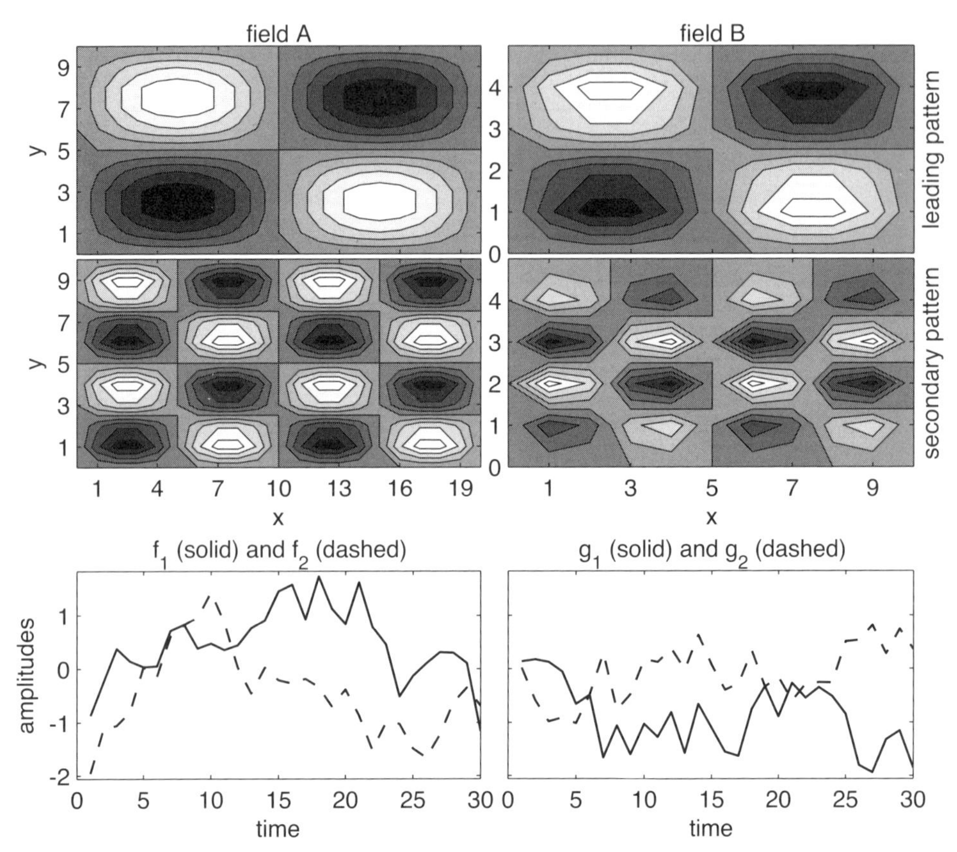

Figure 12.1. The generating patterns of A (left) and B (right), and their time behavior. Uppermost panels are the first pattern, whose time series are shown as a solid line in the corresponding lower panel. The middle panels are the second pattern, whose time series is given as a dashed line in the lowest panels.

The SVD patterns of field A derived from the covariance matrix (fig. 12.3 left) are somewhat similar, but by no means identical, to the generating patterns (fig. 12.1 left). This is a general difference between the individual EOFs of A (which would have looked like the generating patterns), and the SVD of *A and B*. Because we are trying to maximize the explained cross-covariance, not the covariance of A individually, A's SVD patterns need not be the same as A's EOFs. Given the somewhat random time evolution of all involved patterns (governed by the AR(1)), it so happens that the A pattern that best covaries with any pattern in B is almost the first generating pattern, slanted slightly because of interactions between the two generating patterns governed by the specific $f_{1,2}$ realizations.

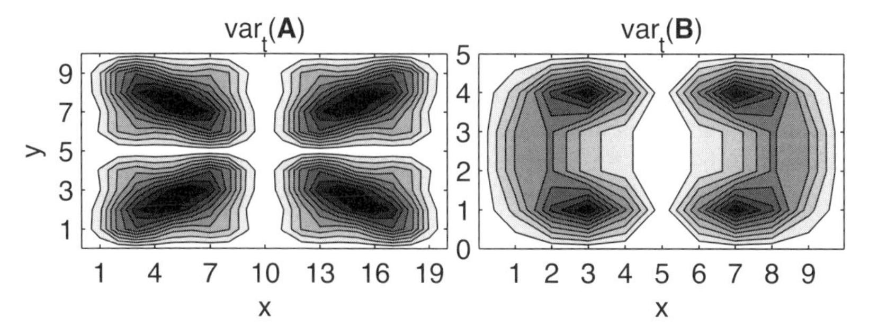

Figure 12.2. The temporal variance of A and B.

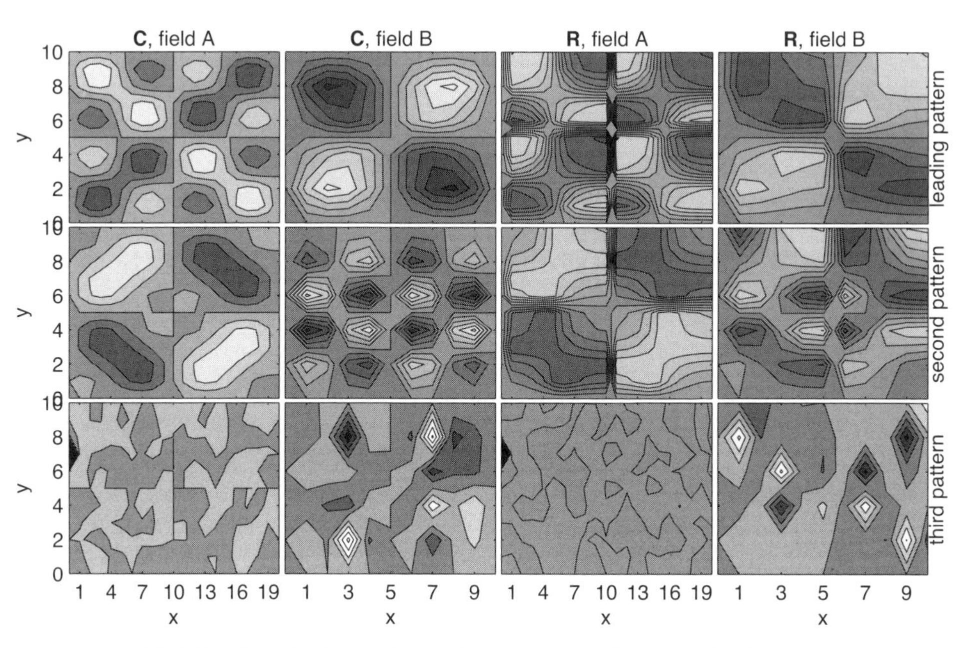

Figure 12.3. The three leading patterns of the cross-covariance matrix of fields A and B. Left (right) columns display the results of analyzing **C**(**R**). The shown modes account for 99.3, 0.7, and essentially 0% of the total cross-covariance and 98.8, 1.2, and ~0% of the total cross-correlation. The percentage of respective variance spanned by the top two patterns of each columns are (from left to right) 92.9 and 7.1; 27.6 and 72.5; 63.2 and 3.2; and 14.2 and 52.2.}

Field A's **R**-based patterns (fig. 12.3 second column from right) are clearly dominated by (leading pattern) or strongly influenced by (second pattern) the singularity that arises where the generating patterns change sign (because there, either $\mathbf{a}_i^T\mathbf{a}_i$ or $\mathbf{b}_j^T\mathbf{b}_j$, or both, vanish in the denominator of the correlation expression.

Field *B*'s **C**-based patterns (fig. 12.3 second column from left), only slightly similar to the generating patterns (fig. 12.1 left), represent some linear combinations of the two. This, again, is a manifestation of the specific time evolution.

All third patterns are unstructured in space, which is not too surprising given that both *A* and *B* are based on two generating patterns only. What is left, as a result, to the third and higher patterns is to attempt to fit noise, resulting in the noisy lower panels.

12.2 A Second Synthetic Example

This example introduces two novelties. First, its generating patterns are not as perfect, comprising a noninteger number of waves in the x and y dimensions. In addition, the temporal evolution is structured rather than random as we have encountered previously.

We consider the generating spatial patterns

$$\mathbf{P}_1 = \cos\left(\frac{2\pi x_a}{11}\right)\cos\left(\frac{2\pi y_a}{19}\right) \tag{12.19}$$

$$\mathbf{P}_2 = 2\exp\left(\frac{-x_a^2}{40}\right)\exp\left(\frac{-(y_a-7)^2}{14}\right) \tag{12.20}$$

for data set A, and

$$\mathbf{S}_1 = \cos\left(\frac{2\pi x_b}{5}\right)\cos\left(\frac{2\pi y_b}{19}\right) \tag{12.21}$$

$$\mathbf{S}_2 = 2\exp\left(\frac{-(x_b-11)^2}{20}\right)\exp\left(\frac{-y_b^2}{7}\right) \tag{12.22}$$

for data set B, with

$$x_i = \begin{cases} 0,1,\ldots,15 & i=a \\ 0,1,\ldots,11 & i=b \end{cases}, \quad y_i = \begin{cases} 0,1,\ldots,7 & i=a \\ 0,1,\ldots,5 & i=b \end{cases}$$

yielding $M_a = 16\cdot 8 = 128$ and $M_b = 12\cdot 6 = 72$.

Figure 12.4 presents the two data sets' generating patterns. Ignoring for a minute the effect of coherent temporal evolution of the two fields on the cross-covariance, the "expected" cross-covariance structure based on the generating patterns alone is

$$\mathbf{C}_{\text{expt}} = (\mathbf{p}_1 + \mathbf{p}_2)(\mathbf{s}_1 + \mathbf{s}_2)^T \in \mathbb{R}M_a \times M_b, \tag{12.23}$$

where, e.g., $\mathbf{p}_2$ is $\mathbf{P}_2$ reshaped into a column vector. Figure 12.5 presents the actual cross-covariance matrix and this "expected" cross-covariance matrix. Their differences are entirely attributable to temporal evolution. (Here "expected" is in quotes because it is only expected ignoring temporal evolution, which is only pedagogically useful.)

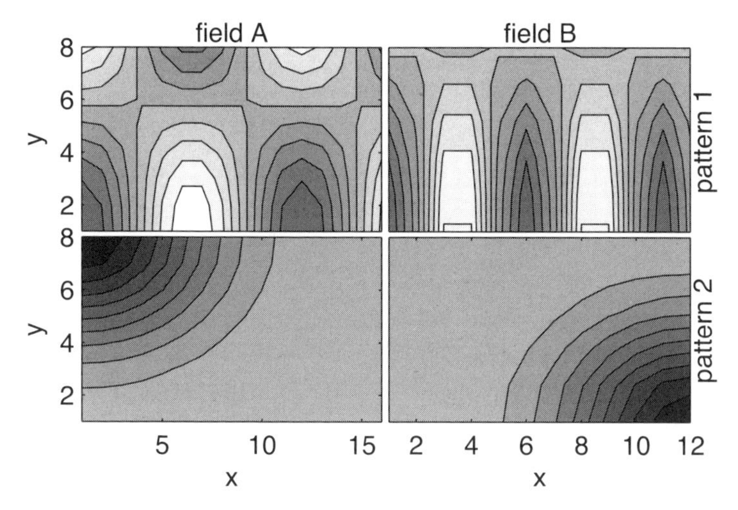

Figure 12.4. The two generating patterns of the second synthetic example.

Both "expected" and actual cross-covariance matrices attain higher values toward the right ($x_2 \geq 9$). This area corresponds to cross-covariance of A's elements with B's right-most columns (high x_2), at the low y_2 region of which (fig. 12.4) B's two generating patterns combine constructively to yield a maximum amplitude. This is most notable in the low range of A's sequential index (below, say, 56), corresponding to A's left-most columns ($x_1 < 6$), where a's two generating patterns combine constructively to yield a maximum amplitude. Within this region, e.g., the column 61 portraits in fig. 12.5's lower panels, cross-covariance is maximized each time the sequential index's value is an integer multiple of 8, i.e., each time it reaches A's upper left corner—where the constructive combination of A's generating patterns is maximized.

The fields evolve temporally according to

$$\mathbf{A} = \cos\left(\frac{2\pi t_i}{N_t}\right)\mathbf{P}_1 + \sin\left(\frac{2\pi t_i}{N_t}\right)\mathbf{P}_2 \tag{12.24}$$

$$\mathbf{B} = \cos\left(\frac{2\pi t_i}{N_t}\right)\mathbf{S}_1 + \sin\left(\frac{2\pi t_i}{N_t}\right)\mathbf{S}_2, \tag{12.25}$$

with $t = 1, 2, \ldots, N_t$ and $N_t = 40$. This choice dictates that the pairs ($\mathbf{P}_2$, $\mathbf{S}_1$) and ($\mathbf{P}_1$, $\mathbf{S}_2$) are in mutual temporal quadrature; when the first member of the pair dominates A, the pair's second member is entirely absent from B, and vice versa. Thus, what matters most is the cross-covariance of ($\mathbf{P}_1$, $\mathbf{S}_1$) and ($\mathbf{P}_2$, $\mathbf{S}_2$). Because both members of the latter pair are not bounded from above like the trigonometric functions, interactions of ($\mathbf{P}_2$, $\mathbf{S}_2$) are particularly important in the vicinity of ($x_a < 3, y_a > 6$) and ($x_b > 9, y_b < 4$). In these regions (in the vicinity of

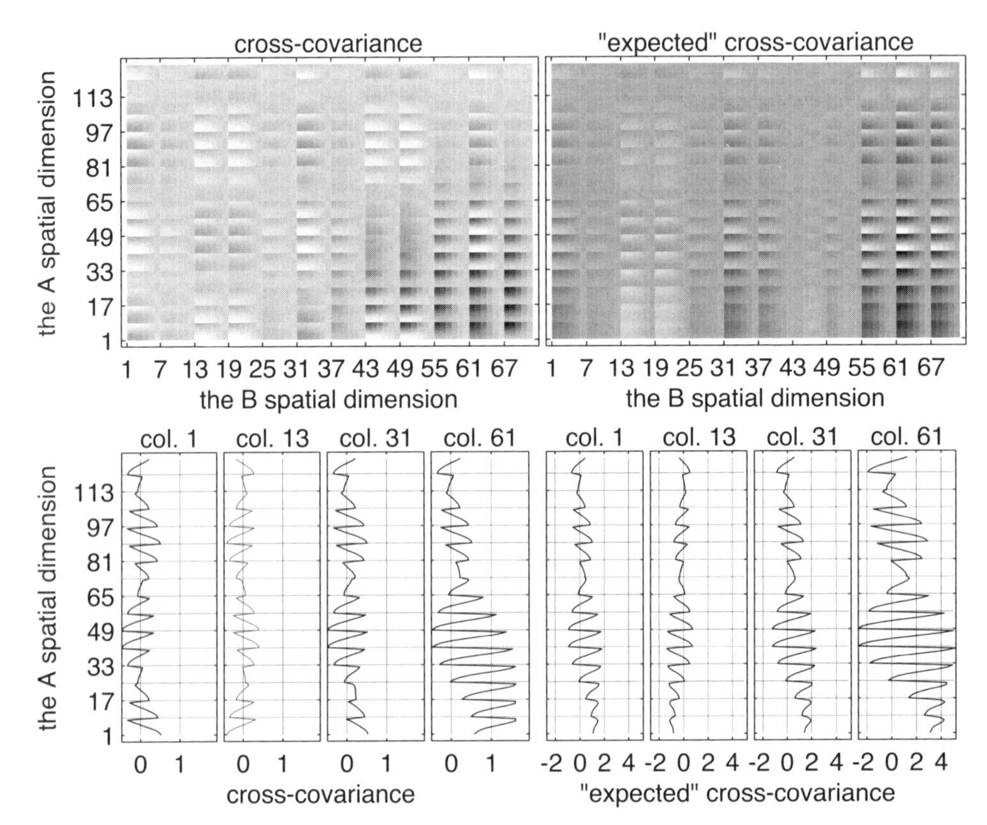

Figure 12. 5. Initial results of the second synthetic example. The upper left panel shows the cross-covariance matrix. The upper right panel shows the "expected" cross-covariance absent contribution due to temporal structure, $(\mathbf{p}_1 + \mathbf{p}_2)(\mathbf{s}_1 + \mathbf{s}_2)^T$, where, e.g., $\mathbf{p}_2$ is $\mathbf{P}_2$ reshaped into a column vector. The lower panels offer up–down cross sections along **C** at the specified column indices. The tick marks along all axes are at the relevant *y* length apart, e.g., the tick mark interval along the vertical axes of the top panels, 8, corresponds to *A*'s *y* dimension, while the interval along the same panels' horizontal axes, 6, corresponds to *B*'s *y* dimension.

the lower right corner of the cross-covariance plots), $(\mathbf{P}_2, \mathbf{S}_2)$ indeed dominate the differences between "expected" and actual cross-covariances. Conversely, where the exponentials of $(\mathbf{P}_2, \mathbf{S}_2)$ are very small (at the confluence of $x_a > 10$ and $x_b < 6$), the cosines dominate the differences between "expected" and actual cross-covariances.

The empirical modes are shown in fig. 12.6. The previously discussed dominance of the cross-covariance of *A*'s upper left corner and *B*'s lower right corner is abundantly clear. The leading mode of **A** is clearly completely dominated by the upper left corner, where $-\mathbf{P}_1$ and $\mathbf{P}_2$ have a positive superposition. As

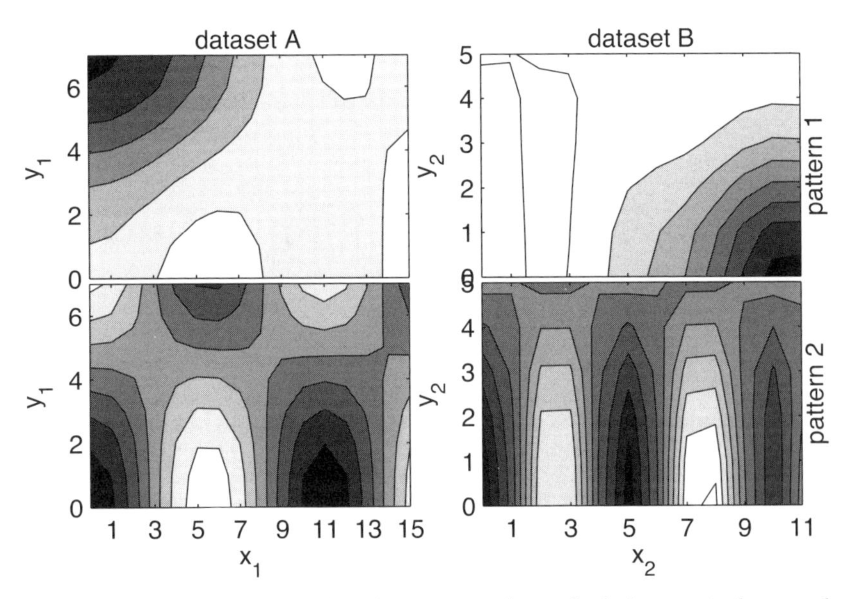

Figure 12.6. The first and second modes corresponding to both data sets in the second synthetic example.

expected, the second mode is mostly the cosine pattern, $\mathbf{P}_1$. Similarly, **B**'s leading mode is primarily focused on the lower right corner.

12.3 A Real Data Example

In this example, we analyze monthly mean anomalies of sea surface temperatures (SSTAs[1]) and 850 mb geopotential height, ϕ_{850}.[2] Both data sets are spatially delimited to 80°–35°W, 18°–60°N, and both comprise 344 monthly means, from November 1981 to June 2010. The ϕ_{850} set is analyzed on its original 2.5° × 2.5° (longitude by latitude) grid, but the SSTA set is degraded, by box averaging, from the original 1° × 1° resolution to the more manageable 2° × 2°. After calculating anomalies and removing time means, the SSTA and ϕ_{850} data sets are in variables `St` and `Ph`. To eliminate the land grid points in the SSTA set, we execute

[1] From a source we have already used, Reynolds, R. W., N. A. Rayner, T. M. Smith, D. C. Stokes, and W. Wang (2002) An improved in situ and satellite SST analysis for climate. *J. Clim.*, **15**, 1609–1625; obtained from iridl.ldeo.columbia.edu/expert/SOURCES/.NOAA/.NCEP/.EMC/.CMB/GLOBAL/.Reyn_SmithOIv2/.

[2] Also from a source we have already used, the NOAA NCEP-NCAR global Reanalysis, www.esrl.noaa.gov/psd/data/reanalysis/reanalysis.shtml.

```
io = find(~isnan(St(:,1))); St = St(io,:);
```

where the vector `io` holds the indices into the full domain, considered sequentially, of all ocean points, and subsequently `St` comprises ocean points only. Next, we obtain the cross-covariance matrix,

```
C = St*Ph'/(Nt-1);,
```

calculate all singular values,

```
sval = svd(C);
```

(when the function `svd` receives only one output argument, in this case `sval`, it returns only all the singular values, but no singular vectors), and compute the percentage cross-covariance spanned, `sval = 100*sval(1:40)/sum(sval);`. The results are summarized in fig. 12.7.

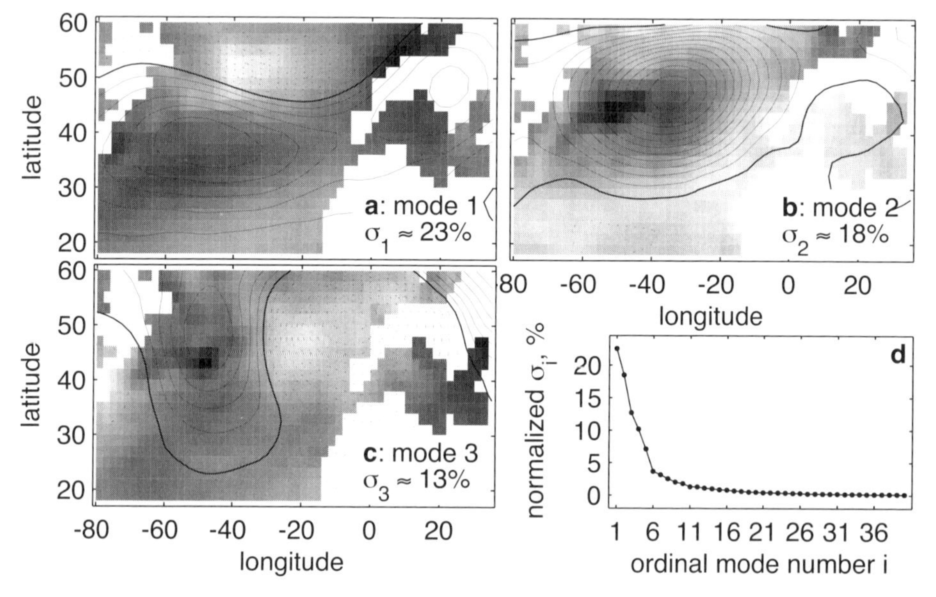

Figure 12.7. Results of the North Atlantic SSTA–ϕ_{850} example. Panels a–c show modes 1–3, spanning the shown percent cross-covariance. Shading presents the SSTA pattern ($\mathbf{u}_i$), and contours the ϕ_{850} patterns ($\mathbf{v}_i$). Contour units are arbitrary, but the interval is uniform throughout, with the zero contour emphasized. Negative contours are dotted. Panel d shows the leading 40 elements of the singular spectrum, $\sigma_{[1,40]}$.

12.4 EOFs as a Prefilter to SVD

In some cases a plain joint SVD analysis may yield disappointing results, with no clear dominance of the leading modes (a slow and steady falloff of the singular spectrum, with power distribution skewed only slightly in favor of lower modes), and no clear physical interpretation of the leading modes' spatial patterns or projection time series. While this can be a manifestation of the two data sets truly having no coherent cross-covariability, it is also possible that the apparent failure stems from the data sets' contamination by high enough noise levels to overwhelm the SVD machinery. In this case, prefiltering the individual participating sets offers an opportunity that may rescue the analysis.

If prefiltering is deemed necessary, it can be achieved by numerous routes. Projection on a complete Fourier or other harmonic bases, and subsequent truncation, is always possible, but makes more physical sense—and offers more economy because the power is spectrally concentrated—when the data clearly exhibit cyclic behavior. Simple temporal or spatial convolution-based smoothing, or an expressly designed digital filter, are also potentially beneficial avenues. The purpose of this section, however, is to describe a natural and simple prefiltering procedure based on the individual data sets' truncated EOF sets.

Suppose the two data sets A and B have each been subjected to individual EOF analysis, which yielded their covariance eigenvalues ($\mathbf{\Lambda}$ for A and $\mathbf{\Delta}$ for B), EOFs ($\mathbf{E}$ for A and $\mathbf{F}$ for B), and projection time series ($\mathbf{P}$ for A and $\mathbf{Q}$ for B), and that suitable truncation points for both eigenspectra have been identified (c and d for A and B),

$$\mathbf{A} \mapsto \begin{cases} \mathbf{\Lambda} \rightarrow \mathbf{\Lambda}_c \in \mathbb{R}^{c \times c} \\ \mathbf{E} \rightarrow \mathbf{E}_c \in \mathbb{R}^{M_a \times c}, \\ \mathbf{P} \rightarrow \mathbf{P}_c \in \mathbb{R}^{N_t \times c} \end{cases} \qquad \mathbf{B} \mapsto \begin{cases} \mathbf{\Delta} \rightarrow \mathbf{\Delta}_c \in \mathbb{R}^{d \times d} \\ \mathbf{F} \rightarrow \mathbf{F}_d \in \mathbb{R}^{M_b \times d} \\ \mathbf{Q} \rightarrow \mathbf{Q}_d \in \mathbb{R}^{N_t \times d} \end{cases}, \tag{12.26}$$

where N_t is the time dimension A and B share, so that $\mathbf{A} \approx \mathbf{E}_c \mathbf{\Lambda}_c^{\frac{1}{2}} \mathbf{P}_c^T$ and $\mathbf{B} \approx \mathbf{F}_d \mathbf{\Delta}_d^{\frac{1}{2}} \mathbf{Q}_d^T$. If this is successful (a surprisingly common outcome), $\mathbf{P}_c$ and $\mathbf{Q}_d$ offer significantly reduced dimensional, imperfect but adequate, representations of A's and B's time behaviors from which much of the noise has been eliminated.

Then we can form the alternative cross-covariance matrix

$$\tilde{\mathbf{C}} = \frac{1}{N_t - 1} \mathbf{P}_c^T \mathbf{Q}_d = \mathbf{\Lambda}_c^{-\frac{1}{2}} \mathbf{E}_c^T \mathbf{C} \mathbf{F}_d \mathbf{\Delta}_d^{-\frac{1}{2}} \in \mathbb{R}^{c \times d} \tag{12.27}$$

that is often dramatically smaller than $\mathbf{C} = (Nt - 1)^{-1} \mathbf{A} \mathbf{B}^T$ and that—if the original A and B are heavily noise contaminated—is based on far less noise. We decompose this $\tilde{\mathbf{C}}$ as before,

$$\tilde{\mathbf{C}} = \mathbf{U} \mathbf{\Sigma} \mathbf{V}^T, \tag{12.28}$$

where $\mathbf{U} \in \mathbb{R}^{c\times c}$ and $\mathbf{V} \in \mathbb{R}^{d\times d}$ hold complete orthonormal spanning sets for *A*'s and *B*'s retained EOF spaces, respectively, and $\boldsymbol{\Sigma}$'s diagonal holds, as usual, the singular values. The modes of this expansion for $\tilde{\mathbf{C}}$ can then be readily transformed back to physical spaces using, for singular mode *i*,

$$\mathbf{g}_i = \mathbf{E}_c\mathbf{u}_i \in \mathbb{R}^{M_a\times 1} \quad \text{and} \quad \mathbf{h}_i = \mathbf{F}_d\mathbf{v}_i \in \mathbb{R}^{M_b\times 1} \tag{12.29}$$

with the same physical interpretation as that of the SVD modes obtained using $\mathbf{C}$ rather than $\tilde{\mathbf{C}}$. That is, e.g., $\mathbf{g}_1$ plays the same role as $\mathbf{u}_1$ in the more traditional, $\mathbf{C}$-based, SVD, from which *A*'s percent covariance spanned and projection time series are derived, and $\mathbf{h}_3$ plays the same role as $\mathbf{v}_3$ in a $\mathbf{C}$-based SVD, from which *B*'s percent covariance spanned and projection time series are derived.

12.5 SUMMARY

The joint SVD analysis of two data sets (that share their time dimension *N* but need not share their space dimensions) identifies patterns of covariability of the two data sets and arranges those patterns is descending order of importance. With the sets identified as $\mathbf{A} \in \mathbb{R}^{M_a\times N}$ and $\mathbf{B} \in \mathbb{R}^{M_b\times N}$ (eqs. 12.1 and 12.2), we form an $M_a \times M_b$ covariability matrix ($\mathbf{C}$, eqs. 12.3 and 12.4, or its correlation-based equivalent, $\mathbf{R}$) and decompose it (eq. 12.5) using the SVD algorithm for a rectangular matrix,

$$\mathbf{C} \text{ or } \mathbf{R} = \mathbf{U}\boldsymbol{\Sigma}\mathbf{V}^T,$$

where $\boldsymbol{\Sigma}$'s diagonal elements $[1, q]$ are the singular values (with q denoting the rank of the chosen covariability matrix), and $\mathbf{U}$ and $\mathbf{V}$ contain the left and right singular vectors. This yields the two complete orthonormal sets

$$\text{span}\left[\{\hat{\mathbf{u}}_i\}_{i=1}^{M_a}\right] = \mathbb{R}^{M_a} \quad \text{and} \quad \text{span}\left[\{\hat{\mathbf{v}}_i\}_{i=1}^{M_b}\right] = \mathbb{R}^{M_b}$$

in terms of which any *A* or *B* state ($\mathbf{a}_k$ or $\mathbf{b}_l$ for any *k* or *l*) can be represented.

The SVD pattern choice is extremely useful; $\hat{\mathbf{u}}_1$ is the *A* spatial pattern that covaries most with any *B* spatial pattern, and the *B* spatial pattern most likely to accompany $\hat{\mathbf{u}}_1$ is $\hat{\mathbf{v}}_1$. If you plot $\hat{\mathbf{u}}_1$ and $\hat{\mathbf{v}}_1$ side by side, you can legitimately say that when *B* closely tracks $\hat{\mathbf{v}}_1$, *A* is most likely to closely track $\hat{\mathbf{u}}_1$, or vice versa. Beyond the leading mode, $\hat{\mathbf{u}}_i$ is the *A* spatial pattern orthogonal to $\hat{\mathbf{u}}_{1\le j<i}$ that covaries most with any *B* spatial pattern orthogonal to $\hat{\mathbf{v}}_{1\le j<i}$. And, as before, the *B* spatial pattern most likely to accompany $\hat{\mathbf{u}}_i$ is $\hat{\mathbf{v}}_i$. For all *i*, the total (*A*, *B*) that the cross-covariance fraction pair ($\hat{\mathbf{u}}_i$, $\hat{\mathbf{v}}_i$) accounts for is $\sigma_i / \sum_{i=1}^{q} \sigma_i$.

Note that $\hat{\mathbf{u}}_1$ is *not* the most dominant *A* pattern; that honor is exclusively reserved for *A*'s leading *individual* EOF. Thus, $\{\hat{\mathbf{u}}_i\}_{i=1}^{M_a}$ do *not* optimally span *A*'s variance, but rather *A*'s covariability with *B*.

Time is reintroduced into the problem by projecting the full data sets on their respective leading cross-covariance patterns, $\mathbf{f}_i = \mathbf{A}^T\mathbf{u}_i$ and $\mathbf{g}_i = \mathbf{B}^T\mathbf{v}_i$ (eqs. 12.6–12.9), and the fractions of *individual* variability spanned by mode *i* is

$$\mathbf{u}_i^T \mathbf{A}\mathbf{A}^T \mathbf{u}_i / \text{trace}(\mathbf{A}\mathbf{A}^T) \quad \text{and} \quad \mathbf{v}_i^T \mathbf{B}\mathbf{B}^T \mathbf{b}_i / \text{trace}(\mathbf{B}\mathbf{B}^T)$$

for data sets A and B, respectively. These fractions are important because the possibility exists that while $\sigma_i / \sum_i \sigma_i$ is large (the fraction of the total (A, B) cross-covariance spanned by mode i is significant), this cross-covariability actually accounts for very little in the individual variability of either A, B, or both. Consequently, the above fractions are crucial irrespective of the relative size of the corresponding singular value.

THIRTEEN

Suggested Homework

13.1 Homework 1, Corresponding to Chapter 3

13.1.1 Assignment

For

$$\mathbf{A} = \begin{pmatrix} 3 & -3 & -1 & 7 & 7 & -2 \\ 13 & 14 & 12 & 10 & 7 & 0 \\ 2 & 14 & 10 & 11 & 11 & 6 \\ 9 & -2 & 2 & 4 & 14 & -4 \end{pmatrix},$$

$$\mathbf{B} = \begin{pmatrix} 2 & 4 & -2 & -7 \\ -3 & 3 & 1 & -5 \\ 6 & 10 & 1 & -8 \\ 3 & 4 & 4 & 6 \\ 9 & 9 & 5 & 12 \end{pmatrix},$$

and

$$\mathbf{C} = \begin{pmatrix} 2 & 0 & 3 & 2 \\ 1 & 5 & -1 & 6 \\ 2 & 1 & 4 & 6 \end{pmatrix},$$

please provide

(a) M, N
(b) rank q; is the matrix full rank? rank deficient?
(c) $\mathbf{U}$ and all necessary $\mathbf{E}_i$s
(d) $\text{basis}[\mathcal{R}(\cdot)] \subseteq \mathbb{R}^M$
(e) $\text{basis}[(\mathcal{N}(\cdot^T)] \subseteq \mathbb{R}^M$
(f) $\text{basis}[\mathcal{R}(\cdot^T)] \subseteq \mathbb{R}^N$
(g) $\text{basis}[(\mathcal{N}(\cdot)] \subseteq \mathbb{R}^N$

and show that

(h) $\mathcal{R}(\cdot) \perp \mathcal{N}(\cdot^T)$
(i) $\mathcal{R}(\cdot^T) \perp \mathcal{N}(\cdot)$
(j) $\mathcal{R}(\cdot) + \mathcal{N}(\cdot^T) = \mathbb{R}^M$
(k) $\mathcal{R}(\cdot^T) + \mathcal{N}(\cdot) = \mathbb{R}^N$

13.1.2 Answers

In many of the tests below, we need to distinguish values that are practically but not identically zero. For this purpose, `format('long')` is useful, as it provides a far better numerical resolution, e.g.,

```
>> format('long'); disp(2/3)
   0.666666666666667
>> format('short'); disp(2/3)
   0.6667
```

(a) To get M and N, use `[Ma,Na=size(A)`, `[Mb,Nb=size(B)`, etc.
(b) To get the rank, we have many options. One is to simply issue `qA = rank(A)`, `qB = rank(B)`, etc. Or, more insightfully, we can use Matlab to automate generating $\mathbf{U}$, saving along the way $\mathbf{E}^a = \prod \mathbf{E}_i^a$ with which $\mathbf{E}^a\mathbf{A} = \mathbf{U}^a$ and so on for $\mathbf{B}$ and $\mathbf{C}$. The code to do this, using $\mathbf{B}$ as an example, is

```
Ub = B;
for i = 1:min(Mb,Nb)
  E = eye(Mb);
  E(i,i) = 1/Ub(i,i); % make u_{ii} = 1
  E(i+1:Mb,i) = Ub(i+1:Mb,i);
  for j = i+1:Mb; E(j,j) = -Ub(i,i); end
  Ub = E*Ub;
  eval(['Eb' int2str(i) ' = E;']);
end
disp(['diag(Ub) = ' int2str(diag(Ub)')])
```

Once we have $\mathbf{U}$ (`Ub` in the code), its nontrivial pivots index basis[$\mathcal{R}(\mathbf{B})$] and their number is the original matrix' rank, e.g.,

```
d  = diag(Ub);
i  = find(d);
qB = length(i);
```

where `i` holds the indices of $\mathbf{B}$'s columns that form basis[$\mathcal{R}(\mathbf{B})$].
(c) To get a basis for the column space, e.g., basis[$\mathcal{R}(\mathbf{B})$] for $\mathbf{B}$, locate the columns in $\mathbf{U}^b$ with nonzero pivots and the corresponding columns back in $\mathbf{B}$ span $\mathcal{R}(\mathbf{B})$. Below, I do this, while also employing Gram-Schmidt to orthonormalize the set:

```
i = find( diag(Ub)==1 );
RB = B(:,i);  % span B's column space
```

```
for i = 1:qB  % loop on range vectors
  p = zeros(Mb,1);
  for j = 1:i-1
    p = p + (RB(:,i)'*RB(:,j))*RB(:,j);
  end
  RB(:,i) = RB(:,i) - p;
  RB(:,i) = RB(:,i)/norm(RB(:,i));
end
disp(RB'*RB) % must produce I if it worked
```

This produces the $M_B \times q_B$ `RB` whose orthonormal columns span $\mathcal{R}(\mathbf{B})$.

(d) To span $\mathcal{R}(\mathbf{B})$'s complementary $\mathbb{R}^M$ space, $\mathcal{N}(\mathbf{B}^T)$ (**B**'s left null space), I first reduce $\mathbf{B}^T$ to a **U** using

```
Ubt = B';
for i = 1:min(Mb,Nb)
  E = eye(Nb);
  E(i,i) = 1/Ubt(i,i);
  E(i+1:Nb,i) = Ubt(i+1:Nb,i);
  for j = i+1:Nb; E(j,j) = -Ubt(i,i); end
  Ubt = E*Ubt;
end
disp(['diag(Ubt) = ' int2str(diag(Ubt)')])
```

Since for the example **B** the rank is 4 while `Mb` = 5, $\mathcal{N}(\mathbf{B}^T)$ must be a single vector. So there's a single free variable to set to 1. Let's do this manually first (later I will automate this, and elaborate on how it works):

```
nlB = zeros(Mb,1);
nlB(Mb  ) = 1;
nlB(Mb-1) = -Ubt(Nb  ,Nb+1 :Mb)*nlB(Nb+1:Mb);
nlB(Mb-2) = -Ubt(Nb-1,Nb   :Mb)*nlB(Nb   :Mb);
nlB(Mb-3) = -Ubt(Nb-2,Nb-1 :Mb)*nlB(Nb-1 :Mb);
nlB(Mb-4) = -Ubt(Nb-3,Nb-2 :Mb)*nlB(Nb-2 :Mb);
nlB = nlB/norm(nlB);
```

This produces the $M_B \times (M_B - q_B)$ `nlB` whose single unit norm column spans $\mathcal{N}(\mathbf{B}^T)$, which can be verified by ascertaining that `B'*nlB` vanishes to within machine accuracy.

(e) To show in one fell swoop that the column and left null spaces are mutually orthonormal and jointly equal $\mathbb{R}^{M_B}$, I place the spanning sets of $\mathcal{R}(\mathbf{B})$ and $\mathcal{N}(\mathbf{B}^T)$ in the single $M_B \times M_B$ matrix RMB and convince myself that $\text{RMB}^T\,\text{RMB} = \mathbf{I}$

```
RMB = [ RB nlB ];
disp(RMB'*RMB);
```

Since this gives $\mathbf{I}_{M_B}$, we are set.

(f) To span **A**'s row space $\mathcal{R}(\mathbf{A}^T)$, recall that **A** and its **U** share the same row space, so it is simply **U**'s rows, Gram-Schmidt orthonormalized:

```
RAT = Ua(qA:-1:1,:)';
for i = 1:qA % loop on R(A^T) spanning vectors
  p = zeros(Na,1);
  for j = 1:i-1
    p = p + (RAT(:,j)'*RAT(:,i))*RAT(:,j);
  end
  RAT(:,i) = RAT(:,i) - p;
  RAT(:,i) = RAT(:,i)/norm(RAT(:,i));
end
p = RAT'*RAT;
disp(diag(p))
p = p - diag(diag(p));
p = max(abs(p(:)));
disp('max(abs(off-diag[R(A^T)^T R(A^T)])) = ')
disp([' = ' num2str(p)])
```

The 16 orders of magnitude difference between the diagonal and off-diagonal elements should convince you that this RAT is indeed orthonormal.

(g) For the above set to successfully span $\mathcal{R}(\mathbf{A}^T)$, every row of **A** must be representable as a linear combination of RAT's columns. Since those columns are orthonormal, it's easy: obtain the coefficients α_i (projections) which render each of **A**'s (transposed) rows a linear combination of RAT's columns $\hat{\mathbf{r}}_i$, $\alpha_i = \mathbf{a}_i^T \hat{\mathbf{r}}_j$, subtract the sum of these contributions from $\mathbf{a}_i$, and show that the result $\mathbf{a}_i - \sum_i \alpha_i \mathbf{r}_i$ is effectively zero, or, in finite precision arithmetic, vanishingly small compared to $\mathbf{a}_i$, $\|\mathbf{a}_i - \sum_i \alpha_i \mathbf{r}_i\| / \|\mathbf{a}_i\| \sim \epsilon$, where $\epsilon \sim \mathcal{O}(10^{-15})$ (with $\mathcal{O}$ denoting "of the order of") is the machine's numerical resolution. On a particular platform, eps can be obtained by issuing eps in the Matlab window. This logic is employed in the following code:

```
for i = 1:Ma    % loop on A's rows
  p = A(i,:)';  % the examined row
```

```
    c = zeros(qA,1);
    for j = 1:qA % loop on R(A') spanning set
      c(j) = p'*RAT(:,j);
    end
    r = RAT*c - p;
    j = ['row ' int2str(i) ' of A: '];
    disp([ j num2str(norm(r)/norm(p))])
  end
```

(h) Our final space is the null space, the set of all vectors that, for **B** as an example, satisfy $\mathbf{B}\mathbf{n}_B = \mathbf{0} \in \mathbb{R}^{M_B}$. This is done using

```
NB = [];
for i = qB+1:Nb % loop on n's free parameters
  n = zeros(Nb,1);
  n(i) = 1;
  for j = qB:-1:1 % loop up Ub's rows
    n(j) = -Ub(j,j+1:Nb)*n(j+1:Nb);
  end
  NB = [ NB n/norm(n) ];
end
```

This is how this works (dropping matrix identifiers for brevity). Recall that because the right-hand-side vector is homogeneous, $\mathbf{A}\mathbf{n} = 0$ is entirely equivalent to and, in particular, shares the solution $\mathbf{n}$, with the more elegant and readily tractable $\mathbf{U}\mathbf{n} = 0$, on which we focus. Let's visualize this for various combinations of (q, M, N):

$$q = N < M, \quad \mathbf{U} = \begin{pmatrix} u_{11} & u_{12} & u_{13} & \cdots & \cdots & u_{1N} \\ 0 & u_{22} & u_{23} & \cdots & \cdots & u_{2N} \\ 0 & 0 & u_{33} & \cdots & \cdots & u_{3N} \\ & & & \ddots & & \\ 0 & 0 & \cdots & \cdots & 0 & u_{qq} \\ 0 & 0 & \cdots & \cdots & 0 & 0 \\ & & & \vdots & & \\ 0 & 0 & \cdots & \cdots & 0 & 0 \end{pmatrix}, \tag{13.1}$$

with $M - N = M - q$ bottom zero rows;

$$q = M < N, \quad \mathbf{U} = \begin{pmatrix} u_{11} & u_{12} & u_{13} & \cdots & \cdots & u_{1N} \\ 0 & u_{22} & u_{23} & \cdots & \cdots & u_{2N} \\ 0 & 0 & u_{33} & \cdots & \cdots & u_{3N} \\ & & & \ddots & & \\ 0 & 0 & \cdots & u_{qq} & \cdots & u_{qN} \end{pmatrix}, \tag{13.2}$$

with the bottom, qth or Mth, row containing $N - M = N - q$ nonzero entries to the right of the diagonal, MM or qq, term; and

$$q < (M, N), \quad \mathbf{U} = \begin{pmatrix} u_{11} & u_{12} & u_{13} & \cdots & \cdots & u_{1N} \\ 0 & u_{22} & u_{23} & \cdots & \cdots & u_{2N} \\ 0 & 0 & u_{33} & \cdots & \cdots & u_{3N} \\ & & & \ddots & & \\ 0 & \cdots & 0 & u_{qq} & \cdots & u_{qN} \\ 0 & \cdots & 0 & \cdots & 0 & 0 \\ & & & \vdots & & \\ 0 & \cdots & 0 & \cdots & 0 & 0 \end{pmatrix}, \tag{13.3}$$

with $M - q$ bottom zero rows, and a bottom, qth, nontrivial row containing $N - q$ nonzero entries to the right of diagonal term qq.

What these situations share is the following: (1) the bottom nontrivial row is always the qth; (2) that bottom row always contains right-of-the-diagonal elements in positions $(q, q+1)$ to (q, N), obviously the empty set when $q \geq N$. Back-substitution always starts at the bottom, qth, row. The null space spanning set comprises $N - q$ $\mathbb{R}^N$ vectors, so if $q \geq N$, the null space is just (0) and there is no need to find a spanning set for it. When $q < N$, the $N - q$ $\mathcal{N}(\mathbf{A})$ spanning vectors have the general form

$$\mathbf{n} = \begin{pmatrix} \alpha_1 \\ \alpha_2 \\ \vdots \\ \alpha_q \\ p_1 \\ p_2 \\ \vdots \\ p_{N-q} \end{pmatrix}, \tag{13.4}$$

where the α_is are solved for, while the p_i parameters are all set to zero except one. We vary this p_i over $i = 1$ to $N - q$, setting one of them to 1 and all the others to zero. That is,

$$\mathbf{n}_1 = \begin{pmatrix} \alpha_1 \\ \alpha_2 \\ \vdots \\ \alpha_q \\ 0 \\ 0 \\ \vdots \\ 0 \\ 1 \end{pmatrix}, \mathbf{n}_2 = \begin{pmatrix} \alpha_1 \\ \alpha_2 \\ \vdots \\ \alpha_q \\ 0 \\ \vdots \\ 0 \\ 1 \\ 0 \end{pmatrix}, \ldots, \mathbf{n}_{N-q} = \begin{pmatrix} \alpha_1 \\ \alpha_2 \\ \vdots \\ \alpha_q \\ 1 \\ 0 \\ 0 \\ \vdots \\ 0 \end{pmatrix}, \tag{13.5}$$

with the α_is solved for by back-substitution. The inner product of **U**'s qth row and $\mathbf{n}_1$ is

$$\begin{pmatrix} 0 & 0 & \cdots & 0 & u_{qq} & \cdots & u_{qN} \end{pmatrix} \begin{pmatrix} \alpha_1 \\ \alpha_2 \\ \vdots \\ \alpha_q \\ 0 \\ 0 \\ \vdots \\ 0 \\ 1 \end{pmatrix} = 0 \tag{13.6}$$

or

$$\alpha_q u_{qq} + u_{qN} = 0 \implies \alpha_q = -\frac{u_{qN}}{u_{qq}}. \tag{13.7}$$

If, in addition, we constructed **U** so that $u_{ii} = 1 \ \forall i$, as I have done in the code and above, this reduces to

$$\alpha_q = -u_{qN}, \tag{13.8}$$

with which the shown algorithm for **n** starts.

With this result, the inner product of the next row up, row $q-1$, and $\mathbf{n}_1$ is

$$\begin{pmatrix} 0 & \cdots & 0 & u_{q-1q-1} & u_{q-1q} & \cdots & u_{q-1N} \end{pmatrix} \begin{pmatrix} \alpha_1 \\ \alpha_2 \\ \vdots \\ \alpha_{q-1} \\ -u_{qN} \\ 0 \\ 0 \\ \vdots \\ 0 \\ 1 \\ 0 \end{pmatrix} = 0, \tag{13.9}$$

which is

$$\alpha_{q-1} u_{q-1q-1} - u_{q-1q} u_{qN} + u_{q-1N-1} = 0 \tag{13.10}$$

or, rearranging and noting again that $u_{q-1q-1} = 1$,

$$\alpha_{q-1} = u_{q-1q} u_{qN} - u_{q-1N-1}. \tag{13.11}$$

This procedure continues recursively (the algorithm's inner, `j`-dependent, loop) until all elements of $\mathbf{n}_1$ are filled. Then, in the algorithm's outer, `i`-dependent, loop, this continues for subsequent $\mathbf{n}_i$s, until $\mathcal{N}(\mathbf{A})$ is fully spanned.

13.2 Homework 2, Corresponding to Chapter 3

13.2.1 *Assignment*

For the sets

$$\mathcal{A} = \left\{ \begin{pmatrix} 1 \\ 0 \\ -1 \\ 1 \end{pmatrix}, \begin{pmatrix} 0 \\ 1 \\ 1 \\ 0 \end{pmatrix}, \begin{pmatrix} -1 \\ -1 \\ 0 \\ 1 \end{pmatrix} \right\},$$

$$\mathcal{B} = \left\{ \begin{pmatrix} 2 \\ 1 \\ -1 \\ 1 \\ 0 \end{pmatrix}, \begin{pmatrix} 0 \\ -1 \\ 0 \\ 2 \\ -1 \end{pmatrix}, \begin{pmatrix} -1 \\ 0 \\ 0 \\ -1 \\ 1 \end{pmatrix}, \begin{pmatrix} 1 \\ 2 \\ 1 \\ 1 \\ 1 \end{pmatrix}, \begin{pmatrix} 1 \\ 1 \\ -2 \\ 0 \\ 1 \end{pmatrix} \right\},$$

and

$$\mathcal{C} = \left\{ \begin{pmatrix} 2 \\ 0 \\ -1 \end{pmatrix}, \begin{pmatrix} 0 \\ 1 \\ 0 \end{pmatrix}, \begin{pmatrix} -1 \\ -1 \\ 2 \end{pmatrix} \right\}$$

use Matlab or Octave to

(a) Determine whether the set elements are linearly independent, and, if linear dependence is detected, the linearly independent subset.
(b) Devise an orthonormal basis spanned by the linearly independent subset.
(c) Devise orthonormal bases for the complementary sets $\mathcal{A}^\perp$, $\mathcal{B}^\perp$, and $\mathcal{C}^\perp$ satisfying

$$\mathcal{A} \perp \mathcal{A}^\perp, \quad \mathcal{B} \perp \mathcal{B}^\perp, \quad \text{and} \quad \mathcal{C} \perp \mathcal{C}^\perp.$$

13.2.2 *Answers*

13.2.2.1 Generating $\mathbf{U}$

To determine linear independence of the sets' elements, using $\mathcal{B}$ as an example, the approach is to

- place $\mathcal{B}$'s elements (individual member vectors) in a matrix, call it $\mathbf{B} \in \mathbb{R}^{M \times N}$
- build the necessary elementary matrices $\mathbf{E}_{b_i}$
- get $\mathbf{E}_b \equiv \prod_i \mathbf{E}_{b_i}$
- get $\mathbf{U}_b = \mathbf{E}_b \mathbf{B}$
- solve $\mathbf{U}_b \mathbf{n} = \mathbf{0} \in \mathbb{R}^M$, the back substitutable analog of $\mathbf{Bn} = \mathbf{0} \in \mathbb{R}^M$
- if $\mathbf{n} = \mathbf{0} \in \mathbb{R}^N$, the set members are linearly independent
- if $\mathbf{n} \neq \mathbf{0} \in \mathbb{R}^N$, the set members are linearly dependent

So the first challenge is to obtain the elementary matrices $\mathbf{E}_i$. Recall the code segment of homework 1, using $\mathbf{B}$ as an example.

```
Ub = B;
for i = 1:min(Mb,Nb)
  E = eye(Mb);
  E(i,i) = 1/Ub(i,i);
  E(i+1:Mb,i) = Ub(i+1:Mb,i);
  for j = i+1:Mb; E(j,j) = -Ub(i,i); end
  Ub = E*Ub;
 eval(['Eb' int2str(i) ' = E;']);
end
disp(['diag(Ub) = ' int2str(diag(Ub)')])
```

It will work well for $\mathcal{B}$ and $\mathcal{C}$, but *not* for $\mathcal{A}$. Let me first create a Matlab function that can handle more general situations, and then explain why it works where the previous, simpler, code reproduced above would not. The new function is

```
  function [U,Es] = getU(A);

% function [U,Es] = getU(A);
%
% Obtain elementary matrices E's and U from A
%
% Input: A, a matrix
% Output:
% 1. U , the reduced upper diagonal form of A.
% 2. Es, a cell array whose {1,1} element is
% the overall product of individual E (i.e.,
% inv(Es{1,1})*U = A) and whose elements
% {2:end,2} are the individual needed E's,
% with elements {2:end,1} non empty if
% row swaps are needed.
```

```
%
% Note that inv(E{1}) is **NOT** necessarily
% L. It IS if no row swaps are necessary but
% it is NOT if row swaps ARE needed. Since the
% row swaps are retained in the Es,it's
% trivial to construct a permutation matrix P
% with which P*A=inv(E{1})*A.
%
% Author: Gidon Eshel, Feb. 21st 2010

[M,N] = size(A);
Es = cell(max(M,N)*2,2);
Eall = eye(M);
U = A;
for i = 1:min(M,N)
  E = eye(M);
  S = eye(M);
  if U(i,i)==0 % is a row swap needed?
    j = find(U(i+1:M,i));
    if ~isempty(j)
      S(i+[0 min(j)],:) = S(i+[min(j) 0],:);
      Es{i+1,1} = S;
      U = S*U;
      Eall = S*Eall;
    end
  end
  E(i,i) = 1/U(i,i);
  E(i+1:M,i) = U(i+1:M,i);
  for j = i+1:M; E(j,j) = -U(i,i); end
  U = E*U;
  Eall = E*Eall;
  Es{i+1,2} = E;
end
Es{1,1} = Eall;
Es = Es(1:i+1,:);
```

This code is used in, e.g.,

```
[Ua,Ea] = getU(A);
[Ub,Eb] = getU(B);
[Uc,Ec] = getU(C);
```

to produce

```
>> Ua

Ua =

     1     0    -1
     0     1    -1
     0     0     1
     0     0     0

>> Ub

Ub =

     1.00          0      -0.50       0.50       0.50
        0       1.00      -0.50      -1.50      -0.50
        0          0       1.00      -3.00       3.00
        0          0          0       1.00      -0.20
        0          0          0          0       1.00

>> Uc

Uc =

     1.0000         0      -0.5000
          0    1.0000      -1.0000
          0         0       1.0000

>>
```

This should tell you that rank(**A**, **B**, **C**) = 3, 5, 3. This means that all sets are linearly independent. Of course, the same can be obtained by

```
>> disp([ rank(A) rank(B) rank(C) ])
     3     5     3
```

confirming our results.

How would the code behave when given a rank deficient matrix? Let's test this be issuing

```
>> [U1,E1]=getU(B([2 3],:));
>> [U2,E2]=getU(B(:,[2 3]));
```

which addresses

$$\text{B}([2\ \ 3],:) = \begin{pmatrix} 1 & -1 & 0 & 2 & 1 \\ -1 & 0 & 0 & 1 & -2 \end{pmatrix}$$

and

$$\text{B}(:,[2\ \ 3]) = \begin{pmatrix} 0 & -1 \\ -1 & 0 \\ 0 & 0 \\ 2 & -1 \\ -1 & 1 \end{pmatrix}$$

and yields

```
U1 =

     1    -1     0     2     1
     0     1     0    -3     1
```

and

```
U2 =

     1     0
     0     1
     0     0
     0     0
     0     0
```

This demonstrates that the code works as required under diverse situations.

13.2.2.2 Orthonormalization

To get orthonormal spanning sets, we take the `q` columns of the original matrix for which `U` has nonzero (unit) pivots and orthonormalize them. In the main code, this looks like

```
% get orthonormal spanning sets for
% the column spaces
ja = find(diag(Ua==1));
jb = find(diag(Ub==1));
jc = find(diag(Uc==1));
RA = orthogonalize(A(:,j));
RB = orthogonalize(B(:,j));
RC = orthogonalize(C(:,j));
```

which makes use of

```
  function R = orthogonalize(R)

% function R = orthogonalize(R)
% Take a linearly independent basis R and
% return orthonormal set.
% Do NOT use if you are not sure R's
% columns are linearly independent!
%
% Author: Gidon Eshel, Feb. 21st, 2010

for i = 1:length(R(1,:));
  q = R(:,i);
  p = 0;
  for j = 1:i-1
    p = p + (R(:,j)'*q)'*R(:,j);
  end
  q = q - p;
  R(:,i) = q/norm(q);
end
```

To test the orthonormalization, I use

```
% test orthonormailty
PA  = RA'*RA;   dA = diag(PA);
PB  = RB'*RB;   dB = diag(PB);
PC  = RC'*RC;   dC = diag(PC);
oA  = PA-diag(dA);  oA = oA(:);
oB  = PB-diag(dB);  oB = oB(:);
oC  = PC-diag(dC);  oC = oC(:);
maA = max(abs(oA));
maB = max(abs(oB));
maC = max(abs(oC));
i   = 'on/off diag ratio, ';
disp([i 'A = ' num2str(min(abs(dA))/maA)])
disp([i 'B = ' num2str(min(abs(dB))/maB)])
disp([i 'C = ' num2str(min(abs(dC))/maC)])
```

in the main code, which calculates the worst case of on- to off-diagonal element magnitude ratio. This produces

```
on/off diag ratio, A = 9007199254740992
```

```
on/off diag ratio, B = 211934100111552.7
on/off diag ratio, C = 6004799503160661
```

Since these numbers are all $\mathcal{O}(10^{15})$ or higher, we are set.

13.2.2.3 The Complementary Sets

When $\mathcal{A}$'s member vectors are made into the columns of **A**, then this question asks you to devise a spanning set for **A**'s left null space $\mathcal{N}(\mathbf{A}^T)$, the set of all $\mathbf{n}_{la} \in \mathbb{R}^M$ vectors satisfying $\mathbf{A}^T\mathbf{n}_{la} = \mathbf{0} \in \mathbb{R}^N$.

To get this, we must reduce $\mathbf{A}^T$, $\mathbf{B}^T$, and $\mathbf{C}^T$ to their respective **U**s, which I do using

```
[Ua,j] = getU(A');
[Ub,j] = getU(B');
[Uc,j] = getU(C');
```

(note that since we do not need the **E** matrices here, I just use the dummy variable `j`). To test the code (introduced below), I also added another matrix, `D = B(:,1:3);`. To generate spanning sets for the left null spaces, I use the following, expanded and more general, code.

```
% left null spaces
for iv = 'ABCD'
  eval(['Mh = ' iv ';'])
  eval(['q = q' iv ';'])
  Mh = Mh';
  % U of the transpose (don't need E)
  [U,j] = getU(Mh);
  [M,N] = size(Mh);
  NL = [];
  % remove from U columns corresponding to
  % unknowns that must vanish in all n_l's
  % because they are in a row by themselves
  k = 1:N;
  for i = 1:M
    j = find(U(i,:));
    if length(j)==1; k(j) = -9; end
  end
  U = U(:,find(k>0));
  % remove empty rows from U
  z = [];
  for i = 1:M
    j = find(U(i,:));
```

```
    if isempty(j); U(i,1) = nan; end
  end
  i = find(~isnan(U(:,1)));
  U = U(i,:);
  [Mr,Nr] = size(U);
  qr = length(find(diag(U)));
  % do back substitution
  % loop on free parameters in n_1,
  % also dim[null(D')]
  for i = Mr+1:Nr
    n = zeros(N,1);
    n(i) = 1;
    for j = qr:-1:1 % loop up U's rows
      n(j) = -U(j,j+1:Nr)*n(j+1:Nr);
    end
    NL = [ NL n/norm(n) ];
  end
  if ~isempty(NL)
    eval(['Nl' iv ' = NL;'])
  end
end
```

I will leave it to *you* to show why the above works where the previous code fails as the subsequent homework exercise!

13.3 Homework 3, Corresponding to Chapter 3

13.3.1 Assignment

Please examine the code of homeworks 1 and 2 and explain the differences in the algorithms. Please use example matrices to demonstrate where and under what circumstances the original homework 1's algorithms fail while homework 2's prevails.

13.3.2 Answers

13.3.2.1 A

This exercise is supposed to both teach you a bit about the actual issues, orthonormalization, space spanning, etc., and also to give you some appreciation of code structure and challenges.

If **A** is full rank, getting its **U** is very easy, because all pivots are nonzero and can be divided by, so

```
U = A;
for i = 1:min(M,N)
```

```
  E = eye(M);
  E(i,i) = 1/U(i,i);
  E(i+1:M,i) = U(i+1:M,i);
  for j = i+1:M; E(j,j) = -U(i,i); end
  U = E*U;
  eval(['E' int2str(i) ' = E;']);
end
```

should do the trick.

What if sometime along the reduction we get something like

$$\mathbf{U} = \begin{pmatrix} 1 & 0 & u_{13} & u_{14} & 0 & u_{16} \\ 0 & 0 & 0 & 0 & 0 & 0 \\ 0 & 0 & u_{33} & u_{34} & u_{35} & 0 \end{pmatrix} ??$$

Depending on the overall objective of the row reduction, it may make sense (as long as we keep track of it in the elementary operation matrices) to carry out a row swap, converting the above to

$$\mathbf{U} = \begin{pmatrix} 1 & 0 & u_{13} & u_{14} & 0 & u_{16} \\ 0 & 0 & 1 & u_{24} & u_{25} & 0 \\ 0 & 0 & 0 & 0 & 0 & 0 \end{pmatrix}.$$

One coding approach to such a row swap may be

```
for i = 1:min(M,N)
  E = eye(M);
  if U(i,i)==0 % is a row swap needed?
    j = find(U(i+1:M,i));
    if ~isempty(j)
      S(i+[0 min(j)],:) = S(i+[min(j) 0],:);
      Es{i+1,1} = S;
      U = S*U;
      Eall = S*Eall;
    end
  end
end
```

If we would like to be able to handle input matrices that are rank deficient, one approach may be

```
for i = 1:min(M,N)
  if U(i,i)~=0
    E(i,i) = 1/U(i,i);
    E(i+1:M,i) = U(i+1:M,i);
```

```
    for j = i+1:M; E(j,j) = -U(i,i); end
    U = E*U;
    Eall = E*Eall;
    Es{i+1,2} = E;
  end
end
```

where all the action is contingent on the pivot being nonzero, so when it is not, no action is taken.

Some may prefer all rows' leftmost elements to be zero no matter whether or not they are pivots. This can be achieved by

```
for i = 1:M              % loop on rows
  j = find(U(i,:));      % all non zeros in row i
  if ~isempty(j)         % do nothing if U(i,:) is
                         %     all zeros
    if j(1)~=1           % do nothing if leftmost
                         %     element is already 1
      E = eye(M);
      E(i,i) = 1/U(i,j(1));
      U = E*U;
      Eall = E*Eall;
      Es{i+I,2} = E;
    end
  end
end
```

Note that getU is by no means exhaustive, in the sense that there may be situations in which it will not do the full job. This is normal in programming: a trivially simple program can get hopelessly complicated fast if you start allowing for more and more scenarios that are more and more diverse. Sometimes, it's OK to cut your losses and simply allow for imperfection in the code in the name of streamlined, short, and elegant code. At least this is my view.

13.4 Homework 4, Corresponding to Chapter 4

13.4.1 Assignment

Consider

$$\mathbf{A} = \begin{pmatrix} 600\frac{2}{3} & 299\frac{1}{3} & -299\frac{1}{3} \\ 299\frac{1}{3} & 151\frac{1}{6} & -151\frac{1}{6} \\ -299\frac{1}{3} & -151\frac{1}{6} & 151\frac{1}{6} \end{pmatrix}. \tag{13.12}$$

Please

(a) Obtain by hand the eigenvector/eigenvalue decomposition of **A**; express the eigenvectors in a rational representation, e.g.,

$$\mathbf{e}_1 = \frac{1}{\sqrt{n}}\begin{pmatrix}\alpha\\ \beta\\ \gamma\end{pmatrix},$$

where n, α, β and γ are integers.

(b) Obtain the same decomposition using Matlab or Octave.

(c) Compare the decompositions and explain.

(d) Comment on the eigenspectrum.

(e) Let $\hat{\mathbf{e}}_{1,2,3}$ denote the three eigenvectors; now

- form $\hat{\mathbf{b}}_i = \hat{\mathbf{e}}_i$, $i = [1, 3]$,
- form $\hat{\mathbf{b}}_4$ derived from $\mathbf{b}_4 = \hat{\mathbf{e}}_1 + \hat{\mathbf{e}}_2$
- form $\hat{\mathbf{b}}_5$ derived from $\mathbf{b}_5 = \hat{\mathbf{e}}_1 + \hat{\mathbf{e}}_3$
- form $\hat{\mathbf{b}}_6$ derived from $\mathbf{b}_6 = \hat{\mathbf{e}}_2 + \hat{\mathbf{e}}_3$
- form $\hat{\mathbf{b}}_7$ derived from $\mathbf{b}_7 = \hat{\mathbf{e}}_1 + \hat{\mathbf{e}}_2 + \hat{\mathbf{e}}_3$

(f) For each of the $\hat{\mathbf{b}}_i$s, evaluate

- $\mathbf{p}_i = \mathbf{E}^T\hat{\mathbf{b}}_i$
- $\mathbf{q}_i = \mathbf{A}\hat{\mathbf{b}}_i = \mathbf{E}\boldsymbol{\Lambda}\mathbf{E}^T\hat{\mathbf{b}}_i = \mathbf{E}\boldsymbol{\Lambda}\mathbf{p}$
- $m = \|\mathbf{E}\boldsymbol{\Lambda}\mathbf{E}^T\hat{\mathbf{b}}_i\|$
- $|p_{ij}|/|q_{ij}|$, $j = [1, 3]$ (don't confuse i and j here!)

and report all numbers in a table or plots.

(g) Explain the above results: what's going on, what are the discernible main trends, and why they are the way they are.

13.4.2 Answers

First, apologies for this ugly matrix, which I needed to get a wide eigenvalue disparity—sorry!

Elegance notwithstanding,

$$\mathbf{A} = \begin{pmatrix} 600\frac{2}{3} & 299\frac{1}{3} & -299\frac{1}{3}\\ 299\frac{1}{3} & 151\frac{1}{6} & -150\frac{1}{6}\\ -299\frac{1}{3} & -150\frac{1}{6} & 151\frac{1}{6}\end{pmatrix} \tag{13.13}$$

$$= \frac{1}{6}\begin{pmatrix} 3604 & 1796 & -1796\\ 1796 & 907 & -901\\ -1796 & -901 & 907\end{pmatrix}. \tag{13.14}$$

The characteristic polynomial is obtained by solving

$$\left| \frac{1}{6} \begin{pmatrix} 3604 - 6\lambda & 1796 & -1796 \\ 1796 & 907 - 6\lambda & -901 \\ -1796 & -901 & 907 - 6\lambda \end{pmatrix} \right| = 0, \tag{13.15}$$

which yields

$$\begin{vmatrix} 3604 - 6\lambda & 1796 & -1796 \\ 1796 & 907 - 6\lambda & -901 \\ -1796 & -901 & 907 - 6\lambda \end{vmatrix} = 0 \tag{13.16}$$

or

$$\begin{aligned} &(3604 - 6\lambda)\left[(907 - 6\lambda)^2 - 901^2\right] \\ -\ &1796\left[1796(907 - 6\lambda) - 901 \cdot 1796\right] \\ -\ &1796\left[-1796 \cdot 901 + 1796(907 - 6\lambda)\right] = 0. \end{aligned} \tag{13.17}$$

The latter two terms combine,

$$(3604 - 6\lambda)\left[(907 - 6\lambda)^2 - 901^2\right] - 12 \cdot 1796^2(1 - \lambda) = 0 \tag{13.18}$$

$$(1802 - 3\lambda)\left[(907 - 6\lambda)^2 - 901^2\right] - 6 \cdot 1796^2(1 - \lambda) = 0. \tag{13.19}$$

As far as I know, this doesn't simplify any further, but upon brute force expansion (I used `wolframalpha.com`) becomes

$$-108\lambda^3 + 97524\lambda^2 - 291816\lambda + 194400 = 0, \tag{13.20}$$

with roots

$$\lambda_i = \begin{cases} 900, & i = 1 \\ 2, & i = 2 \\ 1, & i = 3 \end{cases} \tag{13.21}$$

and corresponding eigenvectors

$$\hat{\mathbf{e}}_1 = \frac{1}{\sqrt{6}} \begin{pmatrix} 2 \\ 1 \\ -1 \end{pmatrix}, \hat{\mathbf{e}}_2 = \frac{1}{\sqrt{3}} \begin{pmatrix} 1 \\ -1 \\ 1 \end{pmatrix}, \hat{\mathbf{e}}_2 = \frac{1}{\sqrt{2}} \begin{pmatrix} 0 \\ 1 \\ 1 \end{pmatrix}. \tag{13.22}$$

Using Matlab, this is trivial:

```
[E,D] = eig(A);
```

which yields the eigenvectors as `E`'s columns and the eigenvalues as `D`'s diagonal elements:

```
[E,D] = eig(A)
```

```
E =

     0.0000   -0.5774   -0.8165
    -0.7071    0.5774   -0.4082
    -0.7071   -0.5774    0.4082

D =

     1.0000         0         0
          0    2.0000         0
          0         0  900.0000
```

It is clear that while Matlab uses $\lambda_1 \leq \lambda_2 \leq \lambda_3$, we do the opposite, i.e., $\lambda_3 \leq \lambda_2 \leq \lambda_1$. Also, the rational representation is slightly tricky; while Matlab's `rats` is supposed to solve this, it's not great. But, if you examine simultaneously

```
E,D

E =

     0.0000   -0.5774   -0.8165
    -0.7071    0.5774   -0.4082
    -0.7071   -0.5774    0.4082

D =

     1.0000         0         0
          0    2.0000         0
          0         0  900.0000
```

and

```
1./[sqrt(1:9)]

ans =

     1.0000  0.7071  0.5774  0.5000  0.4472
     0.4082  0.3780  0.3536  0.3333
```

you immediately find the right rational scaling. To get the right column order, we use

```
[D,i] = sort(diag(D),'descend');
D = diag(D);
E = E(:,i);
```

which yields

```
>> D,E

D =

   900     0     0
     0     2     0
     0     0     1

E =

    0.8165    0.5774         0
    0.4082   -0.5774    0.7071
   -0.4082    0.5774    0.7071
```

as required.

The most obvious thing about this eigenspectrum is its dramatic falloff, with the 450-fold drop-off between λ_1 and λ_2 dwarfing the 2-fold drop-off between λ_2 and λ_1.

In the final question, I messed up the assignment a bit. What is most interesting is the ratio of the initial vector's magnitude to that of the outcome, $\|\mathbf{b}_i\|/\|\mathbf{A}\mathbf{b}_i\|$. But let's see what this is:

$$\frac{\|\mathbf{b}_i\|}{\|\mathbf{A}\mathbf{b}_i\|} = \frac{\|\mathbf{b}_i\|}{\|\mathbf{E}\Lambda\mathbf{E}^T\mathbf{b}_i\|} = \frac{\|\mathbf{b}_i\|}{\|\sum_{j=1}^{N}(\lambda_j\mathbf{e}_j^T\mathbf{b}_i)\hat{\mathbf{e}}_j\|} \tag{13.23}$$

$$= \frac{\|\mathbf{b}_i\|}{\sum_{j=1}^{N}|\lambda_j|\,\|(\mathbf{e}_j^T\mathbf{b}_i)\hat{\mathbf{e}}_j\|} = \frac{\|\mathbf{b}_i\|}{\sum_{j=1}^{N}|\lambda_j\hat{\mathbf{e}}_j^T\mathbf{b}_i|\,\|\hat{\mathbf{e}}_j\|}. \tag{13.24}$$

But, since the eigenvectors are normalized, $\|\hat{\mathbf{e}}_j\| = 1, j = [1,3]$, the latter term in the denominator contributes nothing to the product and can be gotten rid of. That is,

$$\frac{\|\mathbf{b}_i\|}{\|\mathbf{A}\mathbf{b}_i\|} = \frac{\|\mathbf{b}_i\|}{\sum_{j=1}^{3}|\lambda_j\hat{\mathbf{e}}_j^T\mathbf{b}_i|}, \tag{13.25}$$

i.e., the quotient is the ratio of the initial vector norm to the weighted mean absolute eigenvalue, where the weighting factor is the projection of the initial vector on the mode.

13.5 Homework 5, Corresponding to Chapter 5

13.5.1 Assignment

Consider

$$\mathbf{A} = \begin{pmatrix} 0 & \alpha & 0.5 & \alpha & 0 & -\alpha & -0.5 & -\alpha & 0 \\ 0 & 0.5 & 0.0 & -0.5 & 0 & 0.5 & 0.0 & -0.5 & 0 \\ 0 & \alpha & -0.5 & \alpha & 0 & -\alpha & 0.5 & -\alpha & 0 \end{pmatrix},$$

with $\alpha = 0.3536$.

(a) Plot **A**'s rows and guess what function(s), with what scaling and at what resolution, they try imperfectly to emulate.
(b) Obtain $\mathbf{C} \equiv \mathbf{A}\mathbf{A}^T$, report it, and explain what this matrix is and why it has the structure and values it does.
(c) Obtain the reduced SVD representation of **A**, explicitly trying to minimize the calculation burden.
(d) Write down the full (rather than reduced) $\mathbf{\Sigma}$ and explain its structure.
(e) Show that **U** fully spans the space its vectors are from; which space is it?

13.5.2 Answers

(a) A's rows are as shown in the figure, where thick solid-dotted curves show the row elements and thin solid curves show finely sampled single-frequency sine waves. While it may not be entirely trivial, it is fairly easy to speculate that the rows are sine functions. The number of crests/troughs and the amplitudes suggests that

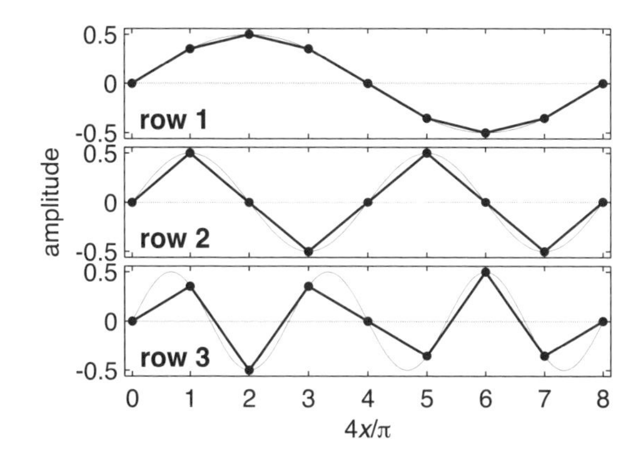

$$\mathbf{A} = \frac{1}{2}\begin{pmatrix} \sin\left(\frac{2\pi x}{L}\right) \\ \sin\left(\frac{4\pi x}{L}\right) \\ \sin\left(\frac{6\pi x}{L}\right) \end{pmatrix}, \quad \text{where} \quad x = \frac{\pi}{4}(0, 1, 2, \ldots, 8)$$

and $L = 2\pi$.

It is clear that while the low-frequency signal (the top row) is very well reproduced by the coarse representation, the high-frequency signal (row 3) is very poorly reproduced. Note also that if you guessed this, it will make sense to use the exact values (the exact sine function values evaluated at each of the x values) instead of the imperfect, low-precision, values in the assignment.

(b) With the high-precision values of the sines, instead of their imperfect printed representation,

$$\mathbf{C} = \mathbf{A}\mathbf{A}^T = \begin{pmatrix} 1 & 0 & 0 \\ 0 & 1 & 0 \\ 0 & 0 & 1 \end{pmatrix},$$

with off-diagonal elements all $\mathcal{O}(10^{\leq -15})$ (of order 10^{-15} or smaller).

(c) Since here $M = 3$ but $N = 9$, it is clear that is is best to eigenanalyze $\mathbf{A}\mathbf{A}^T \in \mathbb{R}^{3\times 3}$ and *not* $\mathbf{A}^T\mathbf{A} \in \mathbb{R}^{9\times 9}$. The characteristic equation of $\mathbf{I}_3$ (arising from the effort to find the null space of $\mathbf{C} - \lambda\mathbf{I}$) is

$$(1-\lambda)^3 = 0,$$

with $\lambda_1 = \lambda_2 = \lambda_3 = 1$. This means that the singular value matrix of $\mathbf{A}$ is

$$\boldsymbol{\Sigma} = \left(\begin{array}{ccc|cccccc} 1 & 0 & 0 & 0 & 0 & 0 & 0 & 0 & 0 \\ 0 & 1 & 0 & 0 & 0 & 0 & 0 & 0 & 0 \\ 0 & 0 & 1 & 0 & 0 & 0 & 0 & 0 & 0 \end{array}\right),$$

where the 3×3 submatrix to the left of the vertical bar is $\boldsymbol{\Sigma}_r$.

To find the eigenvectors, note that while the eigen decomposition of a general symmetric $\mathbf{C}$ is

$$\mathbf{C} = \mathbf{E}\boldsymbol{\Lambda}\mathbf{E}^T,$$

here we have $\boldsymbol{\Lambda} = \mathbf{I}$, so

$$\mathbf{C} = \mathbf{E}\mathbf{E}^T$$

or

$$\mathbf{C}\mathbf{E} = \mathbf{E}.$$

But $\mathbf{C} = \mathbf{I}$ as well, so the above says nothing more than

$$\mathbf{E} = \mathbf{E},$$

and *any* $\mathbb{R}^3$ vector is an eigenvector. Given this unusual freedom, it may make sense to choose

$$\mathbf{E} = \mathbf{I},$$

but that is entirely a matter of taste, which explains the strange results Matlab provides. With our choice of $\mathbf{E} = \mathbf{I}$, then, $\mathbf{A}$'s

$$\mathbf{U} = \mathbf{I}$$

and

$$\mathbf{A} = \mathbf{U}\mathbf{\Sigma}_r\mathbf{V}_r^T = \mathbf{I}\mathbf{I}\mathbf{V}_t^T = \mathbf{V}_r^T$$

so that

$$\mathbf{V} = \mathbf{A}^T,$$

which is trivial to prove right because here

$$\mathbf{U}\mathbf{\Sigma}_r\mathbf{V}^T = \mathbf{I}\mathbf{I}\mathbf{A},$$

which is obviously $\mathbf{A}$.

(d) I already did this above. The 6 columns of zeros that are appended to $\mathbf{\Sigma}_r$'s right to form $\mathbf{\Sigma}$, call them $\mathbf{Z} \in \mathbb{R}^{3\times6}$, multiply $\{\mathbf{v}_i^T\}_{i=4}^9$, the spanning set for $\mathcal{N}(\mathbf{A})$. That is, in general, $\mathbf{x}$ in $\mathbf{Ax}$ has a piece from $\mathbf{A}$'s row and null spaces,

$$\mathbf{x} = \mathbf{x}_r + \mathbf{x}_n,$$

with $\mathbf{A}\mathbf{x}_n = \mathbf{0} \in \mathbb{R}^M$, i.e., $\mathbf{x}_n$ is orthogonal to each of $\mathbf{A}$'s rows. Therefore,

$$\mathbf{Ax} = \mathbf{U}\mathbf{\Sigma}\mathbf{V}^T\mathbf{x}$$

$$= \begin{pmatrix} \vdots & \vdots & \vdots \\ \mathbf{u}_1 & \mathbf{u}_2 & \mathbf{u}_3 \\ \vdots & \vdots & \vdots \end{pmatrix} \begin{pmatrix} \mathbf{I} & \mathbf{Z} \end{pmatrix} \begin{pmatrix} \mathbf{v}_1^T\mathbf{x} \\ \mathbf{v}_2^T\mathbf{x} \\ \mathbf{v}_3^T\mathbf{x} \\ \hline \mathbf{v}_4^T\mathbf{x} \\ \mathbf{v}_5^T\mathbf{x} \\ \mathbf{v}_6^T\mathbf{x} \\ \mathbf{v}_7^T\mathbf{x} \\ \mathbf{v}_8^T\mathbf{x} \\ \mathbf{v}_9^T\mathbf{x} \end{pmatrix}$$

$$= \begin{pmatrix} \vdots & \vdots & \vdots \\ \mathbf{u}_1 & \mathbf{u}_2 & \mathbf{u}_3 \\ \vdots & \vdots & \vdots \end{pmatrix} \begin{pmatrix} \mathbf{I} & \mathbf{Z} \end{pmatrix} \begin{pmatrix} \mathbf{v}_1^T(\mathbf{x}_r + \mathbf{x}_n) \\ \mathbf{v}_2^T(\mathbf{x}_r + \mathbf{x}_n) \\ \mathbf{v}_3^T(\mathbf{x}_r + \mathbf{x}_n) \\ \hline \mathbf{v}_4^T(\mathbf{x}_r + \mathbf{x}_n) \\ \mathbf{v}_5^T(\mathbf{x}_r + \mathbf{x}_n) \\ \mathbf{v}_6^T(\mathbf{x}_r + \mathbf{x}_n) \\ \mathbf{v}_7^T(\mathbf{x}_r + \mathbf{x}_n) \\ \mathbf{v}_8^T(\mathbf{x}_r + \mathbf{x}_n) \\ \mathbf{v}_9^T(\mathbf{x}_r + \mathbf{x}_n) \end{pmatrix}$$

$$= \begin{pmatrix} \vdots & \vdots & \vdots \\ \mathbf{u}_1 & \mathbf{u}_2 & \mathbf{u}_3 \\ \vdots & \vdots & \vdots \end{pmatrix} \begin{pmatrix} \mathbf{I} & \mathbf{Z} \end{pmatrix} \begin{pmatrix} \mathbf{v}_1^T\mathbf{x}_r \\ \mathbf{v}_2^T\mathbf{x}_r \\ \mathbf{v}_3^T\mathbf{x}_r \\ \hline \mathbf{v}_4^T\mathbf{x}_n \\ \mathbf{v}_5^T\mathbf{x}_n \\ \mathbf{v}_6^T\mathbf{x}_n \\ \mathbf{v}_7^T\mathbf{x}_n \\ \mathbf{v}_8^T\mathbf{x}_n \\ \mathbf{v}_9^T\mathbf{x}_n \end{pmatrix},$$

where the last step represents the fact that $\{\mathbf{v}_i\}_{i=1}^3$ span $\mathcal{R}(\mathbf{A}^T)$, to which $\mathbf{x}_n$ is orthogonal, while $\{\mathbf{v}_i\}_{i=4}^9$ span $\mathcal{N}(\mathbf{A})$, to which $\mathbf{x}_r$ is orthogonal.

Now comes the clincher: while, in general, $\mathbf{v}_i^T\mathbf{x}_n \neq 0$ for $i = [4, 9]$, $\mathbf{A}\mathbf{x}_n$ must produce zero no matter what. Making sure this is achieved $\mathbf{Z}$'s role.

(e) $\mathbf{U}$'s columns are supposed to span $\mathbb{R}^M$, in this case $\mathbb{R}^3$. Since $\mathbf{U} = \mathbf{I}$, which has three linearly independent (more than that, in fact; here $\mathbf{U}$'s columns are mutually orthogonal!), they *must* span $\mathbb{R}^3$.

13.6 Homework 6, Corresponding to Chapter 8

13.6.1 Assignment

Consider $\mathbf{A}\mathbf{x}_{1,2,3} = \mathbf{b}_{1,2,3}$ with

$$\mathbf{A} = \begin{pmatrix} \mathbf{a}_1 & \mathbf{a}_2 \end{pmatrix} = \begin{pmatrix} 1 & 2 \\ 3 & 6 \\ 2 & 4.001 \end{pmatrix}, \tag{13.26}$$

$$\mathbf{b}_1 = \begin{pmatrix} 5 \\ 15 \\ 10 \end{pmatrix}, \quad \mathbf{b}_2 = \begin{pmatrix} 5 \\ 15 \\ 10.002 \end{pmatrix}, \quad \text{and} \quad \mathbf{b}_3 = \begin{pmatrix} 0 \\ 0 \\ 0.002 \end{pmatrix}. \tag{13.27}$$

(a) Please obtain the least-squares solutions $\hat{\mathbf{x}}_1$, $\hat{\mathbf{x}}_2$, and $\hat{\mathbf{x}}_3$.
(b) Obtain the reconstructed right-hand sides $\hat{\mathbf{b}}_1$, $\hat{\mathbf{b}}_2$, and $\hat{\mathbf{b}}_3$.
(c) Analyze $\mathbf{A}$ and obtain its fundamental spaces.
(d) Explain the difference between the various solutions and reconstructed right-hand sides based on $\mathbf{A}$'s fundamental spaces.
(e) Explain the general moral of the least-squares failure you unearthed.

13.6.2 Answers

The purpose of this homework is to make you aware of some of the possible pitfalls one can encounter when blindly applying the least-squares solution $\hat{\mathbf{x}} =$

$(\mathbf{A}^T\mathbf{A})^{-1}\mathbf{A}^T\mathbf{b}$ to the problem $\mathbf{Ax} = \mathbf{b}$. To begin with, note that $\mathbf{b}_i \in \mathcal{R}(\mathbf{A})$, $i = [1, 3]$, because $\mathbf{b}_1 = 5\mathbf{a}_1$, $\mathbf{b}_2 = \mathbf{a}_1 + 2\mathbf{a}_2$ and $\mathbf{b}_3 = 2\mathbf{a}_2 - 4\mathbf{a}_1$, so the problems have the exact solutions

$$\mathbf{x}_1 = \begin{pmatrix} 5 \\ 0 \end{pmatrix} \quad \mathbf{x}_2 = \begin{pmatrix} 1 \\ 2 \end{pmatrix}, \quad \text{and} \quad \mathbf{x}_3 = \begin{pmatrix} -4 \\ 2 \end{pmatrix}, \tag{13.28}$$

where, note, there is no need to adorn the least-squares (LS) solution with hats, because they are exact.

Now let's think a bit about the LS solution. Since $\mathbf{A}^T\mathbf{A}$ is square and symmetric, it has a complete set of eigenvectors (which we make $\mathbf{E}$'s columns), and—if it has a full set of nonzero eigenvalues (the nonzero diagonal elements of the eigenvalue matrix $\mathbf{\Lambda}$)—satisfies

$$(\mathbf{A}^T\mathbf{A})^{-1} = (\mathbf{E}\mathbf{\Lambda}\mathbf{E}^T)^{-1}. \tag{13.29}$$

Since $\mathbf{E}$, $\mathbf{E}^T$ and $\mathbf{\Lambda}$ are all square,

$$(\mathbf{E}\mathbf{\Lambda}\mathbf{E}^T)^{-1} = (\mathbf{E}^T)^{-1}\mathbf{\Lambda}^{-1}\mathbf{E}^{-1}. \tag{13.30}$$

This further simplifies due to the fact that, because of $\mathbf{E}$'s columns orthonormality, $\mathbf{E}^{-1} = \mathbf{E}^T$, so that

$$(\mathbf{A}^T\mathbf{A})^{-1} = \mathbf{E}\mathbf{\Lambda}^{-1}\mathbf{E}^T \tag{13.31}$$

and the (in this case exact) LS solution is

$$\hat{\mathbf{x}} = \mathbf{E}\mathbf{\Lambda}^{-1}\mathbf{E}^T\mathbf{A}^T\mathbf{b} = \sum_{i=1}^{N} \left(\frac{\mathbf{e}_i^T(\mathbf{A}^T\mathbf{b})}{\lambda_i} \right) \mathbf{e}_i. \tag{13.32}$$

In the current case, where $N = 2$, this means that

$$\begin{aligned} \hat{\mathbf{x}} &= \left(\frac{\mathbf{e}_1^T(\mathbf{A}^T\mathbf{b})}{\lambda_1} \right) \mathbf{e}_1 + \left(\frac{\mathbf{e}_2^T(\mathbf{A}^T\mathbf{b})}{\lambda_2} \right) \mathbf{e}_2 \\ &= c_1\mathbf{e}_1 + c_2\mathbf{e}_2. \end{aligned} \tag{13.33}$$

Using Matlab, initially things look pretty good, with

$$\begin{aligned} \frac{\|\mathbf{b}_1 - \hat{\mathbf{b}}_1\|}{\|\mathbf{b}_1\|} &\approx 1.2 \cdot 10^{-8}, \\ \frac{\|\mathbf{b}_2 - \hat{\mathbf{b}}_2\|}{\|\mathbf{b}_2\|} &= 0, \quad \text{and} \\ \frac{\|\mathbf{b}_3 - \hat{\mathbf{b}}_3\|}{\|\mathbf{b}_3\|} &\approx 6.6 \cdot 10^{-9}. \end{aligned} \tag{13.34}$$

While these errors are far larger than round-off error which—for 64 bit arithmetic—is typically $\mathcal{O}(10^{-17})$, they are not unheard of in real inversions of noisy data. However, using `format('long')` reveals that while $\hat{\mathbf{x}}_2 = \mathbf{x}_2 = (1, 2)$,

$$\hat{\mathbf{x}}_1 - \begin{pmatrix} 5 \\ 0 \end{pmatrix} = \begin{pmatrix} \mathcal{O}(10^{-8}) \\ 0 \end{pmatrix} \tag{13.35}$$

and, much worse yet,

$$\hat{\mathbf{x}}_3 - \begin{pmatrix} -4 \\ 2 \end{pmatrix} \approx \begin{pmatrix} 8 \\ -4 \end{pmatrix}. \tag{13.36}$$

These errors are definitely not due to normal round-off error. What's the matter with $\hat{\mathbf{x}}_{1,3}$?!

Note that $\mathbf{A}$'s columns, and thus those of

$$\mathbf{A}^T\mathbf{A} \approx \begin{pmatrix} 14.000 & 28.002 \\ 28.002 & 56.008 \end{pmatrix}, \tag{13.37}$$

are *very nearly parallel*, but not quite (or else $\mathbf{A}^T\mathbf{A}$ would have been singular and the entire discussion would have been moot). Most of the action in $\mathbf{A}^T\mathbf{A}$ is along (1, 2), with very little going on in the orthogonal direction $\left(1, -\frac{1}{2}\right)$:

$$\frac{1}{\sqrt{5}}\mathbf{A}^T\mathbf{A}\begin{pmatrix} 1 \\ 2 \end{pmatrix} \approx \begin{pmatrix} 31 \\ 63 \end{pmatrix}, \tag{13.38}$$

while

$$\frac{1}{\sqrt{1.25}}\mathbf{A}^T\mathbf{A}\begin{pmatrix} 1 \\ -1/2 \end{pmatrix} \approx -\begin{pmatrix} 0.001 \\ 0.002 \end{pmatrix}, \tag{13.39}$$

with approximate respective magnitudes of 70 and 0.002 ($70/0.002 \approx 35{,}000$).

Because of this disparity, $|\lambda|_{\max}/|\lambda|_{\min} \sim \mathcal{O}(10^8)$, and the eigenvector associated with the smaller of the two eigenvalues is poorly determined. To show this, let's evaluate the error in the numerical eigenvectors. For mode 1, it is

$$\left\| \mathbf{e}_1 - \frac{1}{\sqrt{1.25}}\begin{pmatrix} 1 \\ -0.5 \end{pmatrix} \right\| \approx 2, \tag{13.40}$$

while for mode 2 it is

$$\left\| \mathbf{e}_2 - \frac{1}{\sqrt{5}}\begin{pmatrix} 1 \\ 2 \end{pmatrix} \right\| \approx 3 \cdot 10^{-5}. \tag{13.41}$$

Thus, $\mathbf{e}_2$ is determined roughly 100,000 times better than $\mathbf{e}_1$, and the latter has the potential to ruin the solution.

Will this potential ruin indeed be realized? The answer is in the magnitude of c_1 of eq. 13.34, and the following table helps address this question:

RHS	$\lvert c_1 \rvert$
$\mathbf{b}_1$	4.5
$\mathbf{b}_2$	$6.4 \cdot 10^{-5}$
$\mathbf{b}_3$	4.5

That is, for $\mathbf{b}_2$, the poorly known mode 1 is scaled by a c_1 whose magnitude is 5 orders of magnitude smaller than in the cases of $\mathbf{b}_1$ and $\mathbf{b}_3$; we fully *expect* $\hat{\mathbf{x}}_2$ to work, and $\hat{\mathbf{x}}_1$ and $\hat{\mathbf{x}}_3$ to fail!

13.7 A Suggested Midterm Exam

13.7.1 Assignment

$$\text{Let}\quad \mathbf{A} = \begin{pmatrix} 1 & 0 & -1 & 2 & 1 & 2 \\ 2 & 1 & -1 & 0 & 1 & 1 \\ 4 & 1 & -3 & 4 & 3 & 5 \end{pmatrix}$$

and

$$\mathbf{B} = \begin{pmatrix} 1 & 0 & 1 \\ 0 & 1 & 0 \\ 0 & 0 & 0 \end{pmatrix}.$$

(a) Enumerate symbolically (by writing down the relevant equations) the 4 fundamental spaces associated with a rank q $M \times N$ matrix, their dimensions, and the various relationships satisfied by those spaces and their parent spaces.
(b) Reduce **A** above to its corresponding **U**. No need to normalize pivots. Report all elementary operations in as many **E**s as needed. Report **A**'s q.
(c) Devise a nice spanning set for $\mathcal{R}(\mathbf{A})$ and explain this space's relationship to a right-hand-side **b** with which $\mathbf{Ax} = \mathbf{b}$ has an exact solution.
(d) Transform the above set to an orthonormal one. Explain briefly what you needed to do.
(e) Devise a nice spanning set for $\mathcal{R}(\mathbf{A}^T)$. Which of **A**'s rows, if any, can you replace so as to render **A** full rank if it isn't already? If you found such a row (or rows), what is the general form of the replacement set?
(f) Devise a nice spanning set for $\mathcal{N}(\mathbf{A})$.
(g) Obtain the general solution of

$$\mathbf{Ax} = \mathbf{b} = \begin{pmatrix} 2 \\ 5 \\ 9 \end{pmatrix},$$

outline its various pieces and their relationships, if any, to any or all of **A**'s fundamental spaces.
(h) If you wanted/needed to, could you devise two solutions $\mathbf{x}_1$ and $\mathbf{x}_2$ to the above system such that $\|\mathbf{x}_1\| = 2\|\mathbf{x}_2\|$?
(i) Find the eigenvalues and eigenvectors of **B**.
(j) Divide $\mathbb{R}^3$ into 2 orthogonal subspaces based on the action of **B**. That is, some $\mathbb{R}^3$ vectors, call them **f**, will have one fate upon being premultiplied

by **B**, while others, call them **g**, will have another fate; give some definition of **f** and **g**, and describe their fates (i.e., **Bf** and **Bg**).

(k) What is **B**'s rank? What is the relationship between **U** and the eigen-decomposition?

13.7.2 *Answers*

(a)

$$\begin{aligned}
\mathcal{N}(\mathbf{A}) &= \{\mathbf{n} \in \mathbb{R}^N : \mathbf{A}\mathbf{n} = \mathbf{0} \in \mathbb{R}^M\} \\
\mathcal{N}(\mathbf{A}^T) &= \{\mathbf{n}_l \in \mathbb{R}^M : \mathbf{A}^T\mathbf{n}_l = \mathbf{0} \in \mathbb{R}^N\} \\
\mathcal{R}(\mathbf{A}) &= \{\mathbf{c} \in \mathbb{R}^M : \mathbf{c} = \mathbf{A}\mathbf{x},\ \mathbf{x} \in \mathbb{R}^N\} \\
\mathcal{R}(\mathbf{A}^T) &= \{r \in \mathbb{R}^N : \mathbf{r} = \mathbf{A}^T\mathbf{y},\ \mathbf{y} \in \mathbb{R}^M\}
\end{aligned}$$

$$\begin{aligned}
\dim[\mathcal{N}(\mathbf{A})] &= N - q \\
\dim[\mathcal{N}(\mathbf{A}^T)] &= M - q \\
\dim[\mathcal{R}(\mathbf{A})] &= q \\
\dim[\mathcal{R}(\mathbf{A}^T)] &= q
\end{aligned}$$

In words:

- **A**'s null space: the set of all $\mathbb{R}^N$ vectors **n** killed by **A** (i.e., mapped by it onto the $\mathbb{R}^M$ zero vector)
- **A**'s left null space: the set of all $\mathbb{R}^M$ vectors $\mathbf{n}_l$ killed by $\mathbf{A}^T$ (i.e., mapped by it onto the $\mathbb{R}^N$ zero vector)
- **A**'s column space: the set of all $\mathbb{R}^M$ vectors **c** that are linear combinations of **A**'s columns
- **A**'s row space: the set of all $\mathbb{R}^N$ vectors **r** that are linear combinations of **A**'s rows

(b) With

$$\mathbf{E}_1 = \begin{pmatrix} 1 & 0 & 0 \\ -2 & 1 & 0 \\ -4 & 0 & 1 \end{pmatrix} \quad \text{and} \quad \mathbf{E}_2 = \begin{pmatrix} 1 & 0 & 0 \\ 0 & 1 & 0 \\ 0 & -1 & 1 \end{pmatrix},$$

$$\mathbf{U} = \mathbf{E}_2\mathbf{E}_1\mathbf{A} = \begin{pmatrix} \boxed{1} & 0 & -1 & 2 & 1 & 2 \\ 0 & \boxed{1} & 1 & -4 & -1 & -3 \\ 0 & 0 & 0 & 0 & 0 & 0 \end{pmatrix},$$

so this **A**'s $q = 2$.

(c)

$$\operatorname{span}[\mathcal{R}(\mathbf{A})] = \{\hat{\mathbf{a}}_1, \hat{\mathbf{a}}_2\} = \left\{ \frac{1}{\sqrt{21}}\begin{pmatrix} 1 \\ 2 \\ 4 \end{pmatrix}, \frac{1}{\sqrt{2}}\begin{pmatrix} 0 \\ 1 \\ 1 \end{pmatrix} \right\}.$$

To show that this set spans $\mathcal{R}(\mathbf{A})$, we need to show that any of $\mathbf{A}$ columns, or linear combinations thereof, can be expressed as a linear combination of the above set. But since most of $\mathbf{A}$ columns are dependent, and only the first (left most) two are independent, all we need to show is that

$$\begin{pmatrix} 1 & 0 \\ 2 & 1 \\ 4 & 1 \end{pmatrix} \begin{pmatrix} \alpha \\ \beta \end{pmatrix},$$

for any (α, β), can be expressed in terms of our spanning set. Since the latter is simply the former normalized, this is a triviality, a condition that is obviously met. If $\mathbf{Ax} = \mathbf{b}$, $\mathbf{b} \in \mathcal{R}(\mathbf{A})$, $\mathbf{b}$ better be from $\mathbf{A}$'s columns space.

(d) First, are the two linearly independent? Since clearly $\hat{\mathbf{a}}_1^T \hat{\mathbf{a}}_2 \neq 0$, they are. With the first one, which is already normalized, we do nothing. For the second,

$$\begin{aligned} \mathbf{q}_2 &= \mathbf{a}_2 - (\mathbf{a}_2^T \hat{\mathbf{q}}_1)\hat{\mathbf{q}}_1 \\ &= \begin{pmatrix} 0 \\ 1 \\ 1 \end{pmatrix} - \frac{\begin{pmatrix} 0 & 1 & 1 \end{pmatrix}}{\sqrt{21}} \begin{pmatrix} 1 \\ 2 \\ 4 \end{pmatrix} \frac{1}{\sqrt{21}} \begin{pmatrix} 1 \\ 2 \\ 4 \end{pmatrix} \\ &= \begin{pmatrix} 0 \\ 1 \\ 1 \end{pmatrix} - \frac{6}{21} \begin{pmatrix} 1 \\ 2 \\ 4 \end{pmatrix} = -\frac{1}{7} \begin{pmatrix} 2 \\ -3 \\ 1 \end{pmatrix}, \end{aligned} \tag{13.42}$$

so

$$\hat{\mathbf{q}}_2 = -\frac{1}{\sqrt{14}} \begin{pmatrix} 2 \\ -3 \\ 1 \end{pmatrix}.$$

(e) Let's reduce $\mathbf{A}^T$ to its $\mathbf{V}$. With

$$\mathbf{F}_1 = \begin{pmatrix} 1 & 0 & 0 & 0 & 0 & 0 \\ 0 & 1 & 0 & 0 & 0 & 0 \\ 1 & 0 & 1 & 0 & 0 & 0 \\ -2 & 0 & 0 & 1 & 0 & 0 \\ -1 & 0 & 0 & 0 & 1 & 0 \\ -2 & 0 & 0 & 0 & 0 & 1 \end{pmatrix},$$

and

$$\mathbf{F}_2 = \begin{pmatrix} 1 & 0 & 0 & 0 & 0 & 0 \\ 0 & 1 & 0 & 0 & 0 & 0 \\ 0 & -1 & 1 & 0 & 0 & 0 \\ 0 & 4 & 0 & 1 & 0 & 0 \\ 0 & 1 & 0 & 0 & 1 & 0 \\ 0 & 3 & 0 & 0 & 0 & 1 \end{pmatrix},$$

$$\mathbf{V} := \mathbf{F}_2\mathbf{F}_1\mathbf{A}^T = \begin{pmatrix} \boxed{1} & 2 & 4 \\ 0 & \boxed{1} & 1 \\ 0 & 0 & 0 \\ 0 & 0 & 0 \\ 0 & 0 & 0 \\ 0 & 0 & 0 \end{pmatrix},$$

so

$$\text{basis}\left[\mathcal{R}(\mathbf{A}^T)\right] = \left\{ \frac{1}{\sqrt{11}} \begin{pmatrix} 1 \\ 0 \\ -1 \\ 2 \\ 1 \\ 2 \end{pmatrix}, \frac{1}{\sqrt{8}} \begin{pmatrix} 2 \\ 1 \\ -1 \\ 0 \\ 1 \\ 1 \end{pmatrix} \right\}.$$

Since $\mathbf{V}$ has nonzero pivots in columns 1 and 2, $\mathbf{A}$'s row to replace is the third. That third row is twice row 1 plus row 2. We have an infinite number of ways to make that not so. It would be most convenient, but by *no means* unique, to make the change in a_{33} so as to get a pivot right away. So any

$$a_{33} \neq 2a_{13} + a_{23}$$

will do.

(f) Revisiting $\mathbf{U}$, recalling that null space vectors are ones satisfying $\mathbf{An} = \mathbf{0} \in \mathbb{R}^M$ and assuming the general structure of null space vectors

$$\mathbf{n} = \begin{pmatrix} \alpha_1 \\ \alpha_2 \\ \alpha_3 \\ \alpha_4 \\ \alpha_5 \\ \alpha_6 \end{pmatrix},$$

the third row is obviously not helpful, so we move to the second, which states that $\alpha_2 = -\alpha_3 + 4\alpha_4 + \alpha_5 + 3\alpha_6$ and the top row, $\alpha_1 = -\alpha_3 + 2\alpha_4 + \alpha_5 + 2\alpha_6$. Then

$$\mathbf{n} = \begin{pmatrix} \alpha_3 - 2\alpha_4 - \alpha_5 - 2\alpha_6 \\ -\alpha_3 + 4\alpha_4 + \alpha_5 + 3\alpha_6 \\ \alpha_3 \\ \alpha_4 \\ \alpha_5 \\ \alpha_6 \end{pmatrix} \qquad (13.43)$$

$$= \alpha_3 \begin{pmatrix} 1 \\ -1 \\ 1 \\ 0 \\ 0 \\ 0 \end{pmatrix} + \alpha_4 \begin{pmatrix} -2 \\ 4 \\ 0 \\ 1 \\ 0 \\ 0 \end{pmatrix} + \alpha_5 \begin{pmatrix} -1 \\ 1 \\ 0 \\ 0 \\ 1 \\ 0 \end{pmatrix} + \alpha_6 \begin{pmatrix} -2 \\ 3 \\ 0 \\ 0 \\ 0 \\ 1 \end{pmatrix},$$

so a basis for **A**'s null space, neither normalized nor orthogonal, is basis[$\mathcal{N}(\mathbf{A})$]

$$= \left\{ \begin{pmatrix} 1 \\ -1 \\ 1 \\ 0 \\ 0 \\ 0 \end{pmatrix}, \begin{pmatrix} -2 \\ 4 \\ 0 \\ 1 \\ 0 \\ 0 \end{pmatrix}, \begin{pmatrix} -1 \\ 1 \\ 0 \\ 0 \\ 1 \\ 0 \end{pmatrix}, \begin{pmatrix} -2 \\ 3 \\ 0 \\ 0 \\ 0 \\ 1 \end{pmatrix} \right\}.$$

(g) The system $\mathbf{Ax} = \mathbf{b}$ is equivalent to $\mathbf{Ux} = \mathbf{d} := \mathbf{E}_2\mathbf{E}_1\mathbf{b}$. Obtaining this modified right-hand side,

$$\mathbf{d} = \begin{pmatrix} 2 \\ 1 \\ 0 \end{pmatrix},$$

is reassuring because the third element vanishes; a solution exists. From the first and second rows of $\mathbf{Ux} = \mathbf{d}$ we get

$$\alpha_1 = 2 + \alpha_3 - 2\alpha_4 - \alpha_5 - 2\alpha_6$$
$$\alpha_2 = 1 - \alpha_3 + 4\alpha_4 + \alpha_5 + 3\alpha_6$$

and thus

$$\mathbf{x} = \begin{pmatrix} 2 + \alpha_3 - 2\alpha_4 - \alpha_5 - 2\alpha_6 \\ 1 - \alpha_3 + 4\alpha_4 + \alpha_5 + 3\alpha_6 \\ \alpha_3 \\ \alpha_4 \\ \alpha_5 \\ \alpha_6 \end{pmatrix} = \mathbf{x}_p + \mathbf{x}_h,$$

where

$$\mathbf{x}_p = \begin{pmatrix} 2 \\ 1 \\ 0 \\ 0 \\ 0 \\ 0 \end{pmatrix}$$

and

$$\mathbf{x}_h = \alpha_3 \begin{pmatrix} 1 \\ -1 \\ 1 \\ 0 \\ 0 \\ 0 \end{pmatrix} + \alpha_4 \begin{pmatrix} -2 \\ 4 \\ 0 \\ 1 \\ 0 \\ 0 \end{pmatrix}$$

$$+ \alpha_5 \begin{pmatrix} -1 \\ 1 \\ 0 \\ 0 \\ 1 \\ 0 \end{pmatrix} + \alpha_6 \begin{pmatrix} -2 \\ 3 \\ 0 \\ 0 \\ 0 \\ 1 \end{pmatrix}. \tag{13.44}$$

Note the general form of the solution. The particular solution, which is unique, is fully determined by the right-hand-side vector **b**. All the indeterminacy is collected into the entirely unconstrained coefficients of the null space vectors, which together make up the homogeneous part of the solution.

(h) Sure I could! Here's how, but note that there are an infinite number of ways of doing this, so my way is by *no means* unique. I am just offering one.

Let's choose, mostly arbitrarily, $\alpha_5 = \frac{1}{2}, \alpha_i = 0, i = 3, 4, 6$, with which

$$\mathbf{x}_1 = \begin{pmatrix} 2 \\ 1 \\ 0 \\ 0 \\ 0 \\ 0 \end{pmatrix} + \frac{1}{2} \begin{pmatrix} -1 \\ 1 \\ 0 \\ 0 \\ 1 \\ 0 \end{pmatrix} = \frac{1}{2} \begin{pmatrix} 3 \\ 3 \\ 0 \\ 0 \\ 1 \\ 0 \end{pmatrix}, \qquad \|\mathbf{x}_1\| = \frac{\sqrt{19}}{2}.$$

We are next looking for an $\mathbf{x}_2$ satisfying

$$\|\mathbf{x}_2\| = \frac{\sqrt{19}}{4} \quad \text{or} \quad \|\mathbf{x}_2\|^2 = \frac{19}{16}.$$

While this vector, in general, comprises a piece from each of the null space vectors, let's restrict our attention to one comprising only $\mathbf{x}_p$ plus some of the null space vector whose coefficient is α_3. Then

$$\mathbf{x}_2 = \begin{pmatrix} 2 \\ 1 \\ 0 \\ 0 \\ 0 \\ 0 \end{pmatrix} + \alpha_3 \begin{pmatrix} 1 \\ -1 \\ 1 \\ 0 \\ 0 \\ 0 \end{pmatrix},$$

and its squared norm is

$$(4 + 2\alpha_3 + \alpha_3^2) + (1 - \alpha_3 + \alpha_3^2) + \alpha_3^2 = \tfrac{19}{16}.$$

Simplifying, we get

$$3\alpha_3^2 + \alpha_3 + 5 - \tfrac{19}{16} = 0,$$

whose roots give the necessary coefficient to satisfy $\|\mathbf{x}_1\| = 2\|\mathbf{x}_2\|$. If they are complex (as they are here, and which they can easily be), the relationship still holds, by definition of the norm, but it may not be exactly what we want…

Can we guarantee a real solution? Yes, again. Here's how, assuming, for simplicity, that only α_3 and α_4 are nonzero, and also that $\alpha_4 = 1$. In this case, The general form of the solution's squared magnitude is

$$\|\mathbf{x}\|^2 = (2 + \alpha_3 - 2)^2 + (1 - \alpha_3 + 4)^2 + \alpha_3^2 + 1 = 3\alpha_3^2 - 5\alpha_3 + 25.$$

For simplicity of notation, let's now call $\mathbf{x}_1$'s α_3 α, and $\mathbf{x}_2$'s α_3 β and note that $\|\mathbf{x}_1\|^2/\|\mathbf{x}_1\|^2 = 4$:

$$\frac{\|\mathbf{x}_1\|^2}{\|\mathbf{x}_2\|^2} = \frac{3\alpha^2 - 5\alpha + 25}{3\beta^2 - 5\beta + 25} = 4.$$

Let's set $\beta = 1$, with which

$$\frac{\|\mathbf{x}_1\|^2}{\|\mathbf{x}_2\|^2} = \frac{3\alpha^2 - 5\alpha + 25}{3 - 5 + 25} = \frac{3\alpha^2 - 5\alpha + 25}{23} = 4$$

$$3\alpha^2 - 5\alpha - 67 = 0.$$

Since this has two real roots ($\sim$5.6 and ~ -3.96), we are set.

(i)

$$\begin{vmatrix} 1-\lambda & 0 & 1 \\ 0 & 1-\lambda & 0 \\ 0 & 0 & -\lambda \end{vmatrix} = -\lambda(1-\lambda)^2 = 0,$$

so

$$\lambda_i = \begin{cases} 1, & i=1 \\ 1, & i=2, \\ 0, & i=3 \end{cases}$$

1 is a repeated eigenvalue with an algebraic multiplicity of 2; would its geometric multiplicity suffice? Let's see, solving (for $\lambda_{1,2} = 1$)

$$\mathbf{B}\mathbf{e}_1 = \begin{pmatrix} 1 & 0 & 1 \\ 0 & 1 & 0 \\ 0 & 0 & 0 \end{pmatrix} \mathbf{e}_1 = \mathbf{e}_1.$$

From the third row it is clear that $e_{13} = 0$, and from the first row,

$$e_{11} + e_{13} = e_{11} = e_{11},$$

which is identically true. The second row is also identically satisfied for any e_{12}. Since neither e_{11} nor e_{12} are constrained and $e_{13} = 0$, the general form of the eigenvectors corresponding to $\lambda = 1$ is

$$\mathbf{e}_1 = \begin{pmatrix} e_{11} \\ e_{12} \\ 0 \end{pmatrix}.$$

Let's set either one once to 1 and once to zero, yielding

$$\hat{\mathbf{e}}_1 = \begin{pmatrix} 1 \\ 0 \\ 0 \end{pmatrix}, \quad \hat{\mathbf{e}}_2 = \begin{pmatrix} 0 \\ 1 \\ 0 \end{pmatrix},$$

so, yes, the geometric multiplicity is as high as it needs to be (as high as the algebraic multiplicity). Good.

What about $\lambda_3 = 0$? Clearly, in this case $e_{32} = 0$, but e_{33} is unconstrained (because now the zero eigenvalue guarantees the third row's vanishing) and, from the first row, $e_{31} = -e_{33}$. Therefore,

$$\mathbf{e}_3 = \begin{pmatrix} e_{31} \\ 0 \\ -e_{31} \end{pmatrix} \implies \hat{\mathbf{e}}_3 = \frac{1}{\sqrt{2}} \begin{pmatrix} 1 \\ 0 \\ -1 \end{pmatrix}.$$

(j) The answer to this is not unique. Subjectively, I think it makes the most sense to split $\mathbb{R}^3$ into those vectors killed by pre multiplication by **B**, $\{\mathbf{f} : \mathbf{B}\mathbf{f} = \mathbf{0} \in \mathbb{R}^3\}$, and a remainder, $\{\mathbf{g} : \mathbf{B}\mathbf{g} \neq \mathbf{0} \in \mathbb{R}^3\}$. Of course, if this is our choice of a split, then $\mathbf{f} = \alpha\hat{\mathbf{e}}_3$ because $\lambda_3 = 0$, i.e., $\hat{\mathbf{e}}_3 = \text{span}[\mathcal{N}(\mathbf{B})]$.

(k) **B**'s rank is $q = 2$. This is revealed by the nonzero pivots in **U** (not that we got it, but in general) and the nonzero eigenvalues. This is the *only* relationship between eigenanalysis and Gaussian elimination that I know of.

13.8 A Suggested Final Exam

(a) Write down the governing equation of the SVD operation. Explain what each term is and point out the dimensions and nature of each participant.
(b) Write the above relations for individual modes. Be sure to include the range of modes over which the modal representation applies.
(c) From the above relations, derive the biorthogonality conditions relating the left and right singular vectors.
(d) Show that the SVD representation of a matrix is a spectral representation.
(e) Explain why the left and right eigenvalue problems share one set of eigenvalues. What are the conditions that set satisfies?
(f) Describe symbolically $\boldsymbol{\Sigma}$ of a rank 1, 3×2 matrix. Point out $\boldsymbol{\Sigma}$'s parts corresponding to the null spaces in the model and data spaces.
(g) Write down the covariance matrix of a $2 \times N$ matrix whose rows are N-element, unit norm time series of perfect sine waves, with the top and bottom series comprising exactly 6 and 3 full waves. What part(s) of the original matrix's SVD can you construct *numerically* with this information? Write down these parts.
(h) How would you get *symbolically* the remainder of that matrix's SVD? What does it represent? What is its general form?
(i) What type of matrix lends itself best to SVD compression? How does it work? At what (accuracy) cost?
(j) What can you say about a time series whose acf is given by $(1, 0, -1, 0, 1, 0, \ldots)$? At what resolution is it sampled?
(k) What can you say about a time series whose acf is given by $(1, 0, 0, 0 \cdot)$?
(l) What can you say about a weekly resolution time series whose acf is given by $(1, 0.65, 0.38, 0.21, 0.09, 0.01 \cdot)$?
(m) Consider a time series y_i, $i = [1, 100]$, measured at t_i, $i = [1, 100]$. Suppose you have a reason to believe y grows exponentially in time. Set up the system that will let you choose the parameters of this growth optimally. Describe and explain the parameters.
(n) Describe how you'd solve the optimization problem. How do the coefficient matrix's fundamental spaces figure into this?
(o) Describe the general split of the right-hand-side vector (**b** in most of our class discussions). How does each of **b**'s pieces affect the solution?
(p) Recast the above solution of the optimization problem in terms of the coefficient matrix's SVD.
(q) How would you obtain a solution to the above problem for the case of a rank-deficient coefficient matrix. What does this procedure amount to?
(r) Write down the general procedure of obtaining the empirical orthogonal functions of an $M \times N$ $\mathbf{A}$.

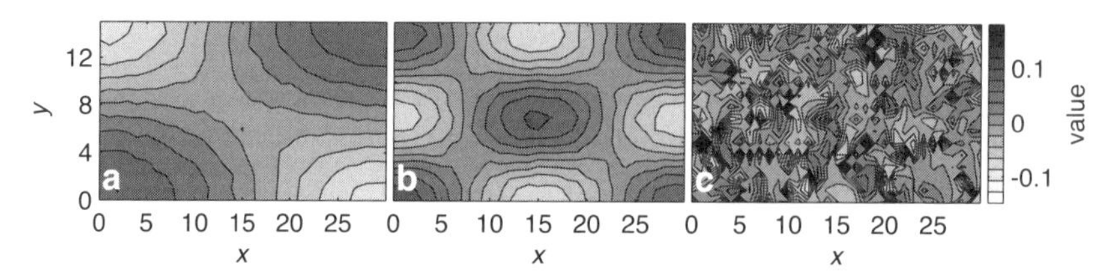

Figure 13.1. Hypothetical EOFs addressed in the exam.

(s) What can you say about a matrix whose three leading EOFs are the fields shown in fig. 13.1 (with panel a showing EOF 1)? What else can you say given $\lambda_1 = 94$, $\lambda_2 = 8$, $\lambda_3 = 0.24$, and $\sum_{i=1}^{N} \lambda_i = 112$?

(t) Describe symbolically the preliminary transformation of a data set depending on two space coordinates and time that will allow you to obtain the data set's EOFs.

Index